Materials Horizons: From Nature to Nanomaterials

Series Editor

Vijay Kumar Thakur, School of Aerospace, Transport and Manufacturing, Cranfield University, Cranfield, UK

Materials are an indispensable part of human civilization since the inception of life on earth. With the passage of time, innumerable new materials have been explored as well as developed and the search for new innovative materials continues briskly. Keeping in mind the immense perspectives of various classes of materials, this series aims at providing a comprehensive collection of works across the breadth of materials research at cutting-edge interface of materials science with physics, chemistry, biology and engineering.

This series covers a galaxy of materials ranging from natural materials to nanomaterials. Some of the topics include but not limited to: biological materials, biomimetic materials, ceramics, composites, coatings, functional materials, glasses, inorganic materials, inorganic-organic hybrids, metals, membranes, magnetic materials, manufacturing of materials, nanomaterials, organic materials and pigments to name a few. The series provides most timely and comprehensive information on advanced synthesis, processing, characterization, manufacturing and applications in a broad range of interdisciplinary fields in science, engineering and technology.

This series accepts both authored and edited works, including textbooks, monographs, reference works, and professional books. The books in this series will provide a deep insight into the state-of-art of Materials Horizons and serve students, academic, government and industrial scientists involved in all aspects of materials research.

Review Process

The proposal for each volume is reviewed by the following:

1. Responsible (in-house) editor
2. One external subject expert
3. One of the editorial board members.

The chapters in each volume are individually reviewed single blind by expert reviewers and the volume editor.

Bibhu Prasad Swain
Editor

Energy Materials

Structure, Properties and Applications

Springer

Editor
Bibhu Prasad Swain
Department of Physics
National Institute of Technology Manipur
Imphal, Manipur, India

ISSN 2524-5384 ISSN 2524-5392 (electronic)
Materials Horizons: From Nature to Nanomaterials
ISBN 978-981-99-3865-0 ISBN 978-981-99-3866-7 (eBook)
https://doi.org/10.1007/978-981-99-3866-7

This Springer imprint is published by the registered company Springer Nature Singapore Pte Ltd.
The registered company address is: 152 Beach Road, #21-01/04 Gateway East, Singapore 189721, Singapore

Preface

Energy Materials: Structure, Properties and Applications is a collection of 11 chapters on significant research that has been carried out to develop energy-related materials to meet worldwide energy demand. Various organic and inorganic materials exhibit strong potential to contribute to the energy demand in various ways, e.g. solar cells, fuel cells, and supercapacitor materials. The charge kinetics at the electrode–electrolyte interface, suitable sealant, and blend of composite electrodes determine the efficiency in a fuel cell. Chapter 1 provides the phenomenon of the electrochemical activity of proton conductors principle different from distinct fuel cells which are categorized on the nature of electrolyte and diffusing charge carriers alongside operating temperature regimes. Chapter 1 also enlightens the physics of structural perturbations and inflexions in charge chemistry. Lower symmetry shifts (distorted structures) although assist unimpeded charge dynamics in cooperative chemical compatibility. It is an attempt at material engineering via heterogeneous impurity substitutions in terms of acceptor dopants.

Electrochemical energy storage is based on two factors that are systems with high energy densities (batteries) or power densities (electrochemical condensers). Chapter 2 focuses on the fundamental properties and synthesis processes of TMNs as electrode materials in EES devices like lithium-ion batteries, sodium-ion batteries, and supercapacitors and also describes their capacitive characteristics.

Various combinations of cathode materials like Li-Fe-phosphate (LFP), Li-Ni-Mn-Co (NCM), Li-Co- Oxide (LCO), Li Ni-Co Al (NCA) LCA, and Li Mn oxide (LMO) are used in lithium-ion batteries (LIBs) based on the type of applications. Modification of electrodes by lattice doping and coatings may play a critical role in improving their electrochemical properties, cycle life, and thermal behaviour. Chapter 3 describes the cathode materials in lithium-ion batteries as energy storage devices.

Dye-sensitized solar cells (DSSC) have evolved as an aspiration for economical solar cells in the era of expensive silicon and thin film-based solar cells. DSSC features low-cost, low-toxic materials, easy fabrication processes, and mainly indoor applications, thus overall escalating its potential and establishing it as efficient, eco-friendly low-weighted solar cells. Chapter 4 describes the overall historical overview, working and mainly the state of the art of each part of DSSC and prospects like graphene-based electrodes, inkjet printer dyes, advanced catalysts, sealants, etc., could greet towards green commercialized DSSC.

The hybrid solar cell is more stable with the combination of inorganic and organic layers materials compared to organic solar cells. Chapter 5 describes the methods to prepare SiNWs, reduced graphene oxide, and reduced graphene oxide P Poly (3,4-ethylene dioxythiophene) polystyrene sulfonate composite for their application in the SiNWs hybrid solar cell.

The perovskite-based materials are crystalline because they have long-range atomic order. However, due to defect distribution, the material's properties are deviations from the ideal arrangement regardless of the temperature or other environmental factors. Chapter 6 gives a detailed description of solid oxide fuel cells (SOFCs) energy conversion and storage devices that have the potential to tackle certain environmental challenges while also reducing resource consumption.

The trends towards the lowering of the solid oxide fuel cell (SOFCs) operating temperature, however, many research activities have been focusing on developing new electrodes, electrolytes, and metallic interconnects; this presents a whole new matrix of possible material interactions. Chapter 7 gives details description of multiphase composite cathode materials such as perovskite magnetite metal oxides, particularly lanthanum magnetite materials for the recent advancements as well as its challenges.

Silicon nanowires and graphene oxide serve as promising candidates for application in photovoltaics due to their excellent properties. Chapter 8 discusses recent studies in the silicon nanowire/graphene heterojunction and their application in solar cell applications.

Oxide-ion conductor solid electrolytes are useful for intermediate-solid oxide fuel cell (IT-SOFCs) applications. Chapter 9 demonstrated details of synthetic Ce-doped IT-SOFCs materials and their properties.

Graphene has the potential to improve the energy and power density of electrochemical energy storage devices such as lithium-ion batteries, supercapacitors, fuel cells, and solar cells. Chapter 10 describes various applications of graphene materials in lithium-ion batteries, supercapacitors, fuel cells, and solar cells.

Recently, cathode materials for lithium-ion batteries (LiBs) have improved in terms of energy and power density, capacity retention over multiple cycles, and safety. Six classes of intercalation compounds including layered and spinel oxides are compounds belonging to the olivine, tavorite, silicate, and borate class. Chapter 11 describes different aspects of improving the overall performance of cathode materials, for instance, improving the conductivity by coating with carbon/increasing the surface area by decreasing the size of particles are found to be advantageous.

Thanks to staff of Springer for making the publication of this book possible. We sincerely hope that it will inspire researchers/scientists to explore the potential in the diverse upcoming research fields.

Imphal, India

Dr. Bibhu Prasad Swain
bpswain@nitmanipur.ac.in

Contents

About the Editor

Bibhu Prasad Swain is currently an associate professor at the department of physics, National Institute of Technology (NIT) Manipur, India. He obtained his B.Sc. Physics (Hons.) from Utkal University, Bhubaneswar, and M.Sc. (Physics), M.Tech. (Materials Science), and Ph.D. from NIT Rourkela, Barkatullah University, and Indian Institute of Technology (IIT), Bombay, respectively. His primary areas of research interests include high bandgap semiconductors thin films for device applications, biocompatibility coating for artificial heart valve coating and stent applications, silicon and carbon-based alloys of nanostructured materials, and titanium nitride-based mechanical challenging coating applications. He has published more than 125 papers in respected national and international journals. Dr. Swain received the Japan Society of Promotion of the Science (JSPS) Fellow at the National Institute of Advanced Industrial Science and Technology, National Research Foundation Fellowship University of Cape Town, and Brain Korea 21 Fellowship Seoul National University. He is also listed in top 2% Indian scientist in the world by Stanford University for the year 2020.

Chapter 1
Proton Conductors: Physics and Technological Advancements for PC-SOFC

D. Vignesh and Ela Rout

1 Introduction

Energy is a ubiquitous entity that dictates countless processes in the universe. Globalization and population growth impose alarming effects on the useable energy sources for sustainability. Rapid energy depletion impels researchers and practitioners to introspect alternate energy sources, storage and conversion technologies. Intermittent, convertible and useable forms of energy demand appropriate technological advancements and fuel cells operate as a valuable contribution to efficient ecological energy production. Distinct fuel cells are categorized based on the electrolyte energy materials and the diffusing charge carriers, respectively [1–3]. However, of the existing fuel cell combinations, solid oxide fuel cells (SOFC), proton-conducting solid oxide fuel cells (PC-SOFC) and polymer electrolyte membrane fuel cells (PEMFC) are widely surveyed due to considerable chemical and electrical efficiency [4, 5]. From the technological prospectus, the electrochemistry in fuel cells is governed by the individual components amongst which electrolyte serves as the core constituents for different charge chemistry. In other words, the physics of energy materials becomes a critical factor to achieve optimized performance. For instance, the high operating temperature (800–1000 °C) of SOFC enforces material degradation and low start-up time and lifetime due to the low thermal, chemical and mechanical stability of the bulk electrolyte [6–8]. Constructive material architecture in terms of proton conductors such as acceptor-doped cerates and zirconates though illustrates improved functional efficiency yet lags in trade-off characteristics. The persistent inverse interplay between chemical stability and proton conductivity under diverse atmospheric constraints curtails their commercial applications and state-of-the-art technological devices [9–11]. As a consequence, it is evident that vast advantages accompany few vital limitations.

D. Vignesh · E. Rout (✉)
Department of Physics, Birla Institute of Technology, Mesra, Ranchi 835215, India
e-mail: elasinha@bitmesra.ac.in

B. P. Swain (ed.), *Energy Materials*, Materials Horizons: From Nature to Nanomaterials,
https://doi.org/10.1007/978-981-99-3866-7_1

Recent research reveals acceptor defects escalate the protonation and chemical stability individually via increased oxygen vacancies and electronegativity of the host composition. However, discrete acceptor characteristics constitute sub-categorical drawbacks in terms of structural phase transition, microstructural impedance and proton trapping effect. Scherban et al. [12] reported structural phase transition in trivalent-doped $BaCeO_3$ (BCO) via Raman spectroscopy. The outcomes display a decline in the number of active modes (24 − 4) due to orthorhombic–tetragonal structural change with increasing Nd^{3+} concentrations. The evolution of Raman active mode at 625 cm^{-1} for 5% Nd^{3+} shifts towards the ideal cubic phase at 10% Nd^{3+}, respectively. High symmetric structures with ideal O–O and B–O distance forbid long-range proton transport due to a large activation barrier and foist delayed proton diffusion (proton trapping effect). Takeuchi et al. [13] in analogy briefed structural phase transitions under distinct Y^{3+} concentrations in $BaCeO_3$ against diverse atmospheres. The outcome highlights the chemical instability of doped $BaCeO_3$ with experimental neutron diffractograms. The structural phase transition from orthorhombic (Pmcn) to rhombohedral (R3c) between $0 \leq x \leq 0.10$ and the co-existence of monoclinic (I2/m) phase under O_2 and 4% H_2 atmospheres between $0.15 \leq x \leq 0.3$ inclines the chemical distress. However, the rational study by Rajendran et al. [14] presents a positive role of tri-doped $BaCeO_3$ and $BaZrO_3$ ($BaCe_{0.5}Zr_{0.2}Y_{0.1}Yb_{0.1}Gd_{0.1}O_3$ (BCZYYbGd)) PCs on the optimal chemical stability against different atmospheres. The results showcase operational stability for 200 h in 50% vol argon (Ar) environment. It is thus conclusive that lower dopant concentration and an overall optimized energy materials forms the backbone for futuristic technologies. The current study highlights the material chemistry and physics of proton conductors for low- and intermediate-temperature fuel cell technology. The study as a result incorporates the role of acceptor substituents for positive and negative impacts on the electrochemical performance with reduced charge kinetics (proton trapping effect). The chapter finally dictates diverse strategies employed to exemplify and mitigate the trade-off parameters.

1.1 Technological Prospects of Proton Conductors

Ceramics are non-metallic, crystalline and inorganic materials with innumerable applications in diverse sectors. Traditional artifacts utilize elementary ceramic materials, while modern-day technologies incorporate complex ceramic matrices. The substrate of which all lies with the befitting choice of the host material. From the material's perspective, PCs and their allied compositions authorize compatible chemical, mechanical, thermal, electrophysical, optical and dielectric characteristics due to supportive microstructures [15–17]. Spectrum proton-conducting material engineering has flourished with the consequent revelation of multiple domains. The physical and chemical characteristics of PCs inevitably rely upon their structural and microstructural features. The underlying platform of solid-state chemistry of distinct energy materials illustrates an interrelation between their properties and structural framework. The PCs in general are either utilized as the host or as a reagent with a distinct class of materials. Novel technologies using standalone and complex proton-conducting oxides are a component of general displacement towards green and sustainable chemistry which assists to the transition towards pollutant-free and reduced carbon footprint with desirable energy sources [18, 19]. In the analogical context, hydrogen constitutes a compelling and economical energy source with abundant natural and anthropogenic synthesis. Besides suitable energy resources, appropriate and green energy materials become an imperative requirement for technological advancements. Proton conductors as a consequence are multifunctional energy materials with desirable materialistic features to serve as smart materials. Recent introspections claim the new gateway of PCs as a potential candidate for isotope separation in the fusion reactor system [20–23], photocatalyst [24, 25], fuel cells [26–29], sensors and actuators for defense applications [30, 31], microwave dielectric resonators [32, 33] and many others, respectively. Figure 1 demonstrates the utilization of smart ceramic proton-conducting composites for advanced applications in diverse domains:

Typical and double perovskite proton conductors offer leverages in the context of fuel cell applications with high electrochemical activity and ionic conductivity [34, 35]. The inherent bond flexibility among the perovskite PCs generates low activation pathways for unimpeded proton transport. The higher operational efficiency of PC-SOFC attributes due to the lighter effective mass of protons over the oxide ions. Acceptor-doped PCs escalate the protonation and the resultant proton conductivity. Gilardi et al. [36] highlight the impact of dopant-host ionic radii difference

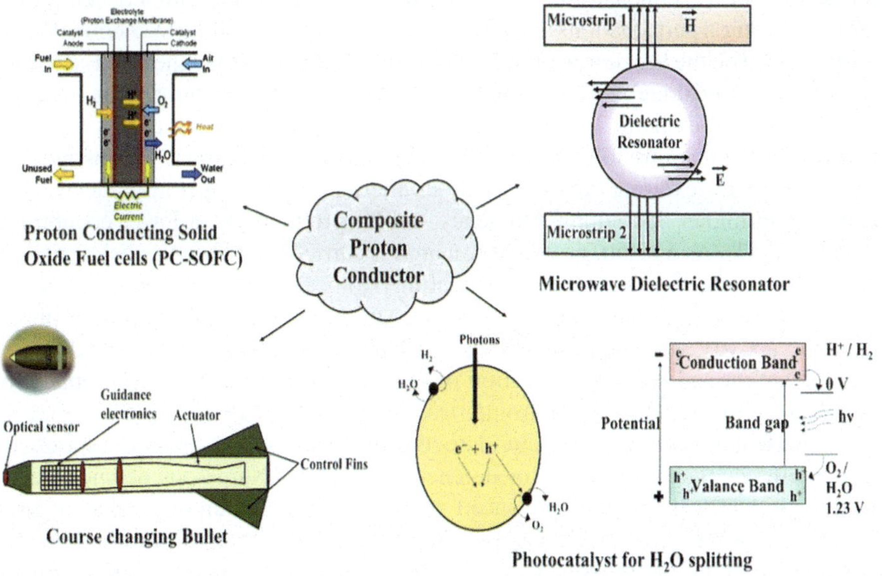

Fig. 1 Multifarious applications of composite proton conductors in diverse domains

on the electrophysical behaviour of the overall composition. The study acquaints the nature of relevant acceptor impurity to accomplish dopant-host compatibility and amplify the resultant proton conductivity against diverse atmospheric constraints. The outcome presents a revamped proton conductivity with acceptor substitutions over their standalone counterpart. Meanwhile, the use of smaller and highly electronegative acceptor substituents (Ga^{3+} ($r = 0.62$ Å), Al^{3+} ($r = 0.53$ Å)) revises the chemical stability due to homogeneous and large grain size distribution while the large acceptor substituents (Y^{3+} ($r = 0.90$ Å), Gd^{3+} ($r = 0.93$ Å)) aid the electrophysical behaviour due to heterogeneous and smaller average grain size. A recent study by Vignesh et al. [9] as a continuum established the concerns over the defect-induced proton trapping effect in $BaCeO_3$ and $BaZrO_3$ PCs. The study dictates diverse factors responsible for delayed proton diffusion through the bulk electrolyte. Besides, the author concludes with an optimal doping range to lie between $x = 0.20$–0.25 to accomplish desirable properties which vary with distinct trivalent impurities.

The ecological energy source is equally complemented by reliable energy conversion technology. In accordance with former discussions, hydrogen forms the basis for compatible fuel and energy source. The usable form of hydrogen among the abundant availability is largely disproportionate. As a result, photocatalytic water splitting into H_2 and O_2 aids a complete utilization of solar radiation to synthesize hydrogen fuel [37–40]. Conventional photocatalyst illustrates a lower absorption spectrum and encounters lower functional efficiency. Yuan et al. [41] examined the photocatalytic response of $BarO_3$ PCs upon UV irradiance. The non-metallic Sn^{4+} doped composition showcases successive excitations and de-excitations with a surge in H_2 and O_2 chemical kinetics. The rate of H_2 productivity enhances with a consequent rise in Sn^{4+} concentration up to a threshold. Elevated doping concentrations induce structural perturbations via multiphase co-existence leading to chemical instability. However, the low-temperature process sets a backdrop to higher photocatalytic activity. On a positive note, the low energy and temperature demand as accessed using the thermodynamic free energy (ΔH and TΔS), respectively, assists extensive plausible permutations for material architecture. On the contrary, exhausting cost and demand for external catalysts with an alternate high-temperature electrolyzer cell present higher operational efficiency and H_2 productivity at lower economical facets [42, 43]. As a result, profound research is carried out to accomplish economically optimized composite proton-conducting photocatalysts with desirable reaction kinetics. However, Gundeboina et al. [25] 2020, highlight the limitation of photocatalytic activity via visible radiations. Stringent absorption spectra, rapid charge recombination and reduced utilization of photoelectrons and holes accomplish the significance to engineer the electronic structure of the host PCs [44–46].

A gradual transit from structural to the mechanical and optical response of PCs overhauls their utilization for defence application. Numerous defence organisations intend to strengthen advanced weaponry and armour to combat security breaches and preserve nation's integrity. Armours and bullets constitute the elemental component among progressive defence technologies. The value of ceramic armour grew steeply post the Vietnam conflict. Breakthroughs in terms of lightweight ceramic and ceramic composites are enforced for modern-day bulletproof vests and shields [47–50]. On the contrary, a tailored heavy ceramic matrix forms the platform for state-of-the-art shields and armours to obstruct major ballistic impacts. Ceramic derivatives such as $BaZrO_3$-based PCs operate as the potential substrate for multilayer composite defense systems. The front face of the armour composes of a rigid ceramic matrix with an energy-absorbing backplate to elevate the mechanical strength of the projectile. The weight distribution of ceramic composites differs among diverse shields [51]. Bulletproof automobiles are designed with heavyweight multilayer proton-conducting composites to combat catastrophes and large ballistic missiles [30]. The approaching projectile is impeded by the hard ceramic shards which pacify the impact. As a result, the ceramic backplate mollifies the overall kinetic energy of the projectile and imposes plastic deformation on the shield. Smart $BaZrO_3$-based proton-conducting composites are thus engineered to cope and counterattack distressing situations with immediate effect. In compliance, Zhang et al. [52] proposed over $(_{0.97-x})(K_{0.48}Na_{0.52})Nb_{0.965}Sb_{0.035-0.03}Bi_{0.5}(K_{0.18}Na_{0.82})_{0.5}ZrO_{3-x}BaZrO_3$ ($x =$ 0–0.06) complex proton-conducting composite for high responsive actuators in smart bullet system. The outcome of the study reveals a high piezoelectric coefficient and bipolar strain (29%) with low hysteresis (13.8%) at a sufficient electric field (4 kV/mm). The high piezoelectric response emerges due to the rhombohedral–orthorhombic–tetragonal (R-O-T) and relaxor-ferroelectric phase transition [53–55].

The aggregate of the following section illustrates the extended implication of $BaZrO_3$-based PCs either as a host or as an additive for the smart material framework. Pragmatic characteristics urge researchers and practitioners to invest their resources in the development of existing technology. Among innumerable applications, discrete sectors such as energy a defence are reported to provide futuristic insight and advancements. Material architecture with suitable acceptor impurity assists to bridge the gap between the trade-off properties in energy conversion devices. The search and hence the anthropogenic productivity of hydrogen fuel can be accomplished with higher chemical kinetics via compatible proton-conducting photocatalyst. The tailored electronic structure of the host material offers the scope to utilize maximum plausible UV–Visible em-radiations to generate photoelectrons and holes with delayed recombination. A shift from structural to optical and mechanical characteristics also proportionally varies the appliance of proton-conducting ceramics and ceramic composites. The enhanced responsivity of optical sensors and actuators with a higher tensile strength of proton-conducting ceramic matrix aids to construct smart course-changing bullet system. Such high-end technology befits to preserve the nation's integrity against unintended security breaches. The cumulative technologies and other future devices impose tedious material-oriented limitations which demand profound investigation, experimentation and understanding. The progressive sections deal with one such technological advancement related to energy conversion devices, viz "the energy materials for proton-conducting solid oxide fuel cells".

1.2 Electrochemistry in Fuel Cell Technology

Fuel cells are an elementary example of energy conversion technology that functions based on electrochemical redox reactions across the electrode–electrolyte interface. The characteristic half-reactions at the relevant electrodes aid to power external circuits utilizing flexible and eco-friendly gaseous fuels. The early insight of fuel cells was conceptualized by Humphry Davy during the nineteenth century with radical progress by Christian Friedrich Schonbein. However, a cluster of experiments conducted by William Grove produced electric current using a gas voltaic battery via electrochemistry between hydrogen (H_2) and oxygen (O_2) gaseous fuels over a platinum catalyst [56]. Progressive research interests in fuel cells grew with distinct prototypes powered by hydrocarbons and ammonia. The continuous energy extraction at the expense of economical fuel sparked the scope of sustainability among diverse researchers and practitioners in the scientific community. Besides technological facets, material chemistry was provided with equivalent priority. The search for optimal energy materials for the fundamental and core constituents of a fuel cell (electrolyte) began much earlier. The present study with $BaCeO_3$ and $BaZrO_3$ proton conductors dates back to the early 1960s. Substantial material engineering has been carried out over the years to meet the desirable characteristics. Meanwhile, polymer electrolyte membrane fuel cells (PEMFC) were targeted for low-temperature and small-scale applications. The competency for combined heat and power systems rapidly increased the reliability of micro-combined heat and power (MCHP) units.

From the perspective of electrode–electrolyte interfacial chemistry, pure and gaseous H_2 become a copious and favourable fuel with reusable by-products. The background of electrochemical reactions at the respective electrodes occurs at the molecular scale with H–H and O–O bond dissociations and a resultant O–H bond association. The H_2O by-product articulates via oxidation of H_2 molecules, while the liberated excessive heat energy associates to the energy gradient between bond dissociation and association process, respectively. The powered external circuit is an outcome of the electron transfer between the concerting molecules. Figure 2 represents the electrochemistry in a fuel cell technology:

A sizeable separation between the gaseous reactants becomes a necessary criterion to accomplish electron transfer to large-length scales. Because of ionic conductivity, proton conductivity was first corroborated due to the encountered isotopic effect of the electrical conductivity of TiO_2 [57]. Meanwhile, Stotz and Wagner expanded the Kröger–Vink notations for point defects to accommodate protonic defects [58]. The pursuit of efficient proton-conducting fuel cells continues with diverse material architecture to subsidize novel challenges regarding proton chemistry and long-range bulk proton diffusion. State-of-the-art acceptor-doped proton conductors showcase compositional-induced structural phase transition under atmospheric constraints. The perturbed proton conductivity and delayed proton diffusion are significant factors that indicate microstructural pitfalls. The impeded proton transfer and dopant-induced structural variability were first reported in the early twentieth century. The persistent

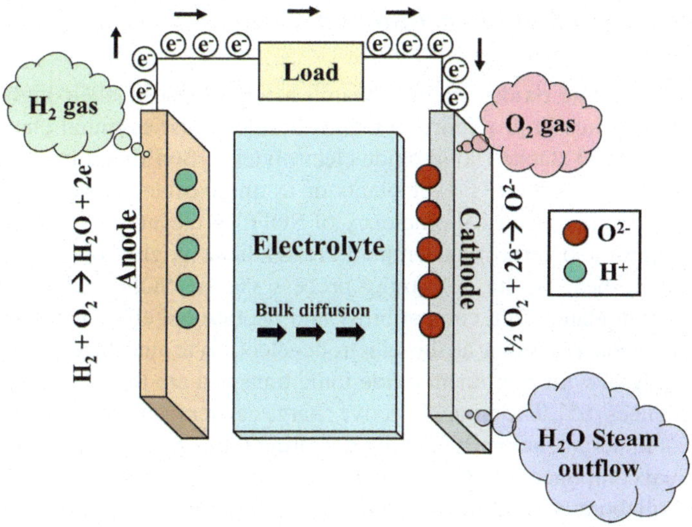

Fig. 2 Schematic representation of fuel cell electrochemistry

trade-off between chemical stability and proton conductivity illustrates a concerning factor among diverse proton conductors to hinder large-scale commercial appliances.

A recent study by Zhang et al. [59] presented the chemical instability in $BaCeO_3$ under reducing atmospheres. The conclusive evidence justifies variable basicity (electron density) around the proton-coordinated oxygen atom and hydration limit of B-site cerium atoms. The attempt of different acceptor substitutions presents distinct supportive strategies. The replacement of host with a more electronegative small-sized trivalent impurity (In^{3+}(1.78) and Ta^{3+} (1.5)) embraces the chemical stability while attenuating long-range proton transport due to dopant-proton coulombic interactions. Low electronegative large-sized acceptor impurity (Y^{3+}(1.22)) on the contrary escalates the proton conductivity at the expense of considerable structural distortions. However, among multiple acceptor impurities, Y^{3+}, Gd^{3+} and Dy^{3+} are cited as more favourable substitutes with the comparable difference between trade-off properties. Apart from acceptor substituents, variable thermodynamical parameters constitute a proton trapping effect. The plausible solution to the afore-limitation is the elevated operational temperatures of the fuel cells. The high-temperature regimes provide additional energy to counterbalance the activation barrier and execute the de-trapping effect. Although the inverse interplay between proton conductivity and chemical stability has invoked novel material modelling constraints, consistent strategies with advanced material processing techniques are introspected to achieve higher operational efficiency and produce smart materials for advanced energy applications.

1.3 SOFC System and Technology

Numerous perovskite oxides excite research activities due to their high electrical conductivity that lies in proximity to metals besides other essential characteristics [60]. Multilayer and stacked solid oxide electrolytes in SOFCs motivate contemporary technology to integrate power plants or as an auxiliary power units in satellites [61, 62]. Higher electrical efficiency of SOFC at elevated temperatures arises due to operating and reforming temperature gradient. Higher operating temperatures contribute heat for the reforming process via the SOFC exhaust heat [60]. SOFCs accommodated high compatibility and operational efficiency through electrochemical redox reactions at the electrode–electrolyte interface [63, 64]. Solid oxide electrolytes with prominent oxide ionic transport are highly compatible with SOFC appliances [65]. The nature of active charge carriers relies completely upon the electrolyte's material chemistry and the operating temperatures. As a consequence, the individual component of SOFC, therefore, becomes essential to meet the requirements for collaborative and ideal functioning. Highly porous cathodes for oxygen reduction reaction (ORR) astride the triple phase boundary (TPB), conductive bulk electrolyte and oxidizing feature of the anode are some of the pre-requisites for electrochemical release of electrons. The electrode–electrolyte interfacial compatibility plays a dominant role on the charge dynamics and fuel-oxidant distinction [66]. Acceptor-doped solid oxide electrolytes are examined to enhance the conductivity of the resultant composition. Nonetheless, ionic conductivity and doping concentrations are monotonically related. The conductivity gradually surges with a consequent rise in dopant concentration up to a threshold and decreases thereafter at higher doping concentrations owing to dopant-induced charge trapping centres and coulombic interactions according to the experimental outcomes of Anjaneya et al. [67] and Steele et al. [68], respectively. In principle, a solid oxide electrolyte is sandwiched between the two electrodes and constitutes bulk oxide ions diffusion. The gaseous H_2 at the anode oxidizes into a surface-adsorbed proton and liberates two electrons. The oxidant at the cathode acquires electrons through the external circuit. The resultant oxide ions diffuse through the bulk electrolyte and form an H_2O by-product. The steam (high-temperature H_2O by-product) is extracted through the water outlet of SOFC [69]. The individual redox reactions at the respective electrodes are illustrated in Eqs. 1 and 2 as follows:

$$\text{Anode: } H_2 + O^{2-} \rightarrow H_2O + 2e \tag{1}$$

$$\text{Cathode: } 1/2O_2 + 2e \rightarrow O_2^- \tag{2}$$

Figure 3 demonstrates a conventional design of planar SOFC.

Besides fuel flexibility, fuel proficiency actuates the electrochemical kinetics across the electrode–electrolyte interface. As a result, hydrogen (H_2) and carbon monoxide (CO) are two radically used gaseous fuel inputs. Although SOFC supports high ionic conductivity, large conducting solid oxide electrolytes establish O_2 leakage

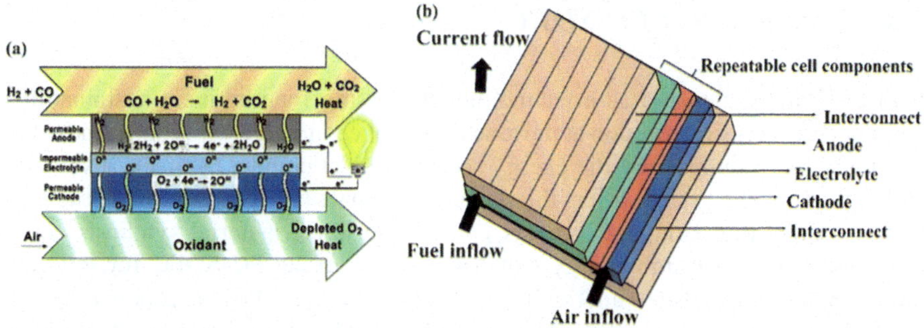

Fig. 3 Schematic representation of SOFC, **a** functioning, **b** cell components

with voltage loss and hence reduce the power output. Zirconium-stabilized electrolytes constitute considerable chemical stability alongside structural phase transitions with monoclinic, tetragonal and cubic as three comprehensive polymorphous phases [61]. Although SOFCs afford a counterbalance for conventional power sources, perpetual technological deficiencies hinder their long run under ambient constraints. The high-temperature SOFCs are advantageous and meantime supervises critical limitations. High operating temperatures as an advantage contribute abundant thermal energy for desirable electrochemistry devoid of external catalyst. On the contrary, high operating temperature also favours electrolyte degradation and longer start-up time. An attempt to curtail the operating temperature perturbs the composition's microstructure with profound grain and grain boundary impedance at low and intermediate temperatures [70]. The thin-film-deposited solid oxide electrolytes are currently exercised for chemically stable oxide ion conductors at intermediate temperatures. This shortens the charge transfer distance and combats the electrolyte's microstructural impedance and interfacial electronic properties.

From dimensional variability, SOFC reigns in different geometries. Planar fuel cell anchors traditional architecture with electrolyte sandwiched between two electrodes. Tubular fuel cells on the other hand allow fuel inflow through the inner tube and oxidant through the outer tubular region [71]. However, the typical drawback with planar SOFCs is fuel-by-product interference. Additionally, the H_2O by-product at the anode motivates a tendency of fuel dilution at the adsorbed anodic surface and obstructs steam reforming reaction [72, 73].

Although theoretical evaluations and proximity differ from experimental circumstances, material purity and processing techniques precisely evaluate the material's structural and dynamical behaviour. In response to such key challenges, technological advancement in terms of proton-conducting solid oxide fuel cell has attracted immense attention which is dealt with in the next section (Sect. 1.4).

1.4 Foundation of PC-SOFC

Fuel cells are recognized as a technological blessing for imminent sustainability. Unimpeded energy conversion at the assessment of economical and eco-friendly fuel flexibility has monopolized the energy sector. Amidst combat between advantages and limitations, PC-SOFC offers the system's compactness and enhanced efficiency by forbidding afterburn requirements [74, 75]. Discrete material synthesis conditions and the use of high-energy exhaust gases advocate PC-SOFC for low- and intermediate-temperature functioning and power harvest. Proton-conducting bulk electrolytes, viz the proton conductors (PCs), display high ionic conductivity at moderate temperatures due to lower activation energy with respect to the solid oxide electrolytes [76, 77]. Established literature claims plausible proton transfer pathways via the lowest possible activation energy for a strong O–H bond network. Bulk diffusion, grain boundary diffusion, diffusion via the open pores in the adsorbed H_2O layer and proton-fortified layer beneath the oxide interface are four predominant proton transfer pathways that devote to the total ionic conductivity. At operating temperatures above 500 °C, the proton conduction follows bulk diffusion due to the narrow bulk impedance offered relative to the grain boundary core. The elemental causality of bulk diffusion emerges due to insufficient H_2O adsorption across the grain boundaries at elevated temperatures. While at intermediate temperatures (< 500 °C), residual open porosity evolves as a competent factor for proton conduction due to enhanced physisorption of water [78]. Meanwhile, proton diffusion is more prominent under the H_2 atmosphere while electron holes and oxygen ions are assertive under wet air atmospheres [79]. These fuel cells operate at lower temperatures due to the higher mobility of protons over oxide ions. However, proton conductivity decreases at elevated temperatures (> 800 °C) due to reduced H_2O solubility in the oxide with a lower protonation. In contrast to solid oxide electrolytes, PCs concede complete pursuit of hydrogen (H_2) composed via H_2O at the cathode to preserve the overall cell's efficiency [78, 80]. Constructive material modelling on barium cerate ($BaCeO_3$) and barium zirconate ($BaZrO_3$) PCs are still in progress to embrace the proton conductivity with optimum chemical stability at reduced operating temperature. However, one of the vital drawbacks of PCs is the synthesis of chemically stable and high-conducting electrolyte material. An illustration of PC-SOFC is shown in Fig. 4.

PC-SOFC showcases long-range proton mobility through the bulk electrolyte via oxygen vacancies at intermediate temperatures (300–700 °C). Protonic defects within PCs emerge via dissociative H_2O absorption at the electrolytic surface [77]. Proton conduction as a result confides over oxygen vacancy concentration which generates reactive sites for ionic diffusion [81]. Grotthuss mechanism or proton jumping is the process by which an 'excess' proton diffuses through the hydrogen bond network of water molecules by the formation of bonds involving neighbouring molecules. The proton transfer between two neighbouring oxygen ions can occur within the same octahedra (intra-octahedral) or between two adjacent octahedra (inter-octahedral)

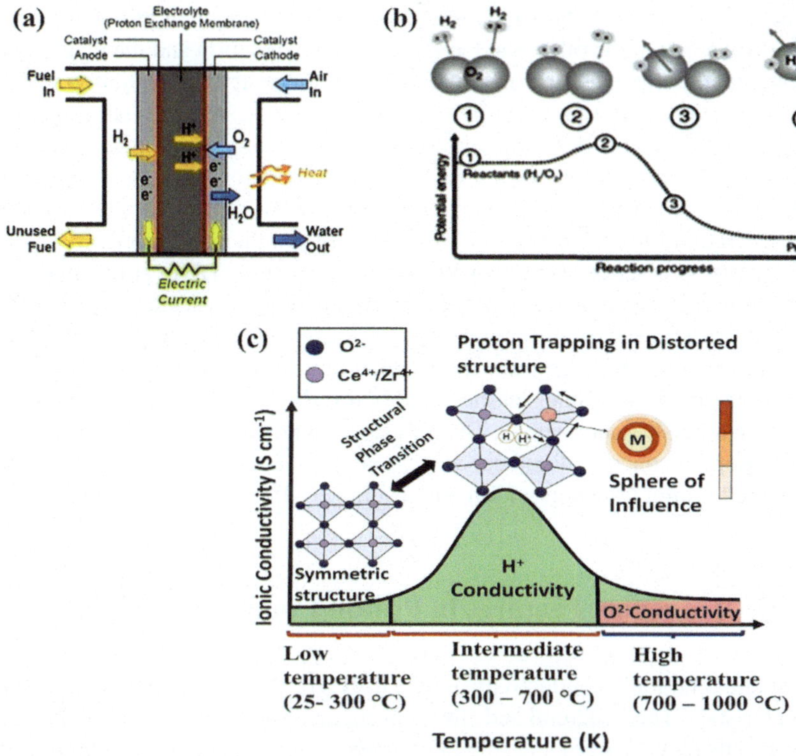

Fig. 4 Representation of, **a** proton-conducting solid oxide fuel cell (PC-SOFC), **b** electrochemistry in PC-SOFC, **c** operating regimes of PC-SOFC

proton transfer in distorted symmetry such as orthorhombic proton conductors. The proton hopping step is foreseen as the rate-limiting step [72, 82].

Additionally, proton diffusion involves rotational and translational propagation around the during the hopping mechanism. The rotation around the proton-coordinated oxygen atoms accompanies by a subsequent hopping to the neighbouring oxygen ion through a hydrogen bond network of the water molecule. In order to migrate, the ions must be associated with hydrogen-bonded clusters; the stronger and more extensive the cluster, the faster is the resultant migration. Stronger hydrogen bonding reduces the O–O distance and eases the closest approach for the proton transfer.

Proton concentration is a function of temperature, under hydration tendency and hydration partial pressure (pH_2O), respectively. Cordial hydration is expressed via the incremental change in exothermic hydration enthalpy (ΔH_{hyd}), feebly negative hydration entropy (ΔS_{hyd}) and negative hydration Gibbs free energy (ΔG_{hyd}) [83]. Doped PCs impact thermodynamical parameters and dictate the dynamical properties of charge carriers. Low electronegative dopants over the replaceable cation enforce more exothermic ΔH_{hyd}, inclining hydration tendency and proton concentration for diffusion. On the other hand, oxygen vacancy location and the magnitude of defect affiliation illustrate the difference in defect chemistry concentration which alters the hydration energy and proton assembly. Besides chemical and structural instability, delayed proton diffusion (proton trapping effect) additionally perturbs long-range proton conduction. High electronegative acceptor dopants constitute a proton trapping effect via a modified activation barrier for proton hopping and the material's hydration limit. Computational evidence of proton trapping effect around acceptor dopants is asserted using density functional theory (DFT) simulations. A plausible solution involves a surge in the number of oxygen vacancies within the host material. Distinct polarity of dopant-oxygen vacancy can accomplice to curtail proton trapping effect [83]. Besides, the trapped proton averts subsequent proton trapping via percolation channels at the cost of perturbed proton mobility owing to proton–proton repulsive interactions. The fundamental knowledge of dopant characteristics and doping concentration becomes essential to obtain desirable protonation and unimpeded proton transport within the host material.

2 Charge Transport Features in Proton Conductors

The partial and total ionic conductivity among different PCs relies upon their chemical characteristics. The structural and chemical characteristics of proton-conducting energy materials play a paramount role in unimpeded long-range proton and proton defect mobility. Doped PCs showcase considerable proton conductivity under wet H_2 atmospheres at intermediate-temperature regimes. The elementary apprehension for the former response is the formation of stable protonic defects via dissociated absorption of H_2O in the presence of oxygen vacancies. Elaborately, the H_2O in the gas phase dissociates into a proton and a hydroxide ion which fills the available oxygen vacancies, while the former charged constituent forms a covalent bond with the neighbouring oxygen lattice. The overall process is demonstrated in Eq. 3:

$$\mathrm{H_2O} + \mathrm{O}_{\mathrm{o}}^{x} + V_{\ddot{\mathrm{O}}} \leftrightarrow 2\mathrm{OH}_{\dot{\mathrm{O}}} \tag{3}$$

where

H_2O = water molecule

O_o^x = neutral oxygen at the respective lattice site

$V_{\ddot{O}}$ = oxygen vacancy

$OH_{\dot{O}}$ = protonic defect.

The protonic defect formation reaction is predicted to be amphoteric where the oxide ion simultaneously displays the acidic and basic characteristics [84]. However, substantial research establishes a higher exothermic enthalpy of hydration reaction for lower electronegative cationic interactions with the oxygen lattice [85]. The implicit oxygen vacancies within a PC generates via stoichiometric variations between host constituents, while the explicit compositional-induced oxygen vacancies emerge to combat the charge imbalance established by acceptor impurities. Ambient hydration allows the oxygen vacancies to be filled by the hydroxide ions (OH^-) due to H_2O dissociation. Diverse kinds of literatures suggest a dual (positive and negative) characteristic of oxygen vacancy. On one hand, the oxygen vacancies aid stable protonic sites for protonation while on the other offers proton trapping sites. The nature of the afore-behaviour depends upon the concentration of the trivalent impurity. Lower B-site acceptor substitutions ($x < 0.25$) stabilizes the electrophysical property of the host PCs with escalated proton conductivity, while the corresponding counterpart of proton trapping effect dominates at higher doping concentrations with attenuated proton diffusion through the bulk electrolyte [9, 10]. The charged protonic defects as shown in Eq. 3 can only diffuse into the bulk oxide when guided by the counter diffusion of oxygen vacancy. This designates an oxide ion conductivity in the dry state and illustrates a rapport for the chemical diffusion of H_2O. The hydration thermodynamics as a result plays a critical role to optimize the resultant proton conductivity. The equilibrium constant K_w for Eq. 3 is shown in Eq. 4:

$$K_w = \frac{[OH_{\dot{O}}]^2}{[V_{\ddot{O}}][O_o^x]pH_2O} \tag{4}$$

where

pH_2O = hydration partial pressure.

The resultant K_w decreases with more electronegative B-site constituents which preserve the following order (Ce → Zr → Sn → Nb → and Ti), respectively. The hydration partial pressure in Eq. 4 constitutes a key element for large-scale protonation within the host PCs. The free proton concentration as a consequence is a function of temperature and relies upon the extent of pH_2O under incomplete hydration. Cordial hydration on the other hand dictates electrochemical spontaneity via superficial variations in hydration enthalpy (ΔH_{hyd}), feebly negative hydration entropy (ΔS_{hyd}) and Gibbs free energy (ΔG_{hyd}) changes, respectively [83]. Bulk proton diffusion under incomplete hydration in PCs is governed by the Grotthuss mechanism. The Grotthuss mechanism otherwise termed proton hopping is the phenomenon by which an excess proton diffuses through the bulk electrolyte via successive hydrogen bonding with neighbouring oxygen atoms [86]. In a partially hydrated PC, the static H_2O molecule shows marked local dynamics. The step-wise protonation occurs in such a way that an initial proton displaces to a neighbouring H_2O molecule due to dissociated H-bond from the former water molecule. A continuum reorientation of the H_2O occurs to subsequently accept the incoming proton and achieve a defined proton transfer trajectory [87].

The Grotthuss mechanism (as shown in Fig. 5) is conventionally a two-step process following a rotation and translation motion of the excess proton via successive BO_6 octahedra [88, 89]. The principal feature among these is the rotational movement of the proton. The rotational diffusion with low activation barrier is predicted to be more rapid than its translation counterpart and constitutes the rate-limiting step

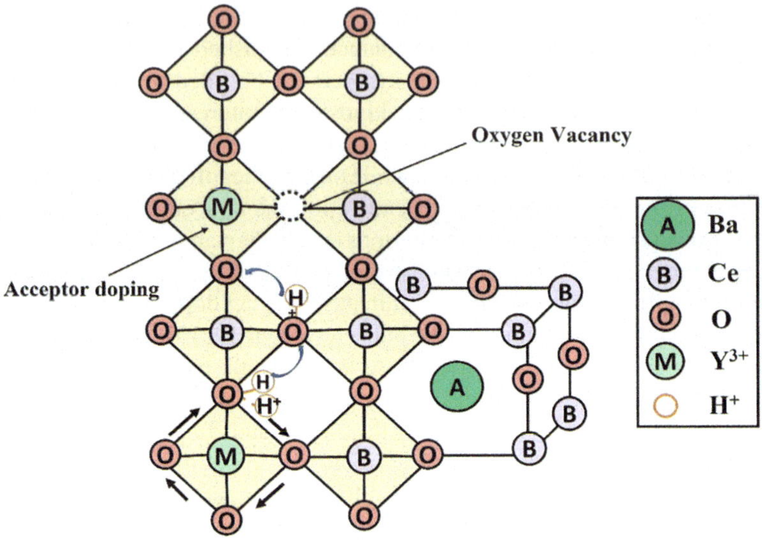

Fig. 5 Protonation and migration between subsequent octahedra

[72, 82]. On the contrary, the red-shifted –OH infrared spectral absorptions illustrate rigid hydrogen-bonded clusters that assist rapid proton transfer reactions over reorientations [76]. The proton diffusion coefficient (*D*) is determined using Eq. 5 shown below:

$$D = \sigma \frac{k_B T}{e^2} \frac{V}{[OH_{\dot{O}}]} \tag{5}$$

where

k_B = Boltzmann constant
E = proton charge
T = operating temperature
V = unit cell volume
σ = proton conductivity.

From the perspective of structural and charge dynamics, structural distortions devote a tremendous effect on the oxygen lattice rearrangement. The inter-octahedral proton transfer pathways become much more favourable in distorted PCs over their intra-octahedral counterpart due to altered O–O bond distance. The proton resides outside the BO_6 octahedra with strong bent hydrogen bonds and aids for proton transfer barrier. The dopant-host coulombic repulsion forms a fundamental factor to the afore-characteristics and forbids linear hydrogen bonds [90]. However, rapid proton diffusion is only plausible with strong hydrogen-bonded cluster. The extent of proton migration increases linearly with the strength of the bonded cluster. A strong cluster curtails the O–O bond distance and establishes low activation proton transfer pathways. Besides, a significant contribution to the activation energy and proton transfer barrier emerges from the B–O bond elongation. As a consequence, H–B bond repulsion may be utilized as an upper threshold for the magnitude of activation enthalpy experienced by participating constituents. Distinct chemical interactions with cations vary the electron densities and hence the proton binding energy of the nearest oxygen atoms. In the case of Y-doped PCs, the activation enthalpy and entropy besides the pre-exponential factor increase with a proportional rise within the host $BaCeO_3$. This constitutes a shallow surge in proton conductivity at elevated doping concentrations. This subsequently favours the antagonistic effects of local ordering with reduced symmetry together with lattice expansion due to dopant-host ionic radius mismatch.

2.1 Standalone and Acceptor-Doped $BaCeO_3$ PCs

Perovskite PCs are a versatile podiums for 3d transition metal catalysts for B-site substitutions. The catalytic activity of PCs emerges due to the exsolved nanoparticles during reduction reactions to escalate the heterogeneous surface catalysis [91]. The early insights of cerium-based PCs began with $BaCeO_3$ and $SrCeO_3$, respectively [92, 93]. Proton assimilation and conduction in doped PCs pursue the Grotthuss mechanism. H_2O vapor contributes interstitial protons to diffuse via the lattice in accordance with the reaction shown in Eq. 6:

$$H_2O + V_{\ddot{O}} \rightarrow O_O^x + 2H_i^+ \quad (6)$$

where

H_2O = steam

$V_{\ddot{O}}$ = oxygen vacancy

O_O^x = lattice oxygen

H_i^+ = interstitial proton.

Protonation and oxygen vacancy within doped PCs are associates with a nonlinear behaviour of acceptor doping concentrations. Facile $Ce^{4+} \rightarrow Ce^{3+}$ reductions of cerates with implicit oxygen vacancies in the Ce^{3+} neighbourhood are responsible for prominent catalytic activity and find their significance as favourable fuel cell components. The amplitude of oxygen vacancy enhances with a consequent rise in the doping concentration up to a threshold beyond which a steep decline accompanies the number of vacancies with a subsequent hike in doping concentrations. Such an attribute emerges due to the A-site substitution tendency of most acceptor substituents at elevated concentrations. Compatible solution energy of acceptor substituents for A and B-site substitutions enforces their dominance with A-site cationic replacement and the resultant oxygen vacancy consumption. Pre and post-humidified protonation and oxygen vacancy dependence are illustrated in Eqs. 7 and 8 as follows:

$$[V_{\ddot{O}}] = 0.5M_B^{3+} \quad (7)$$

$$[H^+] = [M_B^{3+}] \quad (8)$$

where
M_B^{3+} = B-site trivalent impurity.

Standalone and derived $BaCeO_3$-based PCs display adequate proton conductivity against H_2 and hydrated atmospheres over other Ce-based compositions at their operating regimes (400–800 °C) [94, 95]. The higher hydration tendency of Ce^{4+}, the larger cationic radius of Ce and Ba cationic positioning within the lattice and the octahedral tilt of the lower symmetrical structures validate the considerable proton diffusion through the bulk $BaCeO_3$ PCs from the structural viewpoint. Large lattice-free volume for unimpeded proton migration accompanies lattice expansion due to large-sized acceptor substitutions (Y^{3+}, Gd^{3+} and Dy^{3+}) [96]. Supplementary reasons incorporate low proton-dopant inductive effect and grain boundary resistivity towards overall ionic conductivity [97]. Material architecture using organic cations aids layered, multi-structural and chained $BaCeO_3$ derivatives to refine physical characteristics [98, 99]. However, defects are categorized based on dimensionality, purpose and defect characteristics. Among intrinsic, extrinsic and charge transfer, the former two strategies depend on the charged defect evolution within the host crystal lattice. Charge transfer, on the contrary, unfolds due to the commutation of charge carriers among the parent and external constituent [100]. Lattice-contained interstitial protons due to extrinsic doping constitute OH-defects due to affiliations with adjacent oxygen atoms [101]. Besides dimensionality-based defect classification (point (0D), line (1D) and surface (2D) defects), electronic defects (holes (e^+) and electrons (e^-)) occur within the crystal system. Distinct size or charge distribution, lattice vacancies, atomic deviation from the elemental entity for defect genesis. Such doping strategies inflect the compositional chemistry and thermodynamical attributes apart from temperature-induced structural perturbations. Standard compositional stoichiometry ($BaCe_{1-x}M_xO_{3-\delta}$ ($\delta = 0.5$, M = rare-earth impurity)) is preserved to counterbalance the structural and microstructural instability.

Jacobson et al. [102] speculated temperature-dependent structural phase transitions. Standalone $BaCeO_3$ displays orthorhombic (Pmcn, space group # 62) structure at room temperature [13, 103]. The outcomes report three plausible polymorphous phases with inclining temperatures. The primary allotropic phase transition occurs at 290 °C with space group fluctuations (Pmcn – Incn) within the implicit orthorhombic crystal structure. Such space group fluctuations arise due to different crystal orientations and basal plane deviations with respect to pseudo-cubic directions and crystallographic axes. However, structural variations (orthorhombic → rhombohedral) first appear at 400 °C and progress to face-centred higher symmetry (cubic, Pm-3m) at 900 °C, respectively. Such electrochemical energy conversion technologies significantly rely on explicit substitutions to accomplish desirable thermoelectric properties. The outcomes of the study by Matsumoto et al. [104] on the other hand disclose the inverse interplay between chemical stability and proton conduction in acceptor-doped $BaCeO_3$. The trade-off attribute of the host PC makes it vulnerable to reactive atmospheres. As a result, in case of PC-SOFC, acceptor-doped $BaCeO_3$ PCs are heavily investigated and improvised to achieve adequate electrophysical characteristics. Münch et al. [105] in compliance introspected the electrochemistry of different acceptor (M = Y^{3+}, Yb^{3+}, Nd^{3+}, $Sm^{3+}$and Gd^{3+}) doped $BaCeO_3$ ($BaCe_{0.9}M_{0.1}O_{3-\delta}$) PCs. The results display a higher protonic conduction by Y^{3+}-doped $BaCeO_3$ with a considerable proton transport number. The consequence of such a synergetic effect lies in the minimal B/M size difference and hence the compositional-induced structural perturbations. Y^{3+} (1.22) with low electronegativity forbids large dopant-proton coulombic interactions and restricts steep proton traps for long-range proton transport. An analogical study over compositional-induced structural phase transition was conducted by Scherban et al. [106]. The study presented the influence of rising Nd^{3+} concentration upon the phase shifts (lower → higher symmetry). The results claim a decline in the number of Raman active modes (24 to 4) with a linear surge in the doping concentrations. Table 1 dictates a decline in Raman active modes with successive phase transitions:

The pure $BaCeO_3$ illustrates orthorhombic symmetry and shifts towards tetragonal at $x = 0.05$ due to, non-centrosymmetric Ce^{4+} displacement, O–Ce–O stretching and BO_6 antisymmetric bending, respectively. The composition attains the ideal structure (cubic) at an elevated doping range ($x = 0.1$). The study prioritizes the magnitude of structural chemistry (A–O and M–O bond distance) as a consequence of doping concentration. As a result, the dopant characteristics become a prominent pre-requisite prior to dopant substitution.

The overview of assimilated outcomes dictates an inverse interplay between chemical stability and proton conductivity in diverse atmospheric constraints. The consequent attribute emerges due to compositional- and temperature-induced structural phase transition with variable material chemistry. The protonation and oxygen vacancies are the impact of host-dopant charge imbalance and preserves nonlinearity with doping concentration. While B-site substitutions assist oxygen vacancies, A-site occupancy consumes the available oxygen vacancies at elevated doping concentrations. The tendency of large-sized acceptor dopants displays an overlap of solution

Table 1 Decrease in number of Raman active modes with phase transition (table from Scherban et al. [106], *Solid State Ionics* **1993**, Pg 5)

Crystal structure	Factor group	Space group	Z	Predicted fundamental vibrational modes	Number of Raman active modes
Orthorhombic	D_{2h}	D_{2h}^{16}	4	$7A_g + 8A_u + 5B_{1g} + 7B_{2g} + 5B_{3g} + 9B_{1u} + 7B_{2u} + 9B_{3u}$	24
Tetragonal	D_{4h}	D_{4h}^{5}	2	$A_{1g} + A_{2g} + 2A_{1u} + 3A_{2u} + B_{1g} + B_{2g} + 2B_{2u} + E_g + 7E_u$	4
Cubic	O_h	O_h^1	1	$3F_{1u} + F_{2u}$	0

energy for A-site and B-site partial substitutions and introduces electrophysical fluctuations. Hence, large-sized and low electronegative acceptor substituents like Y^{3+}, Gd^{3+} and Dy^{3+} are promising candidates to accomplish desirable outcomes.

2.2 *Standalone and Acceptor-Doped $BaZrO_3$ PCs*

Zriconates are familiar with their high chemical and mechanical integrity. A low Zr–O and Ba–O bond distance in zirconates over their cerate counterpart establishes high chemical stability [76]. However, structural multiplicity in $BaZO_3$ emerges due to rigid Zr–O bonds. Defects and imperfections within the host crystal lattice evolve through the oxygen octahedra and activate self-trapping deviations and redistributions. The synergetic effect imposes suspended charge diffusion through the bulk [80, 107]. The depressed proton conduction in $BaZrO_3$ over $BaCeO_3$ emanates due to the grain-blocking events. Pronounced Zr^{4+} within the host composition introduces the refractive nature of the grain boundary. As a result, the dense microstructure of $BaZrO_3$ appeals to high sintering temperatures (~ 1500–1600 °C). In view of material chemistry, high sintering temperature escalates the Ba vacancy via BaO vaporization [108–111]. Acceptor dopants alongside sintering additives is a typical route to configure material characteristics. $BaZrO_3$ endorses considerable point defects. B-site acceptor substitutions implicitly generate oxygen vacancies and delivers energetically favourable sites for protonation and active charge transport. Preferential order PCs ($BaCeO_3$ > $BaZrO_3$) lie on their respective hydration limit and defect sustainability. Chemical lattice expansion as a consequence of hydration is determined using a dilatometer and high-temperature XRD analysis [112]. Defect incorporation and subsequent hydration impart lattice strain and constitute a structural disorder.

The computational study by Jeong et al. [113] illustrates befitting proton transfer pathways in Y-doped $BaZrO_3$ through low symmetry (orthorhombic) structures. The plausible and preferential proton transfer path depends solely on the rotational and translational activation barrier. The activation energy materializes as a consequence of the hydrogen bond stretching to an optimum distance between the proton donor and acceptor, respectively. Two prominent configurations (obtuse and reflex) within a distorted structure are shown in Fig. 6 as follows.

Obtuse and reflex angle configurations are distinguished based on the magnitude of B–O–B tilt in the proton-coordinated octahedra. Outward stretching of oxygen atoms in the horizontal plane of the proton-coordinated octahedra contributes to obtuse angle configuration. The reflex angle configuration on the contrary corresponds to an inward bending of the former scenario. As a result, distinct proton transfer permutations generate within the low symmetric structures. Plausible proton migration paths among different configurations are displayed in Fig. 7.

From Fig. 7, proton transfer between two distinct configurations occurs via intra/inter-octahedral pathways. In view of structural diversity in low symmetry $BaZrO_3$, intra-octahedral proton transfer demands higher translational barrier over their rotational counterpart due to increased O–O bond distance between horizontal and

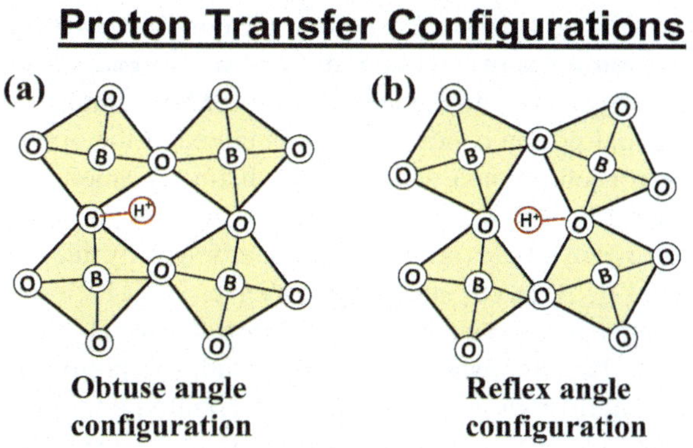

Fig. 6 Different proton transfer configurations, **a** obtuse angle, **b** reflex angle [113]

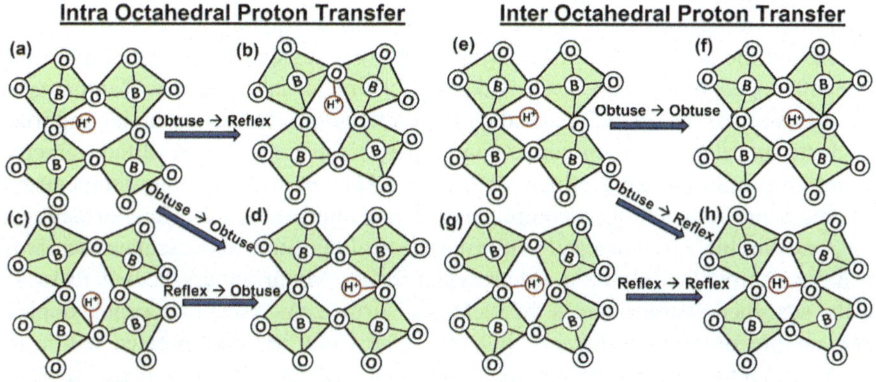

Fig. 7 Intra- and inter-octahedral proton transfer pathways between distinct configurations, **a**, **b** intra-octahedral obtuse to reflex, **a**–**d** intra-octahedral obtuse to obtuse, **c**, **d** intra-octahedral reflex to obtuse, **e**, **f** inter-octahedral obtuse to obtuse, **e**–**h** inter-octahedral obtuse to reflex, **g**, **h** inter-octahedral reflex to reflex [113]

vertical planes within the oxygen octahedra. Proton mobility between two different configurations (obtuse → reflex and vice versa) demands additional activation energy of 0.11 eV due to prolonged pathway. While the inter-octahedral proton transfer befits the proton hopping due to reduced O–O bond distance between the adjacent octahedra. However, the current scenario demonstrates a higher rotational barrier over their translational counterpart to overcome the repulsive effect of the electron cloud between the nearest oxygen atoms in the horizontal plane of the neighbouring oxygen octahedra. Meanwhile, the proton migration between two discrete configurations also demands additional activation energy congruent to the prior case because of an equivalent proton transport mechanism. The proton hopping between obtuse → reflect angle configurations and vice versa within inter-octahedral proton transfer is split into two intra-octahedral proton displacements and follows analogical behaviour relative to the former case. However, among all possible permutations, the inter-octahedral proton transfer between two reflex angle configurations is optimum with the least activation barrier due to the least plausible O–O coordination.

However, potential proton conductivity is obtained at the optimal doping range (at lower doping concentrations) which varies distinctly among different acceptor substituents [114, 115]. The drawback of the elevated doping range is associated with fluctuated charge dynamics around the dopant's vicinity due to the rising inductive effect [116, 117]. A recent study by Draber et al. [117] reported a probable proton transfer pathway in standalone and Y-doped $BaZrO_3$ based on activation energy differences. The simulated outcomes predict the proton trapping effect in acceptor-doped compositions over the undoped counterpart. While the activation energy in pure $BaZrO_3$ hovers between 0.1 and 0.27 eV [117, 118], the Y-doped composition demonstrates double the former values. Such global gradient in activation energy indicates the rise of the proton trapping effect around the dopant's vicinity (in the first closest coordination). Meanwhile, thermal energy at elevated functioning temperatures (> 600 °C) assists proton de-trapping effect [119].

The outline of the section highlights the positives and negatives of pure and acceptor-doped $BaZrO_3$ PCs. Considerable chemical and mechanical integrity in zirconates arises due to the material's implicit structure. However, the lag in proton conductivity relative to their cerate counterpart is the consequence of the refractive nature and grain-blocking effects of $BaZrO_3$. Hence, high sintering temperatures become a pre-requisite to accomplishing dense microstructure. Material degradation, undesirable vacancies and inappropriate sealing might also escalate the ohmic resistance across the electrode–electrolyte interface which hinders the bulk diffusion of the adsorbed charges across the electrode surface. As a result, acceptor dopants and suitable sintering additives (NiO, ZnO and CuO) are incorporated within the host material to lower the processing temperature. Meanwhile, assimilated research confrontations also predict the least activation energy pathways to be highly preferential for optimum proton migration. Compositional-induced structural imperfections develop via oxygen octahedra and constitute self-trapping effects. Thus, the choice of acceptor dopants and their doping concentrations must be regulated to forbid attenuated charge dynamics via proton trapping effect.

2.3 Structural Perspective of $BaCeO_3$ and $BaZrO_3$

Considerable potent of energy conversion by fuel cell technology impels researchers and practitioners to investigate for practical and advanced applications. The perovskite PCs such as $BaCeO_3$ and $BaZrO_3$ display high electrocatalytic activity with extrusive protonic mobility. Besides implicit bond flexibility, active B-sites aids the protonation via oxygen vacancies in doped PCs. Perovskites with distinct cationic pairing and skeletal structures ($A^{3+}B^{3+}O_{2-\delta}$, $A^{+}B^{2+}O_{3-\delta}$ and many others) enhance the protonation feasibility. The fundamental perovskites are additionally pondered as Re–O structures due to corner-sharing BO_6 octahedra and cationic (A-site) interstices [120]. The individual purely cationic constituents establish compositional neutrality with homogeneous stoichiometry. Stoichiometric divergence constitutes novel perovskite oxides. For instance, double perovskites afford B/B′ cationic rearrangement to accomplish essential characteristics [77]. Complex and perovskite composites are contemplated with disordered and partially consumed oxygen sublattices [121, 122]. The ionic defects within the parent compositions significantly alter symmetric atomic positioning and bonding characteristics which is statistically denoted using Goldschmidt's tolerance parameter (t) as shown in Eq. 9 [123]:

$$t = \frac{r_A + r_O}{\sqrt{2}(r_B + r_O)} \tag{9}$$

where

r_A = A-site cationic radius

r_O = radius of oxygen anion

r_B = B-site cationic radius.

Distorted PCs (with lower symmetry) crystallize in an ilmenite polymorphic structure. The chemical coordination of PCs with d-manifold besides the structural standpoint becomes symbolic to conserve material characteristics. For instance, 4-coordinated Ge atoms in $BaGeO_3$ although substantiates cubic symmetry eventually display a silicon-based structure. However, pressurized material processing strategies favour 6-coordinated Ge atoms within the parent material [60]. Similarly, sixfold coordinated Zr^{4+} and Gd^{3+} demonstrate 0.72 Å and 0.94 Å ionic radii, while ninefold coordination constitutes lattice expansion with 0.89 Å and 1.11 Å ionic radii, respectively. As a result, structural imperfections establish a strong bond with the cationic positions and escalate crystal field effects within the host crystal lattice [124].

In view of material architecture, Muñoz-Garc et al. [125] illustrated the oxygen percolation with different cationic ordering ($Sr^{2+} < Ca^{2+} < Ba^{2+}$ (at the A-site) and $Pr^{3+} < Nd^{3+} < Sm^{3+} < Gd^{3+}$ (at the B-site)). The two-fold effect of acceptor substitutions offers hole concentration and contracts oxygen vacancies due to disproportionate charge distribution. The intermediate operational regimes of PCs responses surface oxygen removal due to rapid electrocatalysis [10]. Matsumoto et al. [104] speculated the interdependency between chemical stability and protonic conduction in trivalent-doped (M = Y^{3+}, Tm^{3+}, Yb^{3+}) $BaCeO_3$. Primary composition illustrates an orthorhombic structure that successively distorts the structure to lower symmetry due to dopant-host radii mismatch (from Sc^{3+} to Y^{3+}). The electrophysical characteristics of doped $BaCeO_3$ and $BaZrO_3$ dictate a strong influence on the dopant ionic radii. Among multiple acceptor substituents, Y^{3+} (r = 0.9 Å)-doped $BaCeO_3$ illustrates adequate ionic conductivity due to low electronegativity and feeble B/M cationic difference. Moreover, the structural attributes of the resultant composition constrict the O–O bond distance between adjacent octahedra and pacifies the activation barrier for proton mobility.

Takeuchi et al. [13] showcase the impact of doping concentration on structural differences in $BaCeO_3$ against diverse atmospheres. The outcomes establish an initial orthorhombic phase at lower doping concentrations $0 \le x \le 0.1$ which successively demonstrates disparity at intermediate and elevated doping ranges. The XRD reports claim a decrease in symmetry (orthorhombic → rhombohedral) between $0.15 \le x \le 0.3$, respectively. However, lattice invariance was preserved at the lower doping concentration ($x \le 0.1$) under H_2 and O_2 annealed atmospheres. However, ambiguities build up at elevated doping concentrations $x > 0.15$ due to incomplete phase formation or multiphase co-existence. The essence of such structural variability is visualized via stray peaks in structural characterizations. Successive peak shifts in XRD diffractograms are an elemental example of lattice variations (compression/elongation). Ideal cubic $BaCeO_3$ display symmetric bonds (A–O and B–O) which deviates gradually with lower symmetry (orthorhombic and rhombohedral) [126]. Lattice disorders in lower symmetric structures propose thermodynamical instability and escalate the gradient in charge chemistry. Meanwhile, Mburu et al. [127] conducted IR spectroscopic study to examine structural modifications in acceptor-doped $BaZrO_3$. The outcomes establish the role of doping concentrations upon structural phase transitions and localized symmetry gradient. Figure 8 demonstrates the IR spectral variations in acceptor-doped $BaZrO_3$ for different doping concentrations:

Subtle variations between dry and humidified exposure account for large protonation in the latter case. Three prominent bands (135, 270 and 530 cm^{-1}) in standalone $BaZrO_3$ emerge due to O–Zr–O, Zr–(BO_6) and Zr–O bending, symmetric and asymmetric stretching (symbolized as υ_1, υ_2, υ_3), respectively. While spectral invariance persists in 'hydrated + doped' samples, while the feeble structural disorder emerges in the υ_2 bands. The high spectral amplitude at $x = 0.3$–0.4 corresponds to octahedral tilts due to hydrated oxygen vacancies. This concludes a profound tendency of $BaZrO_3$ to accommodate considerable protonic defects (OH^-) within the host crystal lattice. In compliance and response to compatible acceptor incorporations, Gd^{3+} potentially overhauls germicidal utilizations as a consequence of well-structured narrow emission bands in the UV region. Intense spectral response

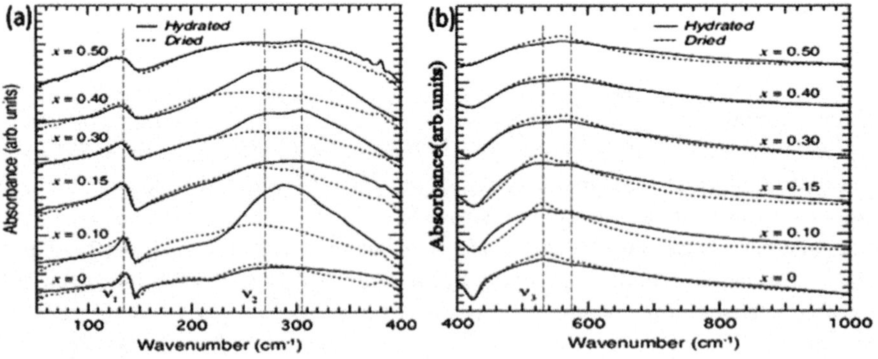

Fig. 8 IR spectra of Y-$BaZrO_3$ in the dry and hydrated sample, **a** range (50–400 cm^{-1}), **b** range (400–1000 cm^{-1}) (figure reproduced with permission from Ref. [127], Copyright 2017, *Journal of Physical Chemistry C* [127])

devotes to the likelihood of large defect centres. Many research confrontations claim the B-site Gd^{3+} occupancy constitutes the former spectral response. A deeper insight suggests a partial A-site and B-site occupancy by acceptor substituents close to Gd^{3+}. However, persistent oxygen vacancies generate highly probable B-site occupancy at lower doping concentrations.

The following section briefs regarding compositional-induced structural disorder which establishes non-linear behaviour with doping concentration beyond the optimal range. Different defect density within the host crystal lattice is responsible for structural imperfections which although assists proton chemistry yet hinders chemical stability. The comparative analysis between $BaCeO_3$ and $BaZrO_3$ claims a prominent structural instability in the latter case due to localized gradient in cation–anion (A–O and B–O) bond distance, respectively. The grounds of the afore-response attributes to the large B/M deviations. This illustrates large protonic defect accommodation by zirconates over their cerates counterpart.

2.4 *Chemical Attributes of Pure and Derived $BaCeO_3$ and $BaZrO_3$ PCs*

PCs are encouraging electrolytes due to the low activation barrier for proton diffusion as a consequence of lower protonic radius over the oxide counterpart [61]. The genesis of proton conductivity demands implicit oxygen vacancies within the host lattice. Proton welcoming cations aggregate protonation and escalate the defect stability and solubility threshold. On the contrary, basic oxides increase the chemical reactivity against acidic gases and acquaint chemical instability. The chemical reactivity of $BaCeO_3$ PCs against acidic environments is shown in Eqs. (10)–(12) below:

$$BaCeO_3 + H_2S \rightarrow BaS + H_2O + CeO_2 \tag{10}$$

$$BaCeO_3 + CO_2 \rightarrow BaCO_3 + CeO_2 \tag{11}$$

$$BaCeO_3 + H_2O \rightarrow Ba(OH)_2 + CeO_2 \tag{12}$$

The following reactions illustrate H_2O dissolution under humidified ambiance which impinges on the resultant electrochemistry. The overall conductivity in $BaCeO_3$ comprises interstitial oxygen and constitutes invariance under humid surroundings. The resultant oxide and hydroxide precipitates arise due to compositional poisoning which establishes depressed protonic conduction. Excessive precipitates favour damped protonation and uplift the oxide ion conduction at escalated temperatures (> 700 °C). However, the synergetic events may be contemplated as the impact of acidic acceptor dopants on the chemical and electrophysical background [128].

Research inquisitions in recent years claim the significant role of interfacial chemistry in the persistent dispute between chemical stability and electrochemical charge dynamics across the electrode–electrolyte interface. Adsorbed charges on the electrode surface experience an ohmic resistance for bulk charge diffusion due to a mismatch of electrode–electrolyte characteristics. Infusion of foreign substituents to tailor-specific material characteristics additionally affects unintended properties and disrupts the electrode–electrolyte chemical compatibility. Such electrolyte impurities evacuate A and B-site cations from the host composition and induce proton-blocking effects [129]. $BaCeO_3$ endures a pragmatic tendency to configure into $BaCO_3$, CeO_2 and $Ba(OH)_2$ against hydrated and CO_2 atmospheres. Comparative dissection between electronegativity and basicity of Zr^{4+} and Ce^{4+} cations fairly dictates a more pronounced basicity with an upraised hydration limit [130]. Besides, the metal–oxygen bond energy gauged via respective formation enthalpies additionally constitutes chemical discrepancies. In compliance with prior discussions, the defining approaches to improve the material's chemical response involve high electronegative acceptor substitutions. Heterovalent impurities like In^{3+}, Al^{3+} and Ga^{3+} significantly stabilize the reactivity of cerates against reducing atmospheres [131–133]. Among different substituents, Zr^{4+} corresponds to a rational choice to enhance the desirable chemical compatibility [79, 134]. High chemical stability zirconates over cerates find their implication as a befitting cationic substituent to combat reactive ambiance. However, Zr^{4+} doping concentrations must be cautiously chosen to avoid poor sinterability and microstructural impedance.

Besides established equilibrium against oxidizing and reducing atmospheres, PCs also interact with the Ni-cermet pair which is vital to accomplish overall electrochemistry and bulk charge diffusion. The significance of the electrolyte-Ni cermet combination is to interpret the Ni-diffusion within the electrolyte lattice. The former consequence is the rapid Ni diffusion across the grain boundary of the resultant PCs. Although, reduced Ni solubility (< 1%) within Ce^{4+} and Zr^{4+} matrix must showcase fairly low negative impacts on the bulk characteristics in accordance with the outcomes of Yang et al. [135] and Yang et al. [136], respectively. Protonation in perovskites not only appears under hydrogen atmospheres, but steam also contributes to the overall conductivity. Adequate defect energy and persistent obstruction to intrinsic disorder constitute high chemical compatibility of $BaZrO_3$. D'Epifanio et al. [137] as a result authorized a comparative analysis regarding the chemical stability of Y-doped $BaCeO_3$ (BCY) and $BaZrO_3$ PCs. The wet chemical (citric-acid sol–gel) synthesis was utilized to accomplish compositional homogeneity with reduced calcination conditions (1100 °C), respectively. The chemical compatibility of the afore-samples was evaluated via their extended exposure to CO_2 atmosphere for about 6 h. The structural stability between $BaCeO_3$ and $BaZrO_3$ also commands chemical coherence. The single phase in Y-doped $BaZrO_3$ (BZY) confronts considerable chemical inactivity against reactive environments, while the $BaCeO_3$ counterpart displays multiphase co-existence due to CeO_2 precipitates. Supplementary reasons explain the incomplete lattice occupancy by Y^{3+} within the lattice and results in phase instability. Figure 9 demonstrates the comparative chemical cohesion between $BaCeO_3$ and $BaZrO_3$ PCs.

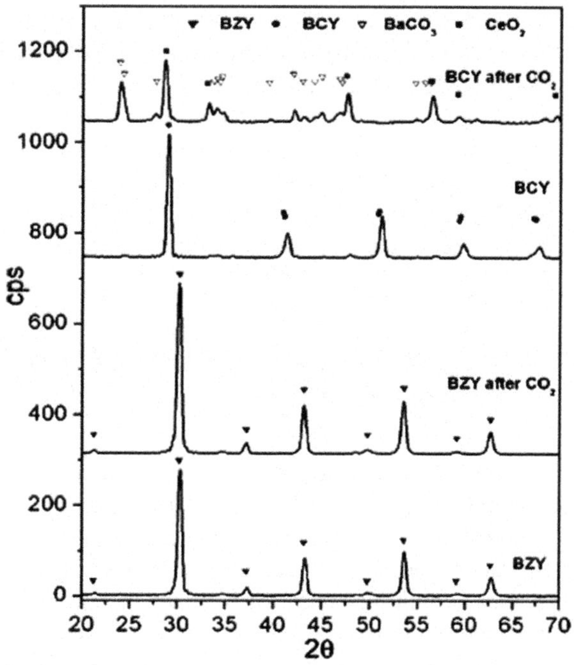

Fig. 9 Chemical stability of BCY and BZY before and after exposure to CO_2 atmosphere (D'Epifanio et al. [137], *Fuel Cells* **2008**, p 4)

Figure 6 displays a surge in the number of stray peaks in Y-doped $BaCeO_3$ PC relative to the standalone composition post-CO_2 exposure. The resultant attribute emerges due to CeO_2 and $BaCO_3$ precipitates besides $Ba(OH)_2$ residue, respectively. However, Y-doped $BaZrO_3$ on the contrary establishes insignificant secondary phases besides a few additional peaks due to $Ba(OH)_2$ residues due to implicit Ba vacancies as a consequence of poor sinterability. The overview of outcomes justifies pronounced chemical stability by zirconates over their cerates counterpart. With an analogical motive, Krug et al. [138] examined the geometrical variations against hydration and dehydration of trivalent-doped PCs. The lattice strain and chemical reactivity was interpreted via ambiguous thermodynamical conformity. Incomplete hydration and consistent heating devote lattice expansion. This constitutes three prominent polymorphic structural phase transitions (I2/m → Imma → R3c) which finally attains ideal crystal structure (Pm-3m) against diverse atmospheres. However, a supportive study of the afore-outcomes by Han et al. [139] underlines the influence of hydration limit over thermoelectric properties of acceptor-doped (Y^{3+} and Yb^{3+}) $BaZrO_3$ PCs. The increased protonation as a consequence of the increase in hydration partial pressure constitutes adequate protonic conduction at the cost of lattice expansion with spatial dilation effect. Stoichiometric irregularity forms the backbone for spatial dilation and isostructural gradient.

The vitality of chemical instability among acceptor-doped $BaCeO_3$ against distinct atmospheres imposes profound concerns. The basicity of Ce^{4+} and acceptor cations lays the platform for unintended chemical activity (in the form of residues and precipitates). While high electronegative acceptor substituents aid chemical cohesion, the contrasting end affects the electrophysical behaviour. Structural phase transitions and reduced charge mobility under distinct atmospheric constraints (such as CO_2, SO_2, N_2 and many other) are all a consequence of low diffusivity of oxygen vacancy due to phonon contributions to the free migration energy. From the perspective of high chemical inertness, zirconates allow Zr^{4+} substituents to decline the chemical reactivity against acidic environments.

2.5 *Electrical Essence of Pure and Derived $BaCeO_3$ and $BaZrO_3$ PCs*

PCs demonstrate considerable proton conductivity (10^{-3} to 10^{-2} S cm^{-1}) with low activation energy (0.3–0.7 eV) at intermediate temperatures (400 °C $\leq T \leq$ 700 °C) [140]. Iwahara et al. [141, 142] disclosed the early instance of proton conductivity among acceptor-doped $BaCeO_3$ and $SrCeO_3$ PCs. Progressive research further extended to multiple other PCs based on their pragmatic characteristics. The classification into single and co-doped PCs was based on chemical, mechanical and electrical features. Protonic conduction is the outcome of bulk proton diffusion through the electrolyte. Ionic conduction among PCs is complex because they can concurrently contribute to proton and oxide ion conductivity. However, the former is prominent at intermediate temperatures (< 600 °C) which declines with a rising dominance by the latter, respectively. The partial and overall protonic conduction relies on numerous factors (temperature, composition, partial pressures and host microstructure) [143–145]. The charge carrier diffusion additionally relies upon the effective mass and the activation energy for long-range transport through the bulk. State-of-the-art composite electrolyte (as shown in Fig. 10) demonstrates the tendency of pronounced mixed ion (H^+ and O_2^-) conductivity despite high activation energy requirement for heavier O_2^- ions. Although the mobility of distinct charge carriers differs to a certain extent, the improved bulk diffusion is attributed to the dense and thin electrolyte composites.

Ionic conduction presents the ionic displacement via defects within the crystal lattice. The amplitude of conduction escalates with temperature. The afore-attribute is subject to microstructural variations. Ionic conduction presents a close alliance with grain size distribution with a pronounced grain boundary width due to grain shrinkage [146]. Standalone and acceptor-derived $BaCeO_3$ PCs offer attractive prospects for the constructive surge in proton conductivity due to structural and electrophysical compatibility [147]. $BaZrO_3$ on the contrary are well known for its extensive chemical stability and mechanical durability. The works of Park et al. [148] illustrated the dominant charge carrier and the resultant conductivity against different atmospheric conditions. The outcomes claim pronounced hole conduction under a humid H_2 atmosphere compared to humid air and prominent oxygen partial pressure (pO_2)

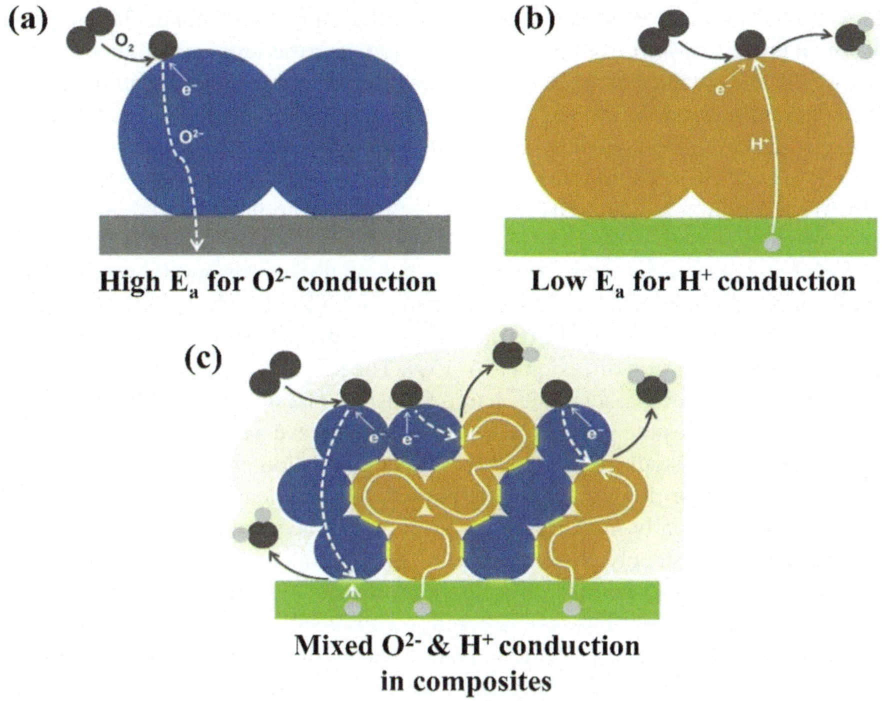

Fig. 10 Conduction attributes of composite electrolyte

between 600 and 800 °C, respectively. However, the conductivity preserved invariance below 600 °C which demonstrates the insignificance of humidified constraints under 600 °C, respectively. From the perspective of proton conductivity, the order of conduction under humid H_2 atmosphere is one order of amplitude more than the dry counterpart. The consequent effect arises due to large protonation in the former atmosphere over the latter. The gradient in proton conductivity as a function of different atmospheric constraints and temperature is a characteristic feature of PCs. An affiliated study by Shin et al. [149] reports abnormal conductivity at low temperatures. The conductivity isotherm displays prominence between 400 and 800 °C which gradually declines with a subsequent decrease in the operating temperatures to 250 °C under humid atmospheres. The outcomes resemble the former study. Since ionic conduction is a function of atmospheric chemistry and temperature, hence the latter study with low protonic conductance is the consequence of decreasing protonation with a consequent decline in the host temperature within the host crystal lattice.

Exner et al. [150] disclosed the role of acceptor substituent upon Ba-based PCs for fuel cell technology. Individual Ba-based PCs doped with ($x = 0.2$) Y^{3+} acceptor was deposited with thin films (5–10 μm) of powdered Ba precursor. Thin-film synthesis procedure aided phase stability with reduced sintering temperature and was devoid of low melting phases. The results dictate suitable proton conductivity (10^{-3} S cm^{-1})

between 400 and 500 °C which deteriorates with rising temperature. The aftermath of which is due to a shift from proton to oxide ion contribution. Supplementary justification for the obtained characteristics is the effect of temperature-dependent structural phase transition. Multiple pieces of literature on the electrophysical behaviour of doped $BaZrO_3$ are conducted with a motive to enhance the ionic conductivity with preserved chemical inertness [151–155]. B-site trivalent cationic substitutions improve the proton conductivity in zirconates relative to their standalone counterpart due to increased oxygen vacancy concentrations and protonation, respectively. However, the obtained conductance yet lags behind their corresponding cerates. The background of which is associated with the hydration limit of the host B-site cations ($Sr^{4+} < Zr^{4+} < Ce^{4+}$). Besides, the refractive microstructure of $BaZrO_3$ prevents unimpeded long-range proton diffusion [156–159]. The nature and dopant concentrations significantly alter the characteristic charge diffusion. The surge in Zr^{4+} content within the host equally rises the sintering temperature to achieve a compact microstructure [160]. The unified study of electrophysical and activation energy fluctuations as a function of acceptor doping concentration indicates enigmatic outcomes. The activation energy gradually increases at lower doping concentrations up to a threshold doping range and steeply declines within the optimal doping range. Amidst the optimal doping concentrations, prominent proton conductivity is achieved among diverse PCs which varies between different acceptor substituents. Beyond the intermediate doping range ($x > 0.25$), the activation barrier steeply increases with a counter effect on the electrophysical behaviour of the overall composition. The incline in the activation barrier and reduced proton mobility is the consequence of active proton-dopant binding due to strong coulombic interactions. The larger the electronegative acceptor substituent, the more rigid is the attractive interaction. Small acceptor substituents with high electronegativity are preferred to account for the material's chemical stability, while larger and low electronegative acceptor dopants on the contrary are opted to escalate the material conductivity response.

Ding et al. [161] annotate the charge chemistry for attenuated proton diffusion. Rising concerns were established about the localized lattice distortions for delayed proton diffusion. The outcomes of the study reveal ambiguous trend over the recent apprehension of threshold doping range. The results devote to comprehend the degree of lattice fluctuations and charge dynamics due to isolated acceptor substituents. As a consequence, the outcome demonstrates a hike in localized symmetry variations due to surge in defect density above the threshold. Such defect cluster as polarons facilitates proton trapping centres for delayed diffusion. A linear trait persists between activation barrier and acceptor concentrations. Figure 11 illustrates proton trapping effect at the expense of structural imperfections:

DFT-based simulations showcase a progressive proton trapping effect with a subsequent surge in the dopant concentration within a confined region of the material and proton. The surge in dopants enforces localized symmetry cohesion and amends the charge chemistry. As a result, the activation barrier subsequently increases and constitutes a proton trapping effect. The synergetic outcomes demand the need to monitor and optimize the doping concentrations below to threshold to configure the

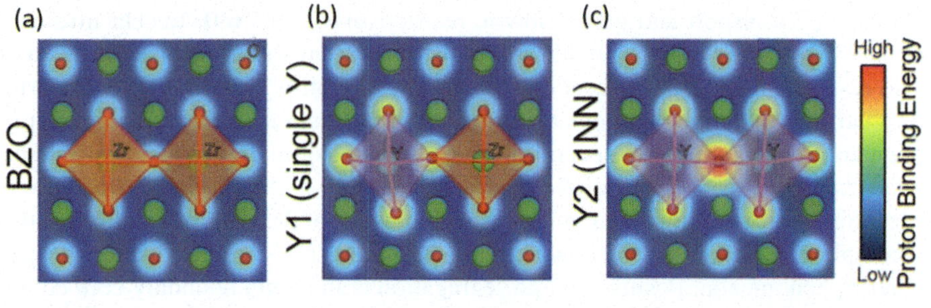

Fig. 11 Schematic illustration of escalating binding energy as a function of dopant concentration contributing proton trapping in Y-doped $BaZrO_3$, **a** undoped $BaZrO_3$, **b** influence of single Y atom towards binding energy, **c** influence of two Y atoms in the vicinity to each other inducing trapping centre amidst them (image from Ding et al. [161], *Chemistry of Materials* **2018**, Pg 4)

trade-off parameters. Extensive studies over Y^{3+} and Gd^{3+} suggest the optimal doping range for the same to waver between $x = 0.125$–0.25, respectively.

The following section summarizes the fluctuations in proton conductivity as a consequence of dopant implanted structural instability. The nonlinear proton conductivity as a function of doping concentration decodes charges dynamical limitations. Rising activation energy from lower to intermediate acceptor concentrations occurs due to an overall increase in the electronegativity of the host composition. A steep decline in the barrier potential at intermediate doping range indicates the optimal acceptor doping concentrations feasible to achieve desirable protonic conduction with minimal structural entropy. Beyond the threshold, electrophysical perturbation inclines to devote proton trapping effect due to surging proton-dopant coulombic interactions and the basicity of the overall composition. As a matter of the influence of chemical surroundings on proton dynamics, humidified ambiance supports considerable protonation, while dry atmospheres display contrary impacts. However, besides a humid atmosphere, adequate thermal energy in terms of operating temperatures (~ 600 °C) is required to achieve long-range proton transport with a de-trapping effect.

2.6 *Microstructural Viewpoint of $BaCeO_3$ and $BaZrO_3$ PCs*

PCs are polycrystalline electroceramics composed of small domains termed grains with the homogeneous crystal structure. However, the grains differ in their orientations with ambiguous symmetry at the contact region (grain boundary region). Symmetry variances introduce potential barrier and obstruct long-range proton diffusion. This validates the microstructural influence upon the total ionic mobility. Thus, microstructural engineering is essential to combat the electrophysical disparity in distinct PCs [162–165]. A conventional technique involves the use of appropriate low melting phases [166, 167]. Among diverse feasible additives, NiO showcases active mass transport and grain growth, respectively [168, 169]. Recent investigations by Medvedev et al. [129] affirm microstructural invariance with NiO additives. The resultant behaviour reflects due to the low solubility limit of Ni^{2+} within the host lattice besides the sedimentation of proton-blocking multiphase across the grain boundaries. A rational introspection for an alternate additive (ZnO and CuO) escalated rapidly [170, 171]. Multiple pieces of literature highlight the competence of acceptor substituents on the thermoelectric behaviour of the host material. Devoid of incomplete phases, low ionic conductivity among $BaZrO_3$ emerges due to the space charge phenomenon. Positive charge segregation at the grain boundary core constitutes a subsequent positive charge depletion at the space charge layer. Analogical attributes are observed with Y-doped $BaCeO_3$ compositions [172, 173]. A deeper insight into the microstructural characteristics of Ti-doped $BaCeO_3$ was carried by Somekawa et al. [174]. The study showcases the chemical instability and electro-physical disparity along the grain boundary of PCs under discrete environments. Significant grain boundary impedance relative to the grain summons a reduced total ionic conductivity. The consequent effect prompts towards temperature-dependent space charge potential gradient. Space charge accumulation astride the grain interface within polycrystalline PCs mollifies the grain boundary conductance. As a result, PCs feature prominent impedance for active charge transport during grain boundary engineering [121]. Such material architecture which curtails the agglomeration force constitutes a defect (acceptor impurities) along the grain boundary core with a unified reduction in the positively charged constituents (protons and oxygen vacancies). The afore-influence has been verified using experimental and computational conducts. Figure 12 demonstrates microstructural chemistry within proton-conducting electrolytes.

Diverse established literature contrary outcomes of charge transport assistance from defect segregation along the grain boundary core. The outcomes of Pasierb et al. [175], De Souza et al. [176] and Nasani et al. [177] claim a positive impact of acceptor dopants such as Yb^{3+} on the reduced space charge potential for unattenuated charge transport. This emphasizes linear reliability with an acceptor doping range. From the macroscopic viewpoint, PCs constitute a gateway for multiphase co-existence due to stoichiometric deviations of the host compositions. This leads to defined levels of defects and charges segregations along the grain boundary. The secondary phase separation differs between individual compositions and their composites [129]. As a result, the nature of microstructural variations proportionally impacts the degree of dopant agglomeration. Ultrafine grains with low dopant segregation constitutes linear surge in the space charge potential. The likelihood of such character is profoundly observed in Ce-based PCs. However, opposing effects of prominent grain boundary conductance in Ce-based PCs accompanies with structural imperfections and reduced space charge potential via secondary and tertiary aftermath [178]. Grain and grain boundary structural disparity is categorized based on the enthalpy of defect formation. Meanwhile, compositions with pronounced Zr^{4+} cations illustrate refractive attributes and resistive grain boundary. As a consequence, a prominent annealing time (> 24 h) with high sintering temperature (> 1400 °C) furnishes dense microstructure [179]. In compliance with prior discussions, poor sinterability constitutes stoichiometric

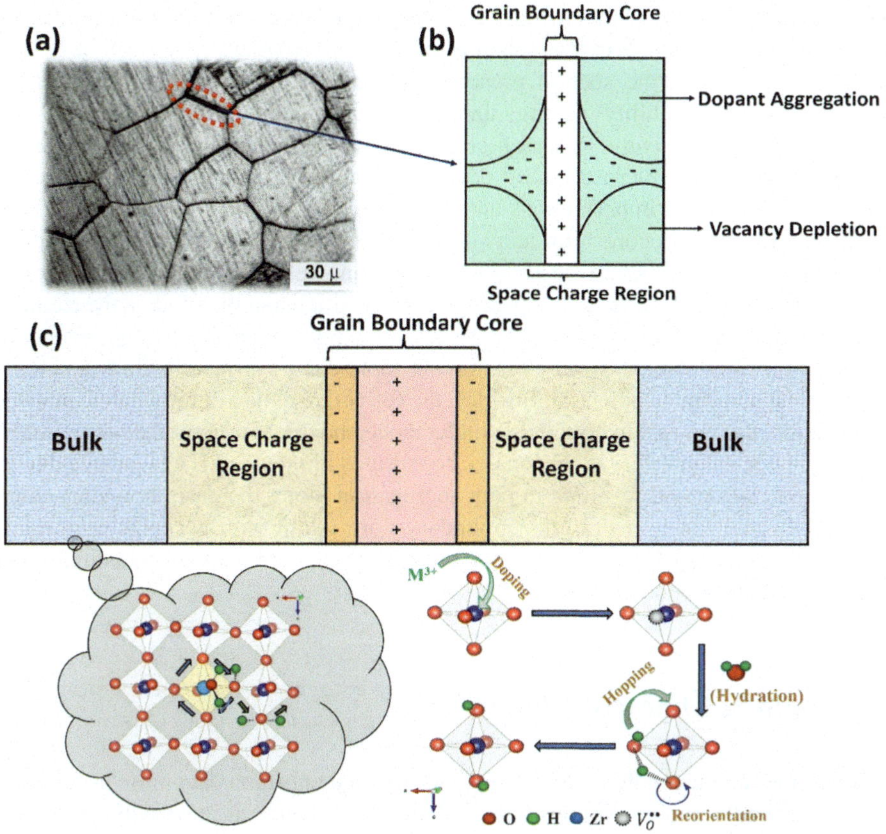

Fig. 12 Microstructural viewpoint of proton conduction, **a** magnified microstructure of polycrystalline proton conductor, **b** magnified section of space charge layer, **c** microstructure of bulk electrolyte with proton-conducting octahedra

divergence due to Ba vaporization and structural uncertainty. The consequent effect ceases compositional neutrality and activates steep potential barriers and proton trapping centres [163].

On the other hand, the thermodynamic properties of the $BaZr_{1-x}Y_xO_{3-\delta}$ (BZY)/NiO system dictate a threshold for Y^{3+} dopant concentrations to prevent BaY_2NiO_5-based proton-blocking phase. Elevated Y^{3+} doping concentrations constitute the Y_2O_3 secondary phase with fractional BaO evaporation. As a result, a lower doping range is instructed to eliminate incomplete and impure phase formations. Besides, an elementary strategy involves the utilization of powered precursors of the host compositions as a protective covering over the pre-sintered pellets to forbid microstructural heterogeneity. Opposingly, excessive Ba^{2+} within the grain boundary core constitutes structural breakdown and activates interactions between gaseous components within the host crystal lattice. The plausible solution to the afore-issues is the synthesis of an acceptor-doped PC with a feeble Ba-deficiency to avoid structural imperfections and also prevent A-site acceptor substitutions.

The overview of the section presents the influence of acceptor defects over microstructural instability among standalone and acceptor-doped $BaCeO_3$ and $BaZrO_3$ PCs, respectively. Research confrontations and outcomes suggest a linear correlation between microstructural dimensions and fluctuations in thermoelectric behaviour. Structural imperfections and symmetry variations across the grain-grain boundary interface accord high activation barrier for attenuated charge transport. Grain boundary architecture via acceptor substituents relies upon two critical parameters. The first is a decline in the space charge potential, and the other is the dilution of impurities. Optimal proton conduction demands appropriate grain dimensions and distribution. While micro-sized grains favour rapid charge diffusion, ultrafine grains uplift the space charge potential. Besides, the refractive nature of zirconates imposes constraints upon the grain boundary conductance and microstructural density. Additionally, poor sinterability of zirconates constitutes unintended Ba vacancies due to BaO vaporization and resultant dopant segregation along the grain boundary core. A proportional surge in the grain boundary core width with acceptor concentrations introduces nonlinearity in thermoelectric characteristics with decreasing Debye length of the host composition.

3 Limiting Characteristics of PCs

Among fuel cell technology, PC-SOFC has attained enormous attention as a potential eco-friendly and continuous energy conversion device at intermediate temperatures [180, 181]. However, practical appliance demands low-temperature operations. Incorporated coolants within the fuel cell framework decline the electrochemical kinetics across the electrode–electrolyte interface. Meanwhile, the afore-implementation affects the economical facets in the large-scale manufacturing. Low-temperature PC-SOFC is hence a plausible solution to the afore-challenge [182, 183]. Low-temperature functioning allows cost-effective material choice and ceramic/non-glass interconnects [184, 185]. The perks of PCs also display lower power density alongside high chemical reactivity and potential barrier for electrochemistry [186, 187]. However, the additional material limitations involve:

1. Grain blocking effects
2. Poor sinterability
3. Large ohmic resistance across electrode–electrolyte interface
4. Large cathode overpotential
5. Steep proton traps.

Researchers and practitioners globally have attempted to resolve the afore-issues via novel electrodes and electrolyte synthesis [188]. Unfortunately, low functioning temperatures of PC-SOFC also illustrates vital limitations. The low temperatures forbid sufficient energy supply for required chemical kinetics across the electrode surface and activation energy for proton hopping. Despite critical drawbacks, researchers aim to accomplish sizeable measures with advanced material architecture and processing techniques [81, 189–193]. A critical limitation among Ce-based PCs is the electronic conduction due to $Ce^{4+} \rightarrow Ce^{3+}$ material chemistry against reducing environments. The consequent effect reduces the cell's Nernstian potential and operational efficiency [194]. Although a feasible measure in terms of a bi-layered electrolyte is fabricated and utilized, it contributes unipolar ionic conduction on the anodic surface as a barrier against Ce reduction [195]. However, the following layered electrolytes demand high sintering conditions and sophisticated technologies. A feeble mismatch between the thermal expansion coefficient of distinct electrolyte layers constitutes delamination and substrate deformation during the sintering process [196]. Thin-film technology with a buffer electrolytic layer is essential to avoid delamination and chemical reactivity of host constituents. As a result, a suitable fundamental material with a plausible material architecture is necessary to replace conventional proton-conducting electrolytes like YSZ and many others [197]. However, the electrolytic thickness pounds as a supplementary drawback due to significant power loss due to ohmic overpotential [198, 199]. Recent works by Lian et al. [198] displayed a high ohmic resistance for thin-film YSZ electrolytes (10–20 μm). Park et al. [199] claimed an increase in a power loss of 0.4 W cm^{-2} with increasing electrolyte thickness (from 2.5 to 8 μm).

Despite the former efforts, chemical instability remains a pronounced concern among PCs. SO_2 and CO_2 poisoning refrain from their utilization under diverse circumstances. Methods of isovalent and aliovalent substitutions are executed to pacify the reactive attributes and uplift the protonic conduction [200]. A recent computational DFT-based study by Vignesh et al. [11] reports the impact of acceptor substituents on the structural disorder. The outcomes cover diverse perspectives ranging from the doping concentration to the proton trapping effect. The simulated results claim a multiphase co-existence under intermediate doping concentration ($x = 0.125$ and 0.375), respectively, due to defect-cluster involvement and electronic factors (unintended cationic displacement from their coordination polyhedra). The study suggests a threshold for material engineering via dopant substitutions ($x < 0.5$) and the resultant oxygen vacancy formation. The amplitude of proton conductivity maximizes at optimal doping range ($x = 0.125$–0.25) for Y-doped $BaCeO_3$ and declines thereafter with the early insight of proton trapping effect at $x = 0.375$, respectively. However, lower B/M deviations refrain from steep proton traps. The study by Polfus et al. [173] addresses a paramount chemical stability against CO_2 atmospheres with the use of Zr-rich PCs. However, reports additionally claim a probable surface reactivity in $BaZrO_3$ due to subsurface Ba deficiency [201, 202]. Grain blocking effects also communicate their hindrance to proton conduction. Localized proton depletion within the space charge region plays a vital role in reduced proton conductivity within zirconates [203]. Oxygen vacancy within bulk phases displays congruent formation energy despite structural and chemical diversity. In a comparative analysis between $BaCeO_3$ and $BaZrO_3$, Lindman et al. [204] claim a high proton formation energy in $BaZrO_3$ concerning $BaCeO_3$ by 0.4 eV. This discloses high exothermic hydration enthalpies by $BaCeO_3$ over $BaZrO_3$ and aids the experimental outcomes. Moreover, the protons within distorted $BaCeO_3$ structures from linear inter-octahedral hydrogen bonds with implicit bond flexibility which is unlikely with the symmetric $BaZrO_3$ structures and large O–Zr–O bonds. Lindman et al. [204] progressively disclosed a pronounced bulk proton transfer within $BaCeO_3$ due to relatively rigid hydrogen bonds and a corresponding lower charge segregation energy with stable protonic sites.

Amidst all material-oriented limitations, material processing and fabrication routes play a critical role in the material and charge chemistry. The conventional solid-state reaction synthesis route establishes the foundation for powered ceramic electrolytes [146, 205]. However, a single-phase complex perovskite composition becomes a tedious method to synthesize. However, alternate wet chemical route (co-precipitate [206], sol–gel [207] and polymeric complexing [208]) assists in complete phase formation for polycrystalline materials. Such synthesis routes additionally aid to minimize the sintering temperature and forbid the residues within the bulk electrolyte [146]. The sol–gel synthesis, on one hand, grants a compositional homogeneity yet on the other side only favours single crystalline ceramic materials [209–212]. Presently, different multilayered, thin-film buffer electrolytes, higher doping strategies, metal–organic frameworks and proton-conducting composites are enterprise to mitigate the trade-off characteristics. The profound material advancements support the material degradation and chemical activity under stringent atmospheric constraints. A detailed study of addressed material advancements is dealt with in the upcoming sections.

3.1 Higher Doping Strategy

Standalone PCs devoid of defects dehydrate and endure structural deviations at elevated temperatures with prominent conductivity loss. This explains the intermediate-temperature operational regimes of PCs [162]. In compliance with the prior section (Sect. 4), chemical, structural and microstructural deploy critical impacts upon the total ionic conductivity of the resultant PCs. Material architecture offers boundless scope for stable electrolytes. However, material improvisations via single acceptor substituents in some way impose drastic constraints over material and charge chemistry. Higher doping strategies and proton-conducting composites assist to bridge the trade-off parameters. While a higher doping strategy involves two or more acceptor substitutions, composites on the other hand are the combination of two or more chemically distinct components [213]. Higher doping techniques incorporate bi, tri and tetra doping combinations with disproportionate distributions to attain optimized composition. Such measures build up the tendency to regulate structural deviations which is counterbalanced by distinctly sized acceptor substituents. A topical (Y^{3+}, Y^{3+} and Gd^{3+}) tri-doped $BaCeZrO_3$ (BCYYbGd) electrolyte with sufficient chemical compatibility was, respectively, coupled with $PrNi_{0.5}Co_{0.5}O_{3-\delta}$ steam and Ni-BCYYbGd hydrogen electrode by Rajendran et al. [14]. The consequent electrolyte displays 200–300 h of operational stability and adequate conductivity (1.1×10^{-2} S cm^{-1} @ 600 °C) against a moisture atmosphere (50 vol.% in Argon). Considerable operational cohesion of the resultant electrolyte empowers adequate protonation under hydrated conditions. Dense and non-porous microstructure on the other hand contributes competent open circuit voltage (adjacent to theoretical predictions). An extrusive novelty of the afore-composition is their ability to execute intermediate-temperature SOFC and SOEC (solid oxide electrolyzer cell) reversible operations. The current density and area-specific resistance (ASR) within SOEC mode gradually differed between 0.70 to −2.40 A cm^{-2} and 0.51–0.15 Ω cm^{-2} between 500–600 °C, respectively. While a peak power density (410 mW cm^{-2}) was obtained under SOFC mode at 600 °C. The result summarizes the chemically stable functioning of the tri-doped electrolyte under high steam constraints. Meanwhile, Kato et al. [214] attempted a partial Ba substitution with Ca and Sr, respectively, to fabricate complex PCs ($Ba_{0.95}Ca_{0.05}Zr_{0.8}Y_{0.2}O_{3-\delta}$ and $Ba_{1-x}Sr_xZr_{0.8}Y_{0.2}O_{3-\delta}$ with ($0.05 \leq x \leq 0.4$)), respectively. The outcome showcases an improved sinterability of the host zirconates with Ca and Sr partial substitutions which on the contrary decreases the ionic transport characteristics with progressive Sr concentrations. The emerged attribute is the consequence of standalone and doped microstructural differences which attenuates the grain boundary conductivity substantially.

Works of Rajendran et al. [215] concerning (Y^{3+}, Pr^{3+} and Gd^{3+}) tri-doped $BaCeZrO_3$ were intended to reform the chemical inertness and operational efficiency of the resultant electrolyte under hydrated atmospheres. The supplementary motive devotes to enhancing the poor sinterability and microstructural disparity due to the presence of Zr^{4+} cations. In accordance with prior discussions, zirconates offer considerable chemical assistance against reactive environments and hence are profoundly incorporated within the host compositions. However, sufficient Zr^{4+} demands high sintering temperatures for a compact microstructure and may lead to material degradation with a reduced lifetime. As a result, the afore-study endeavours to resolve cumulative limitations and contribute an elementary synthesizable complex electrolyte. A preference for 1 wt% ZnO sintering additives was pitched for microstructural densification. Dual samples (with ($Y^{3+} + Gd^{3+} + Pr^{3+}$ ($x = 0.05$)) and ($Y^{3+} + Gd^{3+} + Pr^{3+}$ ($x = 0.1$))) combinations illustrated compositional densification with subsequent shrinkage from 750 to 1250 °C, respectively. Consequent shrinkage post 900 °C showcases the depletion of voids with grain growth (~ 20 μm). The shrinkage rate inclined significantly in the latter composition relative to the former counterpart. This showcases the critical role of Pr^{3+} concentrations over the resultant microstructure with improved sinterability. The grain growth at reduced sintering conditions is the proportional impact of Pr^{3+} concentrations upon the mass diffusion across the grain boundary. Predominant chemical stability offered by Pr^{3+} is owing to its high electronegativity over Gd^{3+}. However, a contrary impact accompanies it in terms of reduced proton conductivity due to steep proton traps with pronounced activation barriers in the closest coordination.

Proton-conducting composite electrolytes on the other hand unite two separately synthesized compositions which pre-demands the optimization of individual compositions. Although the latter insinuates as a complex strategy yet the response of the resultant composites is more desirable than the former case. There may exist different composites ranging from solid oxides to polymers depending upon the centric applications. Complex composites display sub-categorical modifications in terms of different dimensional fillers, core–shell structures and multilayer systems. As a result, Zhu et al. [216] proposed the utilization of co/hybrid ions as nanocomposites for the forefront of fuel cell technology. The outcomes of the study illustrate a promising low-temperature functioning of PC-SOFC with a sizeable increase in ionic conductivity and chemical inertness. The consequent work expanded the scope of hybrid ions for complex PCs [217]. Figure 13 demonstrates a complex proton-conducting composite electrolyte within a fuel cell technology:

A recent study by Xu et al. [218] synthesised a composite proton-conducting electrolyte by coupling Y-doped $BaZrO_3$ (BZY) and Gd-doped CeO_2 (GDC) in distinct weight ratios. Different experimental characterizations were performed on the resultant composite. The electrochemical impedance spectroscopic analysis of total ionic conductivity based on the composition's morphology insists on an increased ionic conductivity with a proportional increase in the activation barrier. The afore-attribute is the aftermath of two discrete grain boundary actions. The first

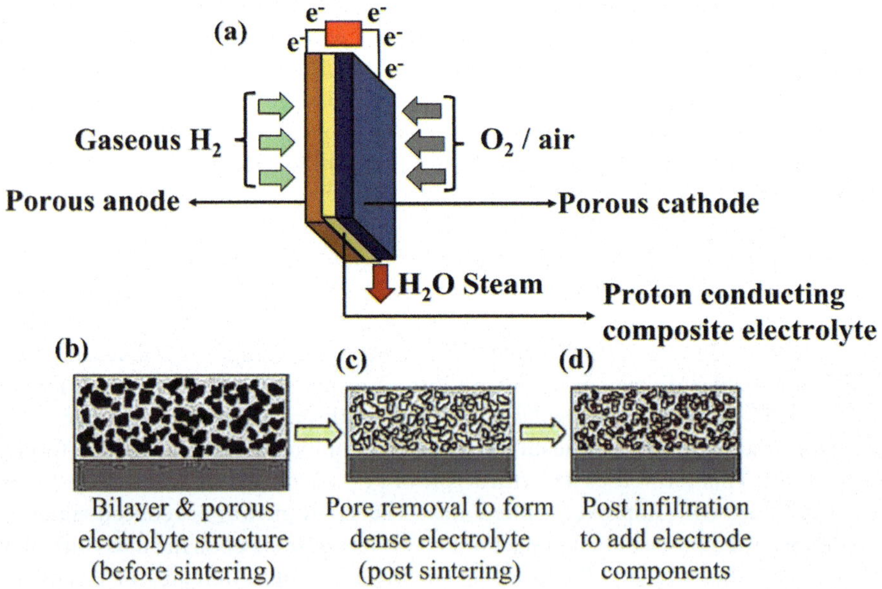

Fig. 13 Proton-conducting solid oxide fuel cell technology with **a** composite electrolyte, **b** porous electrolyte microstructure (before sintering), **c** dense electrolyte (post sintering), **d** infiltered electrolyte with electrode components

of which is the homogeneous grain boundary of the individual compositions (BZY-BZY or GDC-GDC) while the other is the interfacial grain boundary of the coupled constituents (BZY-GDC). The simultaneous increase in conductivity and activation energy was profoundly observed with a subsequent rise in GDC concentration within the composite electrolyte. The effect is significant against humidified H_2 atmosphere with considerable conductivity (0.026 S cm^{-1}) and power density (0.442 W cm^{-2} @ 700 °C), respectively. However, recent research motivates us to unify the higher doping strategies with proton-conducting composites via state-of-the-art processing techniques to model a chemically stable and highly conductive electrolyte.

Despite significant hype for fuel cell technology enhanced among the research community, energy materials such as $BaCeO_3$ and $BaZrO_3$ with pronounced chemical activity and poor proton conductivity destabilize the contributions and interest. A surge and spark of complex material architecture in terms of higher doping strategy and complex electrolytes began to post the introduction of state-of-the-art material processing techniques. The last decade has witnessed a significant uprisal in diverse complex proton conductors and different strategies to accomplish high chemical compatibility with electrophysical attributes. While higher doping strategies (bi/tri/tetra) forbid structural imperfections via counterbalanced B/M ratios. High ionic conductivity is achieved via a cumulative improvement in structural and microstructural disparity. Higher chemical activity by cerates under reactive and reducing atmospheres is rectified with suitable electronegative acceptor substituents alongside optimized Zr^{4+} cations. The electronegative acceptor substituents within the resultant composite revamp the refractive nature of zirconates and introduce the chemical inertness of the overall composition. However, with an increase in total ionic conductivity, a persistent surge in the activation barrier accompanies due to discrete grain boundary effects. A prominent activity is recognized across the interfacial grain boundary of coupling constituents within a composite electrolyte. Concerns rise due to the overall basicity of the host complex with attenuated proton mobility due to the rise in the potential barrier. Besides, acceptor dopants' novel techniques such as buffer thin-film electrolytes mark a potential candidate to combat broader and sub-categorical material limitations.

3.2 Thin-Film Electrolytes and Buffer Layer

One of the major cornerstones in modern-day technological advancement lies in thin-film technology. Thin-film technology has successfully bridged diverse research gaps and has expanded its scope into numerous emerging research areas. A thin film is generally a solid layer sandwiched between two parallel planes and extended infinitely along two directions which are coated upon a compatible substrate with a variable thickness between 1 μm to 10 Å. The attractiveness of thin film emerges due to observable differences from their bulk. The relative difference in the length scales between the third (z) with respect to the other two directions (x, y) categorizes them within 2D materials. Vast differences in the bulk and thin-film characteristics of the same material arise due to the influence of distinct quantum effects at the microscopic scales. As a result, thin films provide a wider view of the microscopic perspective of different materials. Thin films in recent years have established their monopoly among energy materials. Thin-film proton conductors broaden the outlook of new materials for distributed and portable stationary power generation [80]. A decrease in the electrolyte's resistance with a cooperative increase in the ionic conductivity at low temperatures offers adverse advantages among different proton ceramic thin-film electrolytes. A homogeneous thin film is obtained via stacked multiscale porous layers deposited upon an electrolyte membrane surface. Such deposited multilayer membranes aid to achieve stable thin films with reduced microstructural impedance at low- and intermediate-temperature regimes [219, 220]. However, desirable pre-requisites for a thin-film membrane to be utilized for a fuel cell operation are as follows:

1. Chemical compatibility: The designed thin film must satisfy the bonding requisites between electrolyte-membrane assembly and promote rapid electrode kinetics.
2. Stability: The synthesised electrolyte film must display thermal, electrophysical and mechanical stability under the diverse operating ambiance.
3. Transport characteristics: The electrolyte film must accommodate a high hydration tendency without localized drying. The coated film must exhibit low permeability to gaseous reactants to pacify coulombic inefficiency and long-range unattenuated charge transport.
4. Ionic conductivity: Thin-film PCs must illustrate minimal microstructural resistance to exemplify high currents with zero electronic conductivity.
5. Fuel flexibility: Thin-film PCs must display optimal electrochemical response upon the offered fuel flexibility by PC-SOFC at intermediate temperatures.
6. Economic facets: The fabrication of thin-film PCs must be cost-effective for large-scale commercial production and implications [221].

The compatibility among different cell constituents plays the most fundamental entity for a fuel cell's functioning. Substantial reports claim a large ohmic resistance across the electrode–electrolyte interface. Two effective strategies prevail to combat the afore-effects which involve (1) employing a highly conducting electrolyte and (2) reducing the electrolyte thickness significantly. Highly reactive constituent (Ce^{4+}) in standalone and derived $BaCeO_3$ electrolytes hinders their high electrophysical property under diverse atmospheres, while $BaZrO_3$ on the counterpart establishes low proton conductivity due to poor sinterability [222]. Multiphase co-existence and high microstructural resistance additionally constitute low-conducting bulk electrolytes [10, 11].

However, beyond the use of diverse synthesis routes to enrich the ionic conductivity of distinct electrolytes yet the trade-off parameters obstruct the escalation beyond a threshold. The secondary possibility also awaits key limitations. The electrolyte thickness cannot be reduced to less than 10 μm via conventional synthesis routes which additionally demands high operating temperatures ($\approx$ 700 °C) [223]. Several physical deposition techniques avoid high-temperature processing [224,

225]. As a consequence, thin films with electrolyte thickness provide privileges of reduced electrolyte thickness with cost-effective synthesis. In compliance, Hossain et al. [80] incorporated a proton-conducting ($BaZr_{0.9}Y_{0.1}O_{3-\delta}$) nanoparticle within a sulfonated poly(ether ether ketone) polymeric nanocomposite membrane to serve as a low and intermediate operational electrolyte for PC-SOFC. The outcomes reveal a consistent increase in proton conductivity up to a 1.8% volume fraction of nanoparticles and decrease steeply thereafter with a proportional increase in nanoparticle concentration. The emerged attribute devotes to the nanoparticle agglomeration across the microstructural boundary which narrows the proton-conducting pathways. Additional explanations may suggest a rearrangement of hydrogen bonds between the $-SO_3H$ group and the –OH groups of the nanoparticles. The thermal, mechanical and chemical characteristics improved with the incorporated nanoparticles, while the electrophysical property improved due to the presence of Y^{3+} within the overall composition. The acceptor substituent (Y^{3+}) serves as the water intake site to increase the protonation, respectively. The power density as a result inflates from 0.16 to 0.43 W cm^{-2} at lower temperatures (~80 °C). The resultant material properties thus improve with the optimized composite polymer matrix over their standalone polymer counterpart. In acquiescence with the afore-discussion, Fluri et al. [226] reported an elaborative study on the strained $BaZr_{0.8}Y_{0.2}O_{3-\delta}$ thin films epitaxially grown over a MgO substrate with columnar morphology. The strain within the deposited film emerges due to phase instability and the film-substrate crystal lattice mismatch. As a result, a proton-conducting buffer layer ($BaZr_{0.6}Ce_{0.4}O_3$) is additionally incorporated to neutralize the developed strain. The controlled lattice by the buffer layer increases the proton conductivity due to reduced proton-dopant activation energy via tensile in-plane strain. In support of prior studies, Campos Covarrubias et al. [227] examined the strain developed upon the $BaCeO_3$-based thin film on a metallic substrate. The strain on the electron beam deposited film surfaces due to the disparity of thermal expansion coefficient between the film-substrate assembly. However, the presence of amorphous phases with $BaCeO_3$ and CeO_2 residues at lower temperatures (150–300 °C) hinders long-range proton mobility, while proton conductivity surges at intermediate temperatures (450–600 °C) due to the tendency of highest in-plane tensile strain. Low-temperature processing redefines the phase composition and morphology with the formation of a fibrous network. Material architecture at intermediate temperature offers dense homogeneous crack-free electrolyte films with high overall conductivity. Figure 14 illustrates a buffer thin-film layer in PC-SOFC.

Exner et al. [150] devised a novel powder aerosol deposition strategy to produce dense $BaZr_{0.8}Y_{0.2}O_{3-\delta}$ films for high-temperature applications. The novel thin-film fabrication technique contributes homogeneous crack-free dense microstructure with a considerably high mechanical, chemical, thermal and electrical response at high temperatures. The Vicker's microhardness test showcases a surge in the mechanical stability of about 2.4 times over the bulk sintered samples. The electrophysical characteristics profoundly vary at different temperature regimes. Proton conductivity of the thin-film electrolytes is lower than their bulk counterpart at intermediate temperatures (400–700 °C), while the same increases to 2×10^{-2} S cm^{-2} at higher temperatures (800–1000 °C) due to lower crystallite sizes (about 9 times lower than their

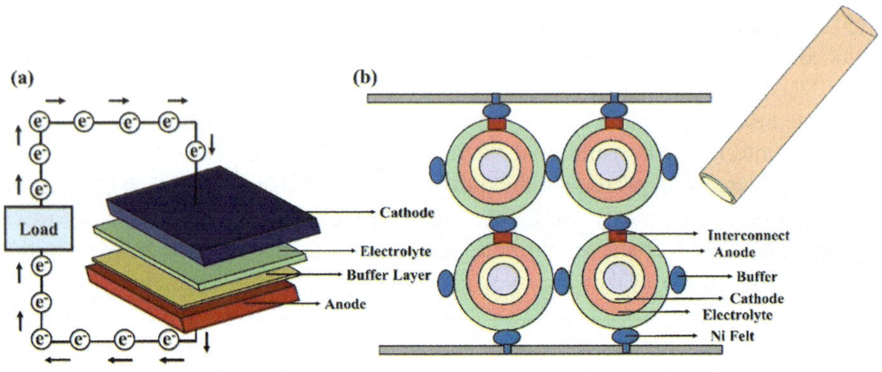

Fig. 14 Use of a thin-film buffer layer as **a** 2D sheet with cell components, **b** tubular geometry with cell components

bulk counterpart) and consequent rise in proton transfer number (~ 0.9). This proves the extended electrochemical activity of the multifunctional proton-conducting thin film for high-temperature applications. Besides numerous advantages, key challenges persist that hinders the operational efficiency of the fuel cell. The utilization of a thin-film proton-conducting electrolyte establishes an interface and contributes chemical and mechanical strain (different from the tensile in-plane strain). This introduces large microstructural resistance for proton migration. Besides, the presence of large acceptor concentrations inculcates proton trapping centres in their close coordination due to inductive effect via electronegativity difference. Thus, multiple factors including dopant-host ionic radius and doping concentration become essential pre-requisite to monitor prior substitution and thin-film synthesis. However, recent pieces of literatures suggest hybrid proton conductors to counterbalance the afore-issues and yield desirable outcomes.

In an overview, adequate structural compatibility between the core film and the substrate becomes essential to achieve optimum conductivity with the absence of undesirable strain that establishes reduced proton transfer due to microstructural impedance and phase instability. The incorporation of a proton-conducting buffer layer additionally pacifies the strain and contributes towards tensile in-plane strain suitable to escalate the resultant proton conductivity of the host composition. Different pieces of literature have also administered numerous deposition techniques beneficial to lower the material processing temperature with high characteristic properties and counter the material degradation issues suffered by their bulk counterpart. The intermediate-temperature-dependent proton conduction by different bulk materials can be extended to higher temperatures with the utilization of a suitable synthesis route.

3.3 Metal–Organic Framework

Metal–organic frameworks (MOFs) comprise porous polycrystalline materials via interconnected metallic clusters and crosslinked organic linkers in a unique arrangement that accommodates the selective adsorption of gaseous constituents [228–230]. Increasing attention in recent years is provided to examining the structural integrity and electrophysical response of MOFs under diverse operating constraints to guarantee their application as the proton-conducting electrolyte in PC-SOFC [77, 231–233]. A diverse range of metal-embedded nodes with variable organic networks has allowed numerous material improvisations over 60,000 distinct MOFs reported structures which is progressively escalating with commercialization motives [234]. The fundamental material architecture begins with a feasible and relevant choice of compositional precursors followed by competent material processing techniques (modified combustion synthesis, Pechini method, slow evaporation and many more), respectively. Fine-tuning of material characteristics is executed via photosynthetic chemical modifications which in compliance with high specific surface area and adequate porosity enhances the surface catalytic activity and the resultant electrochemical response of the cell [235–239].

High charge adsorption and mobility are not only ideal for electrochemistry but even for distinct biological functionality. However, constrained attention over energy conversion devices demands the presence of dense pronounced 1D channels besides sub-micro-sized pores. The adversity of the MOF framework offers two essential categories of proton-conducting electrolytic behaviour. The first of which is the conventional protonation under humified atmospheres with sufficient H_2O intake as steam, while the latter category involves protonic conduction under anhydrous conditions. In principle, the conventional protonation demands feebly porous PCs ($\leq 10\%$) for requisite hydration and subsequent proton migration via the hydrogen-bonded network with the adjacent oxygen octahedra. Such porous MOFs with crosslinked organic frameworks favour the scope of low-temperature proton transfer [240–243]. The anhydrous proton conduction on the contrary traditionally depends on organic heterocyclic compounds which are pronounced charge carriers and favours long-range proton migration. However, the temperature-dependent and microstructural stability of organic constituents provokes critical concerns. As a result, material engineering of highly stable anhydrous proton conductors becomes a calumniatory objective since constrained porosity constitutes a freeway for unimpeded proton transport. Profound investigations to accomplish afore-attributes prompt towards covalent organics [244–247]. Covalent organic frameworks mend the stability criteria via rigid covalent bonding, while the coupled metal constituents within the MOFs assist with the requisite electrical response. A means to achieve free-flow of charge carriers between metal and organic linkage is to strengthen the inter and intralayer interactions which also demonstrates a positive impact on the material's thermodynamical stability [248, 249]. A suitable coupling between basic building blocks within MOFs triggers the proton-loaded channels by preserving the pore distribution on the network [240]. As a result, a metal–organic framework with covalent organics can be designed via the topology-guided origin of polygonal fortitude to obtain dense channels. This additionally favours improvising the pore size and distribution [246, 247].

The works of Tao et al. [250] illustrated a strategy for the material architecture of a covalent organic framework as proton-conducting electrolytes with dense 1-D nanochannels. Such a dense network allows energetically favourable sites on the pore walls to establish a stabilized network between COFs and the proton chain. Additionally, the network further facilitates a proton super-flow via bulk diffusion. The study reveals the three-fold coordinated mechanism for long-range proton transport within anhydrous MOFs. The primary and secondary sequel corresponds to establishing the stability criterion of the MOFs and proton-MOF network on the energetically favourable sites, respectively. The tertiary requisite is to equally improve the density of favourable sites to enhance the charge carrier concentration and the resultant proton conductivity. Figure 15 displays a few energetically favourable sites within a MOF network for active proton transport.

Such a three-fold mechanism is non-degenerate from the proton transport within a biological system which follows a unique molecular mechanism equipped with macropores (~ > 2 nm), respectively. However, the study on a comparative note establishes the superiority of feasible long-range proton transport within MOFs over any other molecular mechanisms due to chemically stable constrained pathways in the form of dense 1D nanochannels. Bao et al. [251] on the other hand disclosed the proton conduction of MOFs and polymer-oriented ligands. Recent events of insoluble and chemically inert MOFs such as Zr-based MOFs open the scope to tune the acidic characteristics of the pores via distinct proton transport mechanisms and defect engineering. The consumption of MOFs with imidazole proposes an isolated

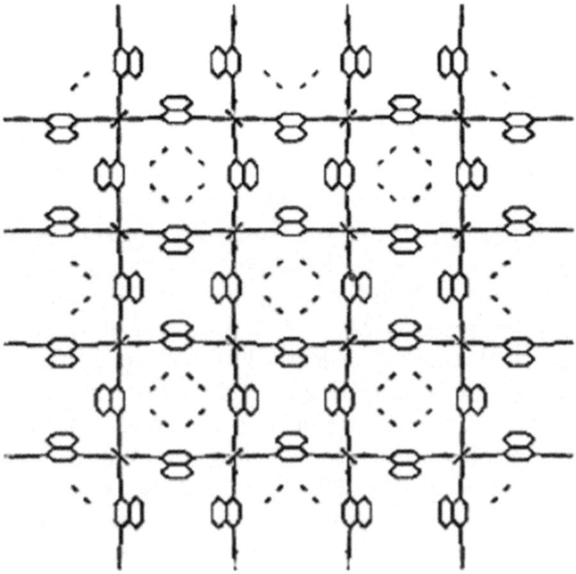

Fig. 15 Energetically favourable sites for proton occupancy within a MOF network

strategy to accomplish proton migration via the Grotthuss mechanism. As a result, Escorihuela et al. [234] claimed the low proton conductivity of imidazole@Al-MOFs between 10^{-5} and 10^{-6} S cm^{-1} at 120 °C, while histamine@Al-MOF reported a pronounced hike up to 10^{-3} S cm^{-1} under congruent conditions (~ 150 °C). The afore-attributes are the consequence of high protonation in histamine over imidazole with favourable proton occupancy sites. While bulk charge diffusion through MOFs is progressively developing, electrode–electrolyte compatibility has been rendered equal priority to enhance the interfacial electrochemistry. This reflects upon the overall cell functionality and efficiency with enhanced power density. The fuel electrode becomes an essential component for considerable fuel utilization and oxidation for active charge carrier generation across the electrolyte. The afore-characteristics are possibly achievable with a complex material design. In principle, the electrode must display both chemical and electrophysical stability under diverse atmospheric exposure with constant surface activity. Meanwhile, the critical issues such as unintended precipitates, rise in polarization resistance and fuel interactions with the electrolyte constituents and the diffusion of porous substituents across the bulk electrolyte must be eliminated with ageing to uplift the operational durability (> 500 h). As a result, a smooth transition towards hybrid and composite cathode materials over their standalone counterpart to ensure their commercial utilization for low- and intermediate-temperature-centric practical applications.

4 Future Research Directions

- Annealing the acceptor-doped proton conductors assists to eliminate the proton trapping effect to the maximum extent.
- The use of a thin-film buffer layer across the electrode–electrolyte interface prohibits the infusion of porous electrode constituents within the electrolyte and the interference of gaseous fuel with the oxidized product on the anodic surface.
- Gas switching within symmetric cells oxidizes and eliminates the carbon and sulphur poisoning for most cerates with active chemical reactivity.
- Controlled hydration partial pressure (pH_2O) compels to optimize device functioning and stability. In a macroscopic view, grain boundary impedance in PCs can be slashed by escalating the grain size and employing nanosized materials with better sinterability or uniform sintering additives.
- Substantial speculation indicates material-oriented research gaps for fuel cell technology. A gradual shift over the years has been visualized from electrolytes to electrodes. The chemical compatibility and operation of the electrode–electrolyte system have been collectively focused and worked upon by multiple research groups.
- Proton-conducting cathode composites appeal material improvisations due to weak link between agglomerated particles and escalating surface diffusion resistance besides charge mobility through porous electrode.

- Chemical and mechanically stable $BaZrO_3$-based PCs have trivially captivated their implications in diverse sectors. Ceramic composites as smart functional materials for advanced applications have subdued energy, defense and communication sectors.
- $BaZrO_3$-modelled materials are recently used for optical applications exploiting wide band gaps (3.8–7.0 eV). Equipped with excellent piezoelectric and ferroelectric properties, novel endeavours are made to vary ferroelectric $BaZrO_3$-based perovskites to ferromagnetic material with progressive metal doping.

5 Conclusion

Proton conductors are utilitarian materials with extensive applications ranging from photocatalysts to fuel cell components. A surge in fuel cell research is foreseen to progress with an ultimatum of sustainable development. Although, fuel cells provide continuous energy conversion with pollutant-free fuel, material choice, synthesis procedures and operating conditions limit the overall cell performance. Hybrid and complex oxides although are a means to extend the functional efficiency; however, the physics of proton conductors is a pre-requisite to suitably engineer such complex ceramics. Among cerates and zirconates, facile reductions of Ce^{4+} to Ce^{3+} are responsible for prominent pronation via oxygen vacancies and photocatalytic activity. On the contrary, the high basicity of cerates increases the chemical reactivity under humified and reducing atmospheres. The refractive microstructure of zirconates on the other hand is the consequence of poor proton conductivity at intermediate temperatures. While acceptor substitutions tailor the material properties, dopant-host inconsistency (ionic radii mismatch and high electronegativity) invites structural phase transitions and charge chemistry variations leading to the proton trapping effect. Recent years have demonstrated sincere attempts in terms of higher doping strategy, thin-film electrolytes and metal–organic frameworks to improve the operational efficiency and eliminate critical drawbacks. Although such material improvisations have showcased a pronounced positivity with respect to former research activities, further optimizations are necessary to boost the resultant electrochemistry.

Acknowledgements The author acknowledges DST-INSPIRE (DST/INSPIRE/03/2021/002004) for financial aid and equivalent support by Birla Institute of Technology Mesra for providing essential resources for the research work.

References

1. Mohammed H, Al-othman A, Nancarrow P et al (2019) Direct hydrocarbon fuel cells: a promising technology for improving. Energy. https://doi.org/10.1016/j.energy.2019.01.105
2. Lyu Y, Xie J, Wang D (2020) Review of cell performance in solid oxide fuel cells. J Mater Sci 55:7184–7207. https://doi.org/10.1007/s10853-020-04497-7

3. Tahir NNM, Baharuddin NA, Samat AA et al (2022) A review on cathode materials for conventional and proton-conducting solid oxide fuel cells. J Alloys Compd 894:162458
4. Chang Y, Qin Y, Yin Y et al (2018) Humidification strategy for polymer electrolyte membrane fuel cells—a review. Appl Energy 230:643–662. https://doi.org/10.1016/j.apenergy.2018.08.125
5. Shi H, Su C, Ran R et al (2020) Electrolyte materials for intermediate-temperature solid oxide fuel cells. Prog Nat Sci Mater Int 30:764. https://doi.org/10.1016/j.pnsc.2020.09.003
6. Shearing PR (2018) Energy Fuels 2. https://doi.org/10.1039/c8se00292d
7. Li G, Gou Y, Qiao J et al (2020) Recent progress of tubular solid oxide fuel cell: from materials to applications. J Power Sources 477:228693. https://doi.org/10.1016/j.jpowsour.2020.228693
8. Junaida A, Aziz A, Akidah N et al (2020) Review of composite cathodes for intermediate-temperature solid oxide fuel cell applications. Ceram Int 46:23314. https://doi.org/10.1016/j.ceramint.2020.06.176
9. Vignesh D, Sonu BK, Rout E (2022) Factors constituting proton trapping in $BaCeO_3$ and $BaZrO_3$ perovskite proton conductors in fuel cell technology: a review. Energy Fuels. https://doi.org/10.1021/acs.energyfuels.2c00650
10. Vignesh D, Rout E (2022) Structural and electrophysical perturbations in Y^{3+} doped $BaCeO_3$ proton conducting electrolyte: a first principles approach. Comput Condens Matter 33:e00763. https://doi.org/10.1016/j.cocom.2022.e00763
11. Vignesh D, Rout E (2022) Analysis of symmetry variation as a function of rare earth dopant concentration in proton conducting solid oxide fuel cells. Mater Today Proc 66:3416–3421. https://doi.org/10.1016/j.matpr.2022.08.373
12. Scherban T, Villeneuve R, Abello L, Lucazeau G (1993) Raman scattering study of acceptor-doped $BaCeO_3$. Solid State Ionics 61:93–98
13. Takeuchi K, Loong C-K, Richardson JW Jr et al (2000) The crystal structures and phase transitions in Y-doped $BaCeO_3$: their dependence on Y concentration and hydrogen doping. Solid State Ionics 138:63–77
14. Rajendran S, Thangavel NK, Ding H et al (2020) Tri-doped $BaCeO_3$–$BaZrO_3$ as a chemically stable electrolyte with high proton-conductivity for intermediate temperature solid oxide electrolysis cells (SOECs). ACS Appl Mater Interfaces 12:38275–38284
15. Sagar R, Gaur MS, Raghav RK (2022) Study of structural, thermal and piezoelectric properties of polyvinylidene fluoride–$BaZrO_3$ nanocomposites. J Therm Anal Calorim 147:10371–10381
16. Rahman MA, Hasan W, Khatun R et al (2023) An ab-initio study to investigate the structural, mechanical, electrical, optical and thermal properties of the $AZrO_3$ (A = Mg, Ca, Sr, Ba, Sn, Cu) compounds. Mater Today Commun 105339
17. Nayak AK, Sasmal A (2023) Recent advance on fundamental properties and synthesis of barium zirconate for proton conducting ceramic fuel cell. J Clean Prod 135827
18. Dawood F, Anda M, Shafiullah GM (2020) Hydrogen production for energy: an overview. Int J Hydrogen Energy 45:3847–3869
19. Zhang Z, Pan S-Y, Li H et al (2020) Recent advances in carbon dioxide utilization. Renew Sustain Energy Rev 125:109799
20. Hibino T, Mizutani K, Iwahara H (1993) H/D isotope effect on electrochemical pumps of hydrogen and water vapor using a proton-conductive solid electrolyte. J Electrochem Soc 140:2588
21. Kawamura Y, Isobe K, Yamanishi T (2007) Mass transfer process of hydrogen via ceramic proton conductor membrane of electrochemical hydrogen pump. Fusion Eng Des 82:113–121
22. Lin K, Chen Q, Gerhardt MR et al (2015) Alkaline quinone flow battery. Science (80-) 349:1529–1532
23. Aristidou N, Eames C, Sanchez-Molina I et al (2017) Fast oxygen diffusion and iodide defects mediate oxygen-induced degradation of perovskite solar cells. Nat Commun 8:15218
24. Kadi MW, Mohamed RM (2019) Synthesis of $BaCeO_3$ nanoneedles and the effect of V, Ag, Au, Pt doping on the visible light hydrogen evolution in the photocatalytic water splitting reaction. J Sol-Gel Sci Technol 91:138–145. https://doi.org/10.1007/s10971-019-05018-y

25. Gundeboina R, Perala V, Muga V (2020) Perovskite material-based photocatalysts. Revolut Perovskite Synth Prop Appl 251–287
26. Mekhilef S, Saidur R, Kamalisarvestani M (2012) Effect of dust, humidity and air velocity on efficiency of photovoltaic cells. Renew Sustain Energy Rev 16:2920–2925
27. Wang J (2015) Barriers of scaling-up fuel cells: cost, durability and reliability. Energy 80:509–521
28. Ioroi T, Siroma Z, Yamazaki S, Yasuda K (2019) Electrocatalysts for PEM fuel cells. Adv Energy Mater 9:1801284
29. Wang Y, Diaz DFR, Chen KS et al (2020) Materials, technological status, and fundamentals of PEM fuel cells—a review. Mater Today 32:178–203
30. Maystrenko AL, Kushch VI, Pashchenko EA et al (2020) Ceramic armour for armoured vehicles against large-calibre bullets. Probl Mechatroniki Uzbroj lotnictwo, inżynieria bezpieczeństwa 11
31. Hart M (2020) Vulnerabilities and challenges of integrating AI into future air force intelligence systems
32. Rizwan M, Aleena S, Shakil M et al (2020) A computational insight of electronic and optical properties of Cd-doped $BaZrO_3$. Chinese J Phys 66:318–326
33. Al Azar S, Al-Zoubi I, Mousa AA et al (2021) Investigation of electronic, optical and thermoelectric properties of perovskite $BaTMO_3$ (TM = Zr, Hf): First principles calculations. J Alloys Compd 887:161361
34. Husain J, Anvarullha M, Raghu N et al. DC Conductivity studies on polyethylene oxide/nickel ferrite composites
35. Sunarso J, Baumann S, Serra JM et al (2008) Mixed ionic–electronic conducting (MIEC) ceramic-based membranes for oxygen separation. J Membr Sci 320:13–41
36. Gilardi E, Fabbri E, Bi L et al (2017) Effect of dopant–host ionic radii mismatch on acceptor-doped barium zirconate microstructure and proton conductivity. J Phys Chem C 121:9739–9747
37. Fu C-F, Wu X, Yang J (2018) Material design for photocatalytic water splitting from a theoretical perspective. Adv Mater 30:1802106
38. Fajrina N, Tahir M (2019) A critical review in strategies to improve photocatalytic water splitting towards hydrogen production. Int J Hydrogen Energy 44:540–577
39. Lin L, Hisatomi T, Chen S et al (2020) Visible-light-driven photocatalytic water splitting: recent progress and challenges. Trends Chem 2:813–824
40. Cao S, Piao L (2020) Considerations for a more accurate evaluation method for photocatalytic water splitting. Angew Chem Int Ed 59:18312–18320
41. Yuan Y, Zhang X, Liu L et al (2008) Synthesis and photocatalytic characterization of a new photocatalyst $BaZrO_3$. Int J Hydrogen Energy 33:5941–5946
42. Jun A, Ju Y-W, Kim G (2015) Solid oxide electrolysis: concluding remarks. Faraday Discuss 182:519–528
43. Jun A, Kim J, Shin J, Kim G (2016) Achieving high efficiency and eliminating degradation in solid oxide electrochemical cells using high oxygen-capacity perovskite. Angew Chem Int Ed 55:12512–12515
44. Luo M, Lu P, Yao W et al (2016) Shape and composition effects on photocatalytic hydrogen production for Pt–Pd alloy cocatalysts. ACS Appl Mater Interfaces 8:20667–20674
45. Qin J, Zeng H (2017) Photocatalysts fabricated by depositing plasmonic Ag nanoparticles on carbon quantum dots/graphitic carbon nitride for broad spectrum photocatalytic hydrogen generation. Appl Catal B Environ 209:161–173
46. Rather RA, Singh S, Pal B (2017) AC_3N_4 surface passivated highly photoactive Au-TiO_2 tubular nanostructure for the efficient H_2 production from water under sunlight irradiation. Appl Catal B Environ 213:9–17
47. Kumar N (2016) Bulletproof vest and its improvement–a review. Int J Sci Dev Res 1:34–39
48. Samir NS, Radwan MA, Sadek MA, Elazab HA (2018) Preparation and characterization of bullet-proof vests based on polyamide fibers. Int J Eng Technol 7:1290–1294

49. Tepeduzu B, Karakuzu R (2019) Ballistic performance of ceramic/composite structures. Ceram Int 45:1651–1660
50. Nurazzi NM, Asyraf MRM, Khalina A et al (2021) A review on natural fiber reinforced polymer composite for bullet proof and ballistic applications. Polymers (Basel) 13:646
51. Reddy PRS, Savio SG, Madhu V (2020) Ceramic composite armour for ballistic protection. Handbook of advanced ceramics and composites: defense, security, aerospace and energy applications, pp 357–402
52. Zhang J, Liu Z, Zhang T et al (2021) High strain response and low hysteresis in $BaZrO_3$-modified KNN-based lead-free relaxor ceramics. J Mater Sci Mater Electron 32:16715–16725
53. Li F, Wu S, Li T et al (2020) Normal-relaxor ferroelectric phase transition induced morphotropic phase boundary accompanied by enhanced piezoelectric and electrostrain properties in strontium modulated $Bi_{0.5}K_{0.5}TiO_3$ lead-free ceramics. J Eur Ceram Soc 40:3918–3927
54. Yin Y, Tang Y, Pan W et al (2021) Relaxor behaviors enhance piezoelectricity in lead-free $BiFeO_3$-$BaTiO_3$ ceramics incorporated with a tiny amount of $Bi(Mg_{1/2}Ti_{1/2})O_3$ near the morphotropic phase boundary. Ceram Int 47:9486–9494
55. Wu L, Zheng T, Wu J (2022) Excellent fatigue resistance in Sb nonstoichiometric KNN-based ceramics by engineering relaxor multiphase state. J Eur Ceram Soc 42:4888–4897
56. Singh K, Kannan R, Thangadurai V (2019) Perspective of perovskite-type oxides for proton conducting solid oxide fuel cells. Solid State Ionics 339:114951
57. Cathcart JV, Perkins RA, Bates JB, Manley LC (1979) Tritium diffusion in rutile (TiO_2). J Appl Phys 50:4110–4119
58. Stotz S, Wagner C (1966) Die löslichkeit von wasserdampf und wasserstoff in festen oxiden. Berichte der Bunsengesellschaft für physikalische Chemie 70:781–788
59. Zhang Z, Chen L, Li Q et al (2018) High performance In, Ta and Y-doped $BaCeO_3$ electrolyte membrane for proton-conducting solid oxide fuel cells. Solid State Ionics 323:25–31
60. Yokokawa H (2009) Overview of intermediate-temperature solid oxide fuel cells. Perovskite oxide solid oxide fuel cells, pp 17–43
61. Choudhury A, Chandra H, Arora A (2013) Application of solid oxide fuel cell technology for power generation—a review. Renew Sustain Energy Rev 20:430–442
62. Tarragó DP, Moreno B, Chinarro Martín E, de Sousa VC (2016) Perovskites used in fuel cells. Pan L, Zhu G (eds) Perovskite Materials: synthesis, characterisation, properties, and applications [recurso eletrônico] [Rijeka, Croatia], chap 21. InTech, pp 619–637
63. Chambi-Rocha A, Cabrera-Domínguez ME, Domínguez-Reyes A (2018) Breathing mode influence on craniofacial development and head posture. J Pediatr (Versão em Port) 94:123–130
64. Li P, Yang W, Tian C et al (2021) Electrochemical performance of $La_2NiO_{4+\delta}$-$Ce_{0.55}La_{0.45}O_{2-\delta}$ as a promising bifunctional oxygen electrode for reversible solid oxide cells. J Adv Ceram 10:328–337
65. Wang W, Medvedev D, Shao Z (2018) Gas humidification impact on the properties and performance of perovskite-type functional materials in proton-conducting solid oxide cells. Adv Funct Mater 28:1802592
66. Horita T, Yokokawa H (2005) Solid oxide fuel cells. In: Materials for energy conversion devices. Elsevier, pp 140–173
67. Anjaneya KC, Nayaka GP, Manjanna J et al (2013) Preparation and characterization of $Ce_{1-x}Gd_xO_{2-\delta}$ ($x = 0.1$–0.3) as solid electrolyte for intermediate temperature SOFC. J Alloys Compd 578:53–59
68. Steele BCH (1994) Oxygen transport and exchange in oxide ceramics. J Power Sources 49:1–14
69. Jaiswal N, Tanwar K, Suman R et al (2019) A brief review on ceria based solid electrolytes for solid oxide fuel cells. J Alloys Compd 781:984–1005
70. Guo X, Waser R (2006) Electrical properties of the grain boundaries of oxygen ion conductors: acceptor-doped zirconia and ceria. Prog Mater Sci 51:151–210

71. Mendonça C, Santos DMF (2021) Towards the commercialization of solid oxide fuel cells: recent advances in materials and integration strategies. Fuels 2:393–419
72. Kreuer K-D (2003) Proton-conducting oxides. Annu Rev Mater Res 33:333–359
73. Fu Y-P, Weng C-S (2014) Effect of rare-earth ions doped in $BaCeO_3$ on chemical stability, mechanical properties, and conductivity properties. Ceram Int 40:10793–10802
74. Li J, Wang C, Wang X, Bi L (2020) Sintering aids for proton-conducting oxides—a double-edged sword? A mini review. Electrochem Commun 112:106672
75. Tarutin A, Kasyanova A, Lyagaeva J et al (2020) Towards high-performance tubular-type protonic ceramic electrolysis cells with all-Ni-based functional electrodes. J Energy Chem 40:65–74
76. Fabbri E, Pergolesi D, Traversa E (2010) Materials challenges toward proton-conducting oxide fuel cells: a critical review. Chem Soc Rev 39:4355–4369
77. Zhang W, Hu YH (2021) Progress in proton-conducting oxides as electrolytes for low-temperature solid oxide fuel cells: from materials to devices. Energy Sci Eng 9:984–1011
78. Meng Y, Gao J, Zhao Z et al (2019) recent progress in low-temperature proton-conducting ceramics. J Mater Sci 54:9291–9312
79. Islam MS, Nolan AM, Wang S et al (2020) A computational study of fast proton diffusion in Brownmillerite $Sr_2Co_2O_5$. Chem Mater 32:5028–5035
80. Hossain MK, Biswas MC, Chanda RK et al (2021) A review on experimental and theoretical studies of perovskite barium zirconate proton conductors. Emerg Mater 4:999–1027. https://doi.org/10.1007/s42247-021-00230-5
81. Hossain S, Abdalla AM, Jamain SNB et al (2017) A review on proton conducting electrolytes for clean energy and intermediate temperature-solid oxide fuel cells. Renew Sustain Energy Rev 79:750–764
82. Kreuer K-D (1996) Proton conductivity: materials and applications. Chem Mater 8:610–641
83. Vera CYR, Ding H, Peterson D et al (2021) A mini-review on proton conduction of $BaZrO_3$-based perovskite electrolytes. J Phys Energy 3:32019
84. Lim D, Kitagawa H (2021) Rational strategies for proton-conductive metal–organic frameworks. Chem Soc Rev https://doi.org/10.1039/d1cs00004g
85. Jing J, Pang J, Chen L et al (2022) Structure, synthesis, properties and solid oxide electrolysis cells application of $Ba(Ce, Zr)O_3$ based proton conducting materials. Chem Eng J 429:132314. https://doi.org/10.1016/j.cej.2021.132314
86. Fischer SA, Gunlycke D (2019) Analysis of correlated dynamics in the Grotthuss mechanism of proton diffusion. J Phys Chem B 123:5536–5544
87. Gupta A, Goswami S, Elahi SM, Konar S (2021) Role of Framework–carrier interactions in proton-conducting crystalline porous materials. Cryst Growth Des. https://doi.org/10.1021/acs.cgd.0c01394
88. Papac M, Stevanović V, Zakutayev A, O'Hayre R (2021) Triple ionic–electronic conducting oxides for next-generation electrochemical devices. Nat Mater 20:301–313
89. Li S, Irvine JTS (2021) Non-stoichiometry, structure and properties of proton-conducting perovskite oxides. Solid State Ionics 361:115571
90. Xu X, Xu Y, Ma J et al (2021) Tailoring electronic structure of perovskite cathode for proton-conducting solid oxide fuel cells with high performance. J Power Sources 489:1–9. https://doi.org/10.1016/j.jpowsour.2021.229486
91. Zhao C, Li Y, Zhang W et al (2020) Heterointerface engineering for enhancing the electrochemical performance of solid oxide cells. Energy Environ Sci 13:53–85
92. Tao Z, Yan L, Qiao J et al (2015) A review of advanced proton-conducting materials for hydrogen separation. Prog Mater Sci 74:1–50
93. Stambouli AB, Traversa E (2002) Solid oxide fuel cells (SOFCs): a review of an environmentally clean and efficient source of energy. Renew Sustain Energy Rev 6:433–455
94. Jamsak W, Assabumrungrat S, Douglas PL et al (2007) Performance of ethanol-fuelled solid oxide fuel cells: proton and oxygen ion conductors. Chem Eng J 133:187–194
95. Medvedev D, Murashkina A, Pikalova E et al (2014) $BaCeO_3$: materials development, properties and application. Prog Mater Sci 60:72–129

96. Lan R, Tao S (2013) Proton-conducting materials as electrolytes for solid oxide fuel cells. Mater High-Temp Fuel Cells 133–158
97. Medvedev DA, Lyagaeva JG, Gorbova EV et al (2016) Advanced materials for SOFC application: strategies for the development of highly conductive and stable solid oxide proton electrolytes. Prog Mater Sci 75:38–79
98. Liu JF, Nowick AS (1992) The incorporation and migration of protons in Nd-doped $BaCeO_3$. Solid State Ionics 50:131–138
99. Saparov B, Mitzi DB (2016) Organic–inorganic perovskites: structural versatility for functional materials design. Chem Rev 116:4558–4596
100. Gao P, Bin Mohd Yusoff AR, Nazeeruddin MK (2018) Dimensionality engineering of hybrid halide perovskite light absorbers. Nat Commun 9:5028
101. Euvrard J, Yan Y, Mitzi DB (2021) Electrical doping in halide perovskites. Nat Rev Mater 6:531–549
102. Jacobson AJ, Tofield BC, Fender BEF (1972) The structures of $BaCeO_3$, $BaPrO_3$ and $BaTbO_3$ by neutron diffraction: lattice parameter relations and ionic radii in O-perovskites. Acta Crystallogr Sect B Struct Crystallogr Cryst Chem 28:956–961
103. Yamanaka S, Fujikane M, Hamaguchi T et al (2003) Thermophysical properties of $BaZrO_3$ and $BaCeO_3$. J Alloys Compd 359:109–113
104. Matsumoto H, Kawasaki Y, Ito N et al (2007) Relation between electrical conductivity and chemical stability of $BaCeO_3$-based proton conductors with different trivalent dopants. Electrochem Solid-State Lett 10:B77
105. Münch W, Seifert G, Kreuer KD, Maier J (1996) A quantum molecular dynamics study of proton conduction phenomena in $BaCeO_3$. Solid State Ionics 86:647–652
106. Scherban T, Villeneuve R, Abello L, Lucazeau G (1993) Raman scattering study of acceptor-doped $BaCeO_3$. Solid State Ionics 61:93–98
107. Zhao F, Chen F (2010) Performance of solid oxide fuel cells based on proton-conducting $BaCe_{0.7}In_{0.3-x}Y_xO_{3-\delta}$ electrolyte. Int J Hydrogen Energy 35:11194–11199
108. Affandi NSM, Zainor ML, Hassan OH et al (2022) Review on the preparation of electrolyte thin films based on cerate-zirconate oxides for electrochemical analysis of anode-supported proton ceramic fuel cells. J Alloys Compd 165434
109. Pİşkİn B, Pİşkİn F (2022) Production and characterization of sputtered Y-doped $BaZrO_3$ for proton conducting oxides. JOM 74:4181–4187
110. Baharuddin NA, Abd Rahman H, Samat AA et al (2023) Perovskite-structured ceramics in solid oxide fuel cell application. In: Perovskite ceramics. Elsevier, pp 221–261
111. Shen H-Z, Guo N, Shen P (2023) Synthesis and densification of $BaZrO_3$ ceramics by reactive cold sintering of $Ba(OH)_2 \cdot 8H_2O$-$Zr(OH)_4$ powders. J Eur Ceram Soc 43:392–400
112. Rizi VS (2019) Ce Pte Us Pt. Mater Res Express 0–12
113. Jeong Y-C, Kim B-K, Kim Y-C (2014) Proton migration in bulk orthorhombic barium cerate using density functional theory. Solid State Ionics 259:1–8
114. Iguchi F, Tsurui T, Sata N et al (2009) The relationship between chemical composition distributions and specific grain boundary conductivity in Y-doped $BaZrO_3$ proton conductors. Solid State Ionics 180:563–568
115. Makagon E, Merkle R, Maier J, Lubomirsky I (2020) Influence of hydration and dopant ionic radius on the elastic properties of $BaZrO_3$. Solid State Ionics 344:115130
116. Yamazaki Y, Blanc F, Okuyama Y et al (2013) Proton trapping in yttrium-doped barium zirconate. Nat Mater 12:647–651
117. Draber FM, Ader C, Arnold JP et al (2020) Nanoscale percolation in doped $BaZrO_3$ for high proton mobility. Nat Mater 19:338–346
118. Bork N, Bonanos N, Rossmeisl J, Vegge T (2010) Simple descriptors for proton-conducting perovskites from density functional theory. Phys Rev B 82:14103
119. Kochetova N, Animitsa I, Medvedev D et al (2016) Recent activity in the development of proton-conducting oxides for high-temperature applications. Rsc Adv 6:73222–73268
120. Sunarso J, Hashim SS, Zhu N, Zhou W (2017) Perovskite oxides applications in high temperature oxygen separation, solid oxide fuel cell and membrane reactor: a review. Prog Energy Combust Sci 61:57–77

121. Alberti G, Casciola M (2001) Solid state protonic conductors, present main applications and future prospects. Solid State Ionics 145:3–16
122. Zhou Y, Guan X, Zhou H et al (2016) Strongly correlated perovskite fuel cells. Nature 534:231–234
123. Levin I, Han M-G, Playford HY et al (2021) Nanoscale-correlated octahedral rotations in $BaZrO_3$. Phys Rev B 104:214109
124. Perrichon A, Jedvik Granhed E, Romanelli G et al (2020) Unraveling the ground-state structure of $BaZrO_3$ by neutron scattering experiments and first-principles calculations. Chem Mater 32:2824–2835
125. Muñoz-García AB, Massaro A, Schiavo E, Pavone M (2020) Tuning perovskite-based oxides for effective electrodes in solid oxide electrochemical cells. In: Solid oxide-based electrochemical devices. Elsevier, pp 1–25
126. Bartel CJ, Sutton C, Goldsmith BR et al (2019) New tolerance factor to predict the stability of perovskite oxides and halides. Sci Adv 5:eaav0693
127. Mburu CW, Gaita SM, Knee CS et al (2017) Influence of yttrium concentration on local structure in $BaZr_{1-x}Y_xO_{3-\delta}$ based proton conductors. J Phys Chem C 121:16174–16181. https://doi.org/10.1021/acs.jpcc.7b05023
128. Medvedev D, Lyagaeva J, Plaksin S et al (2015) Sulfur and carbon tolerance of $BaCeO_3$–$BaZrO_3$ proton-conducting materials. J Power Sources 273:716–723
129. Medvedev DA (2021) Current drawbacks of proton-conducting ceramic materials: how to overcome them for real electrochemical purposes. Curr Opin Green Sustain Chem 32:100549. https://doi.org/10.1016/j.cogsc.2021.100549
130. Zohourian R, Merkle R, Maier J (2017) Proton uptake into the protonic cathode material $BaCo_{0.4}Fe_{0.4}Zr_{0.2}O_{3-\delta}$ and comparison to protonic electrolyte materials. Solid State Ionics 299:64–69
131. Mazzei L, Perrichon A, Mancini A et al (2019) Local structure and vibrational dynamics in indium-doped barium zirconate. J Mater Chem A 7:7360–7372
132. Poolphol P, Muanghlua R, Atiwongsangthong N et al (2019) The study of trivalent-dopants effect on electrical properties of the $BaZr_{0.7}In_{0.3}O_{3-\delta}$ system. Integr Ferroelectr 195:109–118
133. Andreev R, Korona D, Anokhina I, Animitsa I (2022) Proton and oxygen-ion conductivities of hexagonal perovskite $Ba_5In_2Al_2ZrO_{13}$. Materials (Basel) 15:3944
134. Hossain MK, Chanda R, El-Denglawey A et al (2021) Recent progress in barium zirconate proton conductors for electrochemical hydrogen device applications: a review. Ceram Int 47:23725–23748
135. Yang Z-B, Han M-F, Zhu P et al (2011) $Ba_{1-x}Co_{0.9-y}Fe_yNb_{0.1}O_{3-\delta}$ ($x = 0$–0.15, $y = 0$–0.9) as cathode materials for solid oxide fuel cells. Int J Hydrogen Energy 36:9162–9168
136. Yang Z, Liu Y, Zhu T et al (2016) Mechanism analysis of CO_2 corrosion on $Ba_{0.9}Co_{0.7}Fe_{0.2}Nb_{0.1}O_{3-\delta}$ cathode. Int J Hydrogen Energy 41:1997–2001
137. D'Epifanio A, Fabbri E, Di Bartolomeo E et al (2008) Design of $BaZr_{0.8}Y_{0.2}O_{3-\delta}$ protonic conductor to improve the electrochemical performance in intermediate temperature solid oxide fuel cells (IT-SOFCs). Fuel Cells 8:69–76. https://doi.org/10.1002/fuce.200700045
138. Krug F, Schober T, Springer T (1995) In situ measurements of the water uptake in Yb doped $SrCeO_3$. Solid State Ionics 81:111–118
139. Han D, Toyoura K, Uda T (2021) Protonated $BaZr_{0.8}Y_{0.2}O_{3-\delta}$: impact of hydration on electrochemical conductivity and local crystal structure. ACS Appl Energy Mater 4:1666–1676
140. Stevenson DA, Jiang N, Buchanan RM, Henn FEG (1993) Characterization of Gd, Yb and Nd doped barium cerates as proton conductors. Solid State Ionics 62:279–285
141. Iwahara H, Esaka T, Uchida H, Maeda N (1981) Proton conduction in sintered oxides and its application to steam electrolysis for hydrogen production. Solid State Ionics 3:359–363
142. Iwahara H, Uchida H, Ono K, Ogaki K (1988) Proton conduction in sintered oxides based on $BaCeO_3$. J Electrochem Soc 135:529
143. Gu Y-J, Liu Z-G, Ouyang J-H et al (2013) Structure and electrical conductivity of $BaCe_{0.85}Ln_{0.15}O_{3-\delta}$ (Ln = Gd, Y, Yb) ceramics. Electrochim Acta 105:547–553

144. Medvedev DA, Gorbova EV, Demin AK, Tsiakaras P (2014) Conductivity of Gd-doped $BaCeO_3$ protonic conductor in H_2–H_2O–O_2 atmospheres. Int J Hydrogen Energy 39:21547–21552
145. Lim D-K, Lee T-R, Singh B et al (2014) Charge and mass transport properties of $BaCe_{0.45}Zr_{0.4}Y_{0.15}O_{3-\delta}$. J Electrochem Soc 161:F710
146. Shao Z, Zhou W, Zhu Z (2012) Advanced synthesis of materials for intermediate-temperature solid oxide fuel cells. Prog Mater Sci 57:804–874
147. Haugsrud R (2016) High temperature proton conductors-fundamentals and functionalities. Diffusion Foundations, pp 31–79
148. Park J-S, Lee J-H, Lee H-W, Kim B-K (2010) Low temperature sintering of $BaZrO_3$-based proton conductors for intermediate temperature solid oxide fuel cells. Solid State Ionics 181:163–167
149. Shin EK, Anggia E, Park JS (2019) Effects of Al_2O_3 doping in $BaCeO_3$ on chemical stability and electrical conductivity of proton conducting oxides. Solid State Ionics 339:2–7. https://doi.org/10.1016/j.ssi.2019.115007
150. Exner J, Nazarenus T, Kita J, Moos R (2020) Dense Y-doped ion conducting perovskite films of $BaZrO_3$, $BaSnO_3$, and $BaCeO_3$ for SOFC applications produced by powder aerosol deposition at room temperature. Int J Hydrogen Energy 45:10000–10016
151. Fabbri E, Bi L, Tanaka H et al (2011) Chemically stable Pr and Y co-doped barium zirconate electrolytes with high proton conductivity for intermediate-temperature solid oxide fuel cells. Adv Funct Mater 21:158–166
152. Han D, Shinoda K, Sato S et al (2015) Correlation between electroconductive and structural properties of proton conductive acceptor-doped barium zirconate. J Mater Chem A 3:1243–1250
153. Gonçalves MD, Maram PS, Navrotsky A, Muccillo R (2016) Effect of synthesis atmosphere on the proton conductivity of Y-doped barium zirconate solid electrolytes. Ceram Int 42:13689–13696
154. Zhu Z, Sun W, Shi Z, Liu W (2016) Proton-conducting solid oxide fuel cells with yttrium-doped barium zirconate electrolyte films sintered at reduced temperatures. J Alloys Compd 658:716–720
155. Bi L, Shafi SP, Da'as EH, Traversa E (2018) Tailoring the cathode–electrolyte interface with nanoparticles for boosting the solid oxide fuel cell performance of chemically stable proton-conducting electrolytes. Small 14:1801231
156. Demin AK, Tsiakaras PE, Sobyanin VA, Hramova SY (2002) Thermodynamic analysis of a methane fed SOFC system based on a protonic conductor. Solid State Ionics 152:555–560
157. Assabumrungrat S, Sangtongkitcharoen W, Laosiripojana N et al (2005) Effects of electrolyte type and flow pattern on performance of methanol-fuelled solid oxide fuel cells. J Power Sources 148:18–23
158. Sun W, Zhu Z, Shi Z, Liu W (2013) Chemically stable and easily sintered high-temperature proton conductor $BaZr_{0.8}In_{0.2}O_{3-\delta}$ for solid oxide fuel cells. J Power Sources 229:95–101
159. Cervera RB, Oyama Y, Miyoshi S et al (2014) Nanograined Sc-doped $BaZrO_3$ as a proton conducting solid electrolyte for intermediate temperature solid oxide fuel cells (IT-SOFCs). Solid State Ionics 264:1–6
160. Yoo Y, Lim N (2013) Performance and stability of proton conducting solid oxide fuel cells based on yttrium-doped barium cerate-zirconate thin-film electrolyte. J Power Sources 229:48–57
161. Ding J, Balachandran J, Sang X et al (2018) The influence of local distortions on proton mobility in acceptor doped perovskites. Chem Mater 30:4919–4925. https://doi.org/10.1021/acs.chemmater.8b00502
162. Zhu Z, Wang S (2019) Investigation on samarium and yttrium co-doping barium zirconate proton conductors for protonic ceramic fuel cells. Ceram Int 45:19289–19296
163. Loureiro FJA, Nasani N, Reddy GS et al (2019) A review on sintering technology of proton conducting $BaCeO_3$-$BaZrO_3$ perovskite oxide materials for protonic ceramic fuel cells. J Power Sources 438:226991

164. Tao Z, Xu X, Bi L (2021) Density functional theory calculations for cathode materials of proton-conducting solid oxide fuel cells: a mini-review. Electrochem commun 129:107072
165. Sonu BK, Sinha E (2021) Structural, thermal stability and electrical conductivity of zirconium substituted barium cerate ceramics. J Alloys Compd 860:158471
166. Iwahara H, Asakura Y, Katahira K, Tanaka M (2004) Prospect of hydrogen technology using proton-conducting ceramics. Solid State Ionics 168:299–310
167. Bévillon É, Hermet J, Dezanneau G, Geneste G (2014) How dopant size influences the protonic energy landscape in $BaSn_{1-x}M_xO_{3-x/2}$ (M= Ga, Sc, In, Y, Gd, La). J Mater Chem A 2:460–471
168. Shakel Z, Loureiro FJA, Antunes I et al (2022) Tailoring the properties of dense yttrium-doped barium zirconate ceramics with nickel oxide additives by manipulation of the sintering profile. Int J Energy Res 46:21989–22000
169. Li X, Li Z, Zhu Y (2022) Effect of CoO–NiO additives on the microstructure and mechanical properties of microcrystalline corundum abrasives with in-situ formed needle-shaped $LaAl_{11}O_{18}$. Ceram Int 48:33794–33800
170. Leng Z, Huang Z, Zhou X et al (2022) The effect of sintering aids on $BaCe_{0.7}Zr_{0.1}Y_{0.1}Yb_{0.1}O_{3-\delta}$ as the electrolyte of proton-conducting solid oxide electrolysis cells. Int J Hydrogen Energy 47:33861–33871
171. Babar ZUD, Hanif MB, Gao J-T et al (2022) Sintering behavior of $BaCe_{0.7}Zr_{0.1}Y_{0.2}O_{3-\delta}$ electrolyte at 1150° C with the utilization of CuO and Bi_2O_3 as sintering aids and its electrical performance. Int J Hydrogen Energy 47:7403–7414
172. Lindman A, Helgee EE, Wahnstrom G (2017) Comparison of space-charge formation at grain boundaries in proton-conducting $BaZrO_3$ and $BaCeO_3$. Chem Mater 29:7931–7941
173. Polfus JM, Pishahang M, Bredesen R (2018) Influence of Ce^{3+} polarons on grain boundary space-charge in proton conducting Y-doped $BaCeO_3$. Phys Chem Chem Phys 20:16209–16215
174. Somekawa T, Matsuzaki Y, Sugahara M et al (2017) Physicochemical properties of Ba(Zr, Ce)$O_{3-\delta}$-based proton-conducting electrolytes for solid oxide fuel cells in terms of chemical stability and electrochemical performance. Int J Hydrogen Energy 42:16722–16730
175. Pasierb P, Wierzbicka M, Komornicki S, Rekas M (2009) Electrochemical impedance spectroscopy of $BaCeO_3$ modified by Ti and Y. J Power Sources 194:31–37
176. De Souza RA, Dickey EC (2019) The effect of space-charge formation on the grain-boundary energy of an ionic solid. Philos Trans R Soc A 377:20180430
177. Nasani N, Shakel Z, Loureiro FJA et al (2021) Exploring the impact of sintering additives on the densification and conductivity of $BaCe_{0.3}Zr_{0.55}Y_{0.15}O_{3-\delta}$ electrolyte for protonic ceramic fuel cells. J Alloys Compd 862:158640
178. Kim H-W, Seo J, Yu JH et al (2021) Effect of cerium on yttrium-doped barium zirconate with a ZnO sintering aid: grain and grain boundary protonic conduction. Ceram Int 47:32720–32726
179. Hudish G, Manerbino A, Coors WG, Ricote S (2018) Chemical expansion in $BaZr_{0.9-x}Ce_xY_{0.1}O_{3-\delta}$ (x = 0 and 0.2) upon hydration determined by high-temperature X-ray diffraction. J Am Ceram Soc 101:1298–1309
180. Tarancón A, Skinner SJ, Chater RJ et al (2007) Layered perovskites as promising cathodes for intermediate temperature solid oxide fuel cells. J Mater Chem 17:3175–3181
181. Choi S, Yoo S, Kim J et al (2013) Highly efficient and robust cathode materials for low-temperature solid oxide fuel cells: $PrBa_{0.5}Sr_{0.5}Co_{2-x}Fe_xO_{5+\delta}$. Sci Rep 3:2426
182. Haile SM (2003) Fuel cell materials and components. Acta Mater 51:5981–6000
183. Wachsman ED, Lee KT (2011) Lowering the temperature of solid oxide fuel cells. Science 80(334):935–939
184. Zhao Y, Xia C, Jia L et al (2013) Recent progress on solid oxide fuel cell: lowering temperature and utilizing non-hydrogen fuels. Int J Hydrogen Energy 38:16498–16517
185. Kim J, Choi S, Jun A et al (2014) Chemically stable perovskites as cathode materials for solid oxide fuel cells: La-doped $Ba_{0.5}Sr_{0.5}Co_{0.8}Fe_{0.2}O_{3-\delta}$. Chemsuschem 7:1669–1675
186. Bohn HG, Schober T (2000) Electrical conductivity of the high-temperature proton conductor $BaZr_{0.9}Y_{0.1}O_{2.95}$. J Am Ceram Soc 83:768–772

187. Fabbri E, Pergolesi D, Licoccia S, Traversa E (2010) Does the increase in Y-dopant concentration improve the proton conductivity of $BaZr_{1-x}Y_xO_{3-\delta}$ fuel cell electrolytes? Solid State Ionics 181:1043–1051
188. Kim J, Sengodan S, Kim S et al (2019) Proton conducting oxides: a review of materials and applications for renewable energy conversion and storage. Renew Sustain Energy Rev 109:606–618. https://doi.org/10.1016/j.rser.2019.04.042
189. Shao Z, Tadé MO (2016) Intermediate-temperature solid oxide fuel cells. Chem Soc Rev 37:1568
190. Da Silva FS, de Souza TM (2017) Novel materials for solid oxide fuel cell technologies: a literature review. Int J Hydrogen Energy 42:26020–26036
191. Abdalla AM, Hossain S, Azad AT et al (2018) Nanomaterials for solid oxide fuel cells: a review. Renew Sustain Energy Rev 82:353–368
192. Wang F, Lyu Y, Chu D et al (2019) The electrolyte materials for SOFCs of low-intermediate temperature. Mater Sci Technol 35:1551–1562
193. Boldrin P, Brandon NP (2019) Progress and outlook for solid oxide fuel cells for transportation applications. Nat Catal 2:571–577
194. Pikalova EY, Kalinina EG (2021) Approaches to improving efficiency of solid oxide fuel cells based on ceramic membranes with mixed conductivity. Russ Chem Rev 90
195. Prakash BS, Pavitra R, Kumar SS, Aruna ST (2018) Electrolyte bi-layering strategy to improve the performance of an intermediate temperature solid oxide fuel cell: a review. J Power Sources 381:136–155
196. Kalinina E, Pikalova E, Ermakova L, Bogdanovich N (2021) Challenges of formation of thin-film solid electrolyte layers on non-conductive substrates by electrophoretic deposition. Coatings 11:805
197. Pikalova E, Osinkin D, Kalinina E (2022) Direct electrophoretic deposition and characterization of thin-film membranes based on doped $BaCeO_3$ and CeO_2 for anode-supported solid oxide fuel cells. Membranes (Basel) 12:682. https://doi.org/10.3390/membranes12070682
198. Lian Y, Zheng M (2020) Investigation of the electrochemical performance of anode-supported SOFCs under steady-state conditions. Int J Electrochem Sci 15:12475–12490
199. Park B-K, Barnett SA (2020) Boosting solid oxide fuel cell performance via electrolyte thickness reduction and cathode infiltration. J Mater Chem A 8:11626–11631
200. Norman NW, Somalu MR, Muchtar A (2018) A short review on the proton conducting electrolytes for solid oxide fuel cell applications. Akademia Baru 2:115–122
201. Sažinas R, Bernuy-López C, Einarsrud M-A, Grande T (2016) Effect of CO_2 exposure on the chemical stability and mechanical properties of $BaZrO_3$-ceramics. J Am Ceram Soc 99:3685–3695
202. Polfus JM, Yildiz B, Tuller HL, Bredesen R (2018) Adsorption of CO_2 and facile carbonate formation on $BaZrO_3$ surfaces. J Phys Chem C 122:307–314
203. Kim D-H, Kim B-K, Kim Y-C (2012) Energy barriers for proton migration in yttrium-doped barium zirconate super cell with $Σ$5 (310)/[001] tilt grain boundary. Solid State Ionics 213:18–21
204. Lindman A, Helgee EE, Wahnström G (2017) Comparison of space-charge formation at grain boundaries in proton-conducting $BaZrO_3$ and $BaCeO_3$. Chem Mater 29:7931–7941. https://doi.org/10.1021/acs.chemmater.7b02829
205. Ishihara T, Matsuda H, Takita Y (1994) Doped $LaGaO_3$ perovskite type oxide as a new oxide ionic conductor. J Am Chem Soc 116:3801–3803
206. Xi X, Abe H, Kuruma K et al (2014) Novel Co-precipitation method to synthesize NiO–YSZ nanocomposite powder for solid oxide fuel cell. Adv Powder Technol 25:490–494
207. Yang W-D, Chang Y-H, Huang S-H (2005) Influence of molar ratio of citric acid to metal ions on preparation of $La_{0.67}Sr_{0.33}MnO_3$ materials via polymerizable complex process. J Eur Ceram Soc 25:3611–3618
208. Li P, Li F, Deng G et al (2016) Polymerized-complex method for preparation of supported bimetallic alloy and monometallic nanoparticles. Chem Commun 52:2996–2999

209. Galceran M, Pujol MC, Aguiló M, Díaz F (2007) Sol-gel modified Pechini method for obtaining nanocrystalline KRE (WO_4) 2 (RE = Gd and Yb). J Sol-Gel Sci Technol 42:79–88
210. Gao D, Guo R (2010) Yttrium-doped barium zirconate powders synthesized by the gel-casting method. J Am Ceram Soc 93:1572–1575
211. Abdullah NA, Osman N, Hasan S, Hassan OH (2012) Chelating agents role on thermal characteristics and phase formation of modified cerate-zirconate via sol-gel synthesis route. Int J Electrochem Sci 7:9401–9409
212. Zhao H (2015) Research on impact of $BaZr_{0.1}Ce_{0.7}Y_{0.2}O_{3-\delta}$ electrolyte doped with a small amount of Al_2O_3 on the proton conducting solid oxide fuel cell. In: 5th international conference on information engineering for mechanics and materials, pp 1694–1700
213. Sabry F (2022) Multi function structure: future air force systems will become integrated into multi-function material airframes with embedded sensor, and network components. One Billion Knowledgeable
214. Kato K, Han D, Uda T (2019) Transport properties of proton conductive Y-doped $BaHfO_3$ and Ca or Sr-substituted Y-doped $BaZrO_3$
215. Rajendran S, Thangavel NK, Alkatie S et al (2021) Y, Gd, and Pr tri-doped perovskite-type proton conducting electrolytes with improved sinterability and chemical stability. J Alloys Compd 870:159431. https://doi.org/10.1016/j.jallcom.2021.159431
216. Zhu B, Yang XT, Xu J et al (2003) Innovative low temperature SOFCs and advanced materials. J Power Sources 118:47–53
217. Huang J, Gao Z, Mao Z (2010) Effects of salt composition on the electrical properties of samaria-doped ceria/carbonate composite electrolytes for low-temperature SOFCs. Int J Hydrogen Energy 35:4270–4275
218. Xu P, Li Y, Liu T et al (2022) Electrochemical analysis of $BaZr_{0.8}Y_{0.2}O_{3-\delta}$-$Gd_{0.2}Ce_{0.8}O_{2-\delta}$ composite electrolytes by distribution of relaxation time method. Ceram Int 48:12856–12865. https://doi.org/10.1016/j.ceramint.2022.01.157
219. Laali KK, Greves WJ, Correa-Smits SJ et al (2018) Novel fluorinated curcuminoids and their pyrazole and isoxazole derivatives: synthesis, structural studies, computational/docking and in-vitro bioassay. J Fluor Chem 206:82–98
220. Xu X, Bi L, Zhao XS (2018) Highly-conductive proton-conducting electrolyte membranes with a low sintering temperature for solid oxide fuel cells. J Membr Sci 558:17–25
221. Sudhakar YN, Selvakumar M, Bhat DK (2018) Biopolymer electrolytes for fuel cell applications. Biopolym electrolytes, pp 151–166. https://doi.org/10.1016/b978-0-12-813447-4.00005-4
222. Duval SBC, Holtappels P, Vogt UF, Pomjakushina E, Conder K, Stimming U, Graule T (2007) Solid State Ionics 178:1437
223. Liang F, Yang J, Zhao Y et al (2022) A review of thin film electrolytes fabricated by physical vapor deposition for solid oxide fuel cells. Int J Hydrogen Energy
224. Pergolesi D, Fabbri E, Traversa E (2010) Chemically stable anode-supported solid oxide fuel cells based on Y-doped barium zirconate thin films having improved performance. Electrochem commun 12:977–980
225. Bae K, Jang DY, Choi HJ et al (2017) Demonstrating the potential of yttrium-doped barium zirconate electrolyte for high-performance fuel cells. Nat Commun 8:14553
226. Fluri A, Marcolongo A, Roddatis V et al (2017) Enhanced proton conductivity in Y-doped $BaZrO_3$ via strain engineering. Adv Sci 4:1700467
227. Campos Covarrubias MS, Sriubas M, Bockute K et al (2020) Properties of barium cerate thin films formed using E-beam deposition. Crystals 10:1152
228. Eddaoudi M, Kim J, Rosi N et al (2002) Systematic design of pore size and functionality in isoreticular MOFs and their application in methane storage. Science 80(295):469–472
229. Kitagawa S, Kitaura R, Noro S (2004) Functional porous coordination polymers. Angew Chem Int Ed 43:2334–2375
230. Férey G (2008) Hybrid porous solids: past, present, future. Chem Soc Rev 37:191–214
231. Lim D-W, Kitagawa H (2021) Rational strategies for proton-conductive metal–organic frameworks. Chem Soc Rev 50:6349–6368

232. Yu S, Wang Y, Bi L (2022) Tailoring $BaCe_{0.8}Y_{0.2}O_{3-\delta}$ proton-conducting oxide with U ions for an enhanced stability. Ceram Int 48:17987–17993
233. Li J, Yu S, Bi L (2022) Sr-doped $BaZr_{0.5}Fe_{0.5}O_{3-\delta}$ cathode with improved chemical stability and higher performance for proton-conducting solid oxide fuel cells. Ceram Int 48:35642–35648
234. Escorihuela J, Narducci R, Compañ V, Costantino F (2019) Proton conductivity of composite polyelectrolyte membranes with metal-organic frameworks for fuel cell applications. Adv Mater Interfaces 6:1–30. https://doi.org/10.1002/admi.201801146
235. Li J-R, Sculley J, Zhou H-C (2012) Metal–organic frameworks for separations. Chem Rev 112:869–932
236. Gascon J, Corma A, Kapteijn F, Xamena FX (2014) Metal organic framework catalysis: Quo vadis? Acs Catal 4:361–378
237. Wu HB, Lou XW (2017) Metal-organic frameworks and their derived materials for electrochemical energy storage and conversion: promises and challenges. Sci Adv 3:eaap9252
238. Wang Y, Zhao D (2017) Beyond equilibrium: metal–organic frameworks for molecular sieving and kinetic gas separation. Cryst Growth Des 17:2291–2308
239. Zhu L, Liu X-Q, Jiang H-L, Sun L-B (2017) Metal–organic frameworks for heterogeneous basic catalysis. Chem Rev 117:8129–8176
240. Chandra S, Kundu T, Kandambeth S et al (2014) Phosphoric acid loaded azo (-N=N-) based covalent organic framework for proton conduction. J Am Chem Soc 136:6570–6573
241. Yin Y, Li Z, Yang X et al (2016) Enhanced proton conductivity of Nafion composite membrane by incorporating phosphoric acid-loaded covalent organic framework. J Power Sources 332:265–273
242. Yang F, Xu G, Dou Y et al (2017) A flexible metal–organic framework with a high density of sulfonic acid sites for proton conduction. Nat Energy 2:877–883
243. Meng Z, Aykanat A, Mirica KA (2018) Proton conduction in 2D aza-fused covalent organic frameworks. Chem Mater 31:819–825
244. Xu H, Gao J, Jiang D (2015) Stable, crystalline, porous, covalent organic frameworks as a platform for chiral organocatalysts. Nat Chem 7:905–912
245. Waller PJ, Gándara F, Yaghi OM (2015) Chemistry of covalent organic frameworks. Acc Chem Res 48:3053–3063
246. Rodríguez-San-Miguel D, Zamora F (2019) Processing of covalent organic frameworks: an ingredient for a material to succeed. Chem Soc Rev 48:4375–4386
247. Nie H, Schauser NS, Dolinski ND et al (2020) Light-controllable ionic conductivity in a polymeric ionic liquid. Angew Chemie 132:5161–5166
248. Chen X, Addicoat M, Irle S et al (2013) Control of crystallinity and porosity of covalent organic frameworks by managing interlayer interactions based on self-complementary π-electronic force. J Am Chem Soc 135:546–549
249. Chen X, Addicoat M, Jin E et al (2015) Locking covalent organic frameworks with hydrogen bonds: general and remarkable effects on crystalline structure, physical properties, and photochemical activity. J Am Chem Soc 137:3241–3247
250. Tao S, Zhai L, Dinga Wonanke AD et al (2020) Confining H_3PO_4 network in covalent organic frameworks enables proton super flow. Nat Commun 11:8–15. https://doi.org/10.1038/s41467-020-15918-1
251. Bao S-S, Shimizu GKH, Zheng L-M (2019) Proton conductive metal phosphonate frameworks. Coord Chem Rev 378:577–594

Chapter 2
Transition Metal Nitrides as Energy Storage Materials

Aishwarya Madhuri, Sanketa Jena, and Bibhu Prasad Swain

1 Introduction

Society is more concerned about global warming, energy production and energy storage which are the main topics of discussion nowadays. There is only one way to fulfil the energy demand of the escalating global population which is to double the current rate of energy production (14–28 TW) by the year 2050 which is equal to 130,000 TWh yr^{-1} or the equivalent of 10^{10} tons of oil yearly [1]. This increase in the requirement for energy has to be attained without the emission of carbon dioxide or depending only on fossil fuel resources which forced the development of advanced renewable energy resources. As there is a dying reduction in fossil fuels, every economic country should be committed to a shift to electrochemical energy in order to prevent from environmental pollution. Renewable energy sources like wind, tidal, geothermal, solar and biomass energies can be used to reach the goal of lowering greenhouse gas emissions. Supercapacitors, Lithium-ion batteries, Sodium-ion batteries, Lithium–sulphur (Li–S) batteries and Lithium-ion hybrid capacitors (LiHCs) are excellent sources of electrochemical energy storage devices having common configurational features. Rechargeable batteries are superb sources of energy storage and conversion devices. An energy storage device has three main parts: a cathode and anode (electrodes) and an aqueous electrolyte. The electrodes are for collecting charges in EES and are the main important parts for determining the energy storage performance (Fig. 1).

In recent past years, the development of numerous electrode materials for EES devices has been achieved, for example, metal oxides, carbon materials, nitrides, carbides, sulphides, phosphides and alloy elements. Because of their distinctive electronic structure, excellent electrocatalytic activity, large conductivity and great volumetric energy density, group IVB-VIB early transition metal nitrides (TMNs) are

A. Madhuri (✉) · S. Jena · B. P. Swain
Department of Physics, National Institute of Technology, Manipur, Langol, Imphal 795004, India
e-mail: aishwaryamadhuri99@gmail.com

B. P. Swain (ed.), *Energy Materials*, Materials Horizons: From Nature to Nanomaterials,
https://doi.org/10.1007/978-981-99-3866-7_2

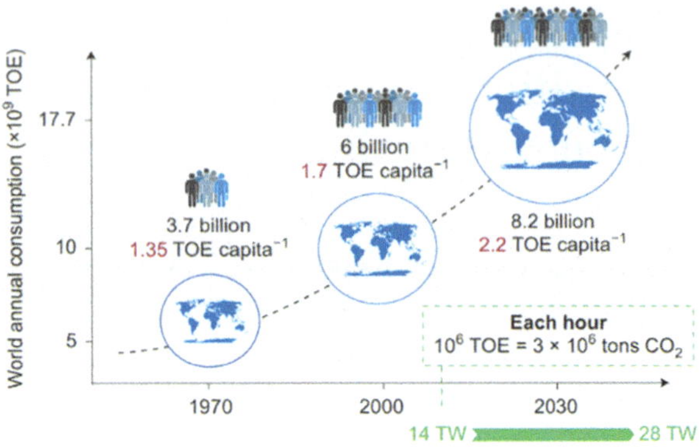

Fig. 1 Past, present and forecast of the energy needs from 1950 to 2050

favorable materials for electrodes in EES devices. In order to improve the power and energy densities as well as life cycle of EES devices, we need to develop the efficiency of the electrode materials. Synthesis and EES characteristics of TMNs with difference in structural morphology that includes nanoparticles, nanorods, nanotubes, nanosheets, nanowires and nanohybrids of vanadium nitride (VN), titanium nitride (TiN), niobium nitride (NbN or Nb_4N_5) and molybdenum nitride (MoN or Mo_2N) have been studied. However, because of the limitation in electrochemically active sites, poor longevity, not-so-high capacitance and brittleness, the practical application of TMNs in EES is hindered. Prevention of agglomeration, improvement in the electrical conductivity, magnification in the electrochemically active sites and further ease of ion transport can be accomplished by using TMN-based nanocomposites to increase the EES performance.

Figure 2 shows the energy density as a function of the power density of energy storage devices like fuel cells, Li-ion, Nickel-metal hybrid (NiMH), Lead-acid batteries and supercapacitors [2]. This plot is known as the "Ragone plot" which is based on the comparison of the performance of different energy storage devices. Fuel cells have low power density but larger energy density, supercapacitors possess high power density but very low energy density and LIBs own average energy and power densities. Hence, a more efficient energy storage device with high power and energy storage is highly desired.

2 Structural and Physical Properties of TMNs

Group IVB-VIB TMNs are interstitial compounds because the atomic size of transition metals is very large having voids and nitrogen being a small atom trapped inside the interstitial sites of the metals [3]. TMNs are normally of cubic, simple hexagonal and hexagonal close-packed structures with the non-metal atoms haphazardly distributed at the interstitial sites. By looking at the neighborhood of the non-metal that is nitrogen in TMNs, the structural elements can be described of interstitial

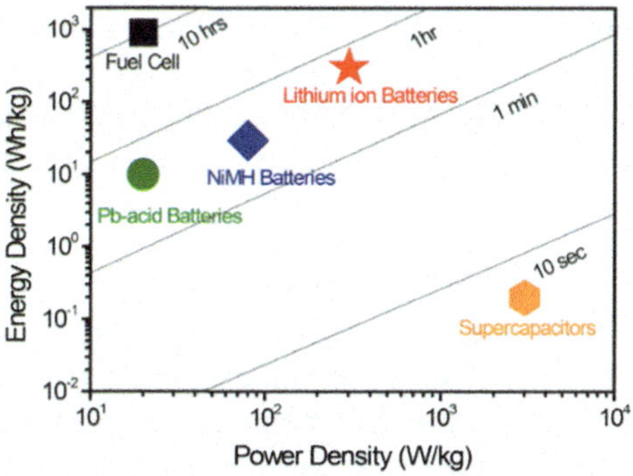

Fig. 2 Ragone plot of the energy storage devices that are commonly available

alloys. The common structural element of group IV and V TM in nitrides is the T_6N octahedron. Sometimes another structural element is observed in TMNs, especially in group V comprising T_5N square pyramids with lower symmetry than octahedrons. Such nitrides have more complicated structures. This square pyramid is also found in complex nitrides like the Z-phases of NbCrN, NbMoN and TaCrN, etc. (Fig. 3).

The fcc structure of δ-nitrides is sabotaged because other structural elements are formed except T_6N octahedra. The destabilization increases with an increase in the group number, and also within a group it increases with an increase in the periodic number. Such as δ-HfN_{1-x}, δ-TiN_{1-x} and δ-ZrN_{1-x} in group IV nitrides crystallize in the fcc structure. At low temperatures, no change in the stoichiometry of these nitrides is known. Group V nitrides also emerge into fcc structure, but δ-VN changes into tetragonal symmetry at 205 K [4], δ-TaN into hexagonal ε-TaN at 1940 °C [5] and δ-NbN into η-NbN below 1320 °C. However, group VI nitrides with fcc symmetry are less stable. CrN with sodium chloride structure is an ordered stoichiometric phase in the low-temperature region and changes into a tetragonal antiferromagnetic compound below 280 K [6]. The chemical stability of TMNs is very large so that they are not attacked instantly by dilute acids except HF and oxidizing acids or alkaline solutions. Their thermal stability is linked with the Gibbs free energy of formation where the stability decreases with increasing the group number. The group IV TMNs can be melted without decomposition while nitrides of

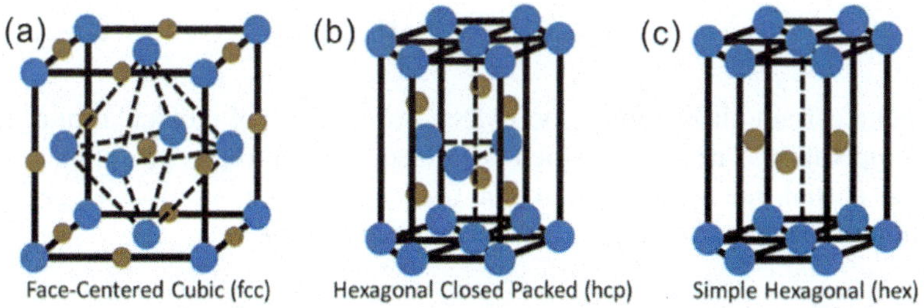

Fig. 3 Crystal structures of TMNs **a** cubic, **b** hexagonal closed packed and **c** simple hexagonal. The blue circles denote transition metal atoms and the brown circles are nitrogen atoms

other groups decompose before the melting point is attained and nitrogen is liberated [7].

The bonding in TMNs is defined as a strange combination of metallic, ionic and covalent properties. Early TMNs with high electrical conductivities, large melting points, hardness and Young's modulus show metallic character. For instance, the electrical conductivity of metallic VN monocrystal nanosheets is about 1.44×10^{-5} $S\ m^{-1}$ at room temperature [8]. The bonding mechanism is reported in various ways to calculate the Density Of States (DOS) in fcc TMNs and also their electron density. Hence, TMNs are prominently used as shielding material on industrial components and also in microelectronic devices to prevent diffusion between two layers of materials. The modification in the metal lattice of parent metal in TMNs is caused due to the formation of the M–N bond which causes expansion of the d-band and therefore results in a higher density of states (DOS) near the fermi level of the metal. TMNs have high compression moduli. Experimentally TMNs of Group IV and period 4 possess moduli less than 300 GPa while the iron-based TMNs possess the maximum compressibility.

The ionic contribution to the binding mechanism of TMNs can be approximated from the charge transfer occurring from the metal atom to the non-metal atom which is around half an electron per atom that contributes to the electrostatic interaction of metal and non-metal (Table 1).

3 Synthesis Methods of Metal Nitrides

TMNs in bulk form can be prepared by the reaction of the powdered or compact form of the metal or the metal hydride with molecular or atomic nitrogen or flowing ammonia:

$$\mathrm{Ti} + \frac{1}{2}\mathrm{N_2} \rightarrow \mathrm{TiN} \tag{1}$$

$$\mathrm{ZrH_2} + \frac{1}{2}\mathrm{N_2} \rightarrow \mathrm{ZrN} + \mathrm{H_2} \tag{2}$$

$$\mathrm{Mo} + \mathrm{NH_3} \rightarrow \mathrm{MoN} + \frac{3}{2}\mathrm{H_2} \tag{3}$$

To get a certain nitride phase for TMNs, the pressure of molecular nitrogen at a given temperature is necessary which is defined as the nitrogen potential; otherwise known as partial free energy of nitrogen. Carbon is used as a reducing agent for the formation of nitrides from oxides, following the reaction;

$$\mathrm{TiO_2} + 2\mathrm{C} + \frac{1}{2}\mathrm{N_2} \rightarrow \mathrm{TiN} + 2\mathrm{CO} \tag{4}$$

Table 1 Elemental properties of some TMNs at room temperature for stoichiometric composition

Parent metal	Metal nitrides	Color	Crystal structure	Lattice parameters (in A°)	Density (in gm/ cm^3)	Melting point (in °C)	Microhardness (in GPa)	Young's modulus (in GPa)	Heat conductivity (in $Wm^{-1}K^{-1}$)	Electrical resistivity (in μ)
Ti	TiN	Golden yellow	Cubic	4.24	5.4	2930	17	420	29	27
Zr	ZrN	Green yellow	Cubic	4.56	7.09	2980	15	460	11	24
Hf	HfN	Dark yellow	Cubic	4.53	13.2	3305	18	380	11	27
V	VN	Brown	Cubic	4.12	6.13	2050	5.7	380	11	65
Nb	NbN	Light yellow	Cubic	4.39	8.47	Decomposes	11	360	3.8	60
Ta	TaN	Gray Yellow	Cubic	4.42	14.97	Decomposes	32		–	–
Cr	CrN	Gray brown	Cubic	4.15	6.14	Decomposes	13	450	11.7	640

This reaction normally carries on with various intermediate products and the following nitride may contain carbon and oxygen.

While preparing nitrides from metal powders, the exothermic heat of formation can be used in the Self-sustaining High-temperature (SHS) Synthesis process. If the heat generated in the exothermic reaction is enough to hit the reaction temperature, then the starting material is transformed into nitrides and the heat is dissipated. The development of zone-annealing techniques was done for the formation of large crystals. The starting material is pressed onto rods which are either solid-state sintered nitride powders or metal powders. A hot section is moved across the rod in a nitrogen atmosphere. In order to obtain large diffusion of nitrogen in the nitrides, a very high temperature must be applied, but it should not exceed the melting temperature of the compound to get rid of variance in melting and decomposition. This technique is used for the formation of crystals of the order of 1 cm^3. Well, there is the possibility of the appearance of some concentration gradients in the compounds. Still, they can be used for solid-state investigations, for example, neutron diffraction.

Oxide precursors are used for the production of nanostructured TMNs which is done via post-NH_3 reduction annealing. During this conversion process, the morphology of the oxides is maintained. Because of the density difference between the oxide precursors and the nitride products, there is volume shrinkage from oxide to nitride and the conversion process produces abundant mesopores. Gao et al. [9] synthesized mesoporous VN nanowires by thermal nitridation of V_2O_5 nanowires that are prepared by a hydrothermal method using V_2O_5 powders and H_2O_2 as precursors as shown in Fig. 4. Xiong et al. [10] also synthesized Mo_3N_2 nanobelts by a temperature programmed reaction of α-MoO_3 nanobelts with ammonia (NH_3) at 750 °C that inherited the structure and morphology of the precursor. Bi et al. [8] prepared VN monocrystal nanosheets by the nitridation of layered $Na_2V_6O_{16}$ nanosheets. Xiao et al. [11] reported the scalable salt-templated fabrication of two-dimensional metallic MoN nanosheets by the reduction of 2D hexagonal MoO_3-coated NaCl in ammonia with the exceeding lateral size of flakes from 20 μm and subnanometer thickness. Well, the production of various TMN nanostructures such as TiN, VN, Nb_4N_5, etc. by the topochemical conversion with different morphologies varies from zero dimension (0D) to three dimension (3D) from transition metal oxides (TMOs) due to their improved morphologies. Carbon-activated TMNs hybrids were done by the mixture of existing TMNs with carbon nanotubes (CNT) or by carbon coating on TMNs. For instance, Xiao et al. [12] synthesized mesoporous VN nanowires/CNT hybrid electrodes by vacuum filtering method using the collegial effects from the high conductivity of CNTs and large electrochemical performance of VN nanowires. Lu et al. [13] prepared uniformly carbon-coated TiN nanowires by hydrothermal treatment using glucose as the carbon precursor ensued by at 800 °C in a nitrogen atmosphere. TMNs/C hybrids can also be fabricated by the deposition of TMNs on carbon. Zhang et al. [14] deposited 3D electrochemically supercapacitive CNT arrays covered by nanocrystalline VN strongly adhered to glassy carbon and polished Inconel substrates by rf magnetron sputtering of the Vanadium target with the nitrogen gas flow. Yang et al. [15] synthesized corn-like TiN nanostructures on Ni plate, stainless steel and carbon fabric by Atomic Layer Deposition

(ALD) of TiO_2 film with $Co_2(OH)_2CO_3$ nanowire precursor that was obtained by a standard hydrothermal method followed by the removal of $Co_2(OH)_2CO_3$ precursor in dilute acid and annealing of TiO_2 in dry NH_3 atmosphere. Haldorai et al. [16] fabricated TiN nanostructures on reduced graphene oxide (rGO) coated with Pt to obtain Pt@TiN/rGO ternary hybrid catalyst for improved electrocatalytic activity followed by the backflow of $Ti(OBu)_4$ precursor and GO followed by the nitridation. Other TMN-based hybrid materials such as TMN-supported nitrides (TMNs/TMNs), TMN-supported TMOs (TMOs/TMNs), TMN-based conducting polymers (CPs/TMNs) can be synthesized by the use of existing TMN as supporting material to deposit with other materials. For example, Peng et al. [17] synthesized coaxial PANI/TiN/PANI nanotube arrays (exhibiting a specific capacitance of 242 mF cm^{-2}) by electrochemical polymerization of polyaniline (PANI) on the inner and outer surfaces as well as nanopores on the walls of the TiN nanotube arrays followed by thermal annealing of anodized TiO_2 in ammonia (Fig. 5).

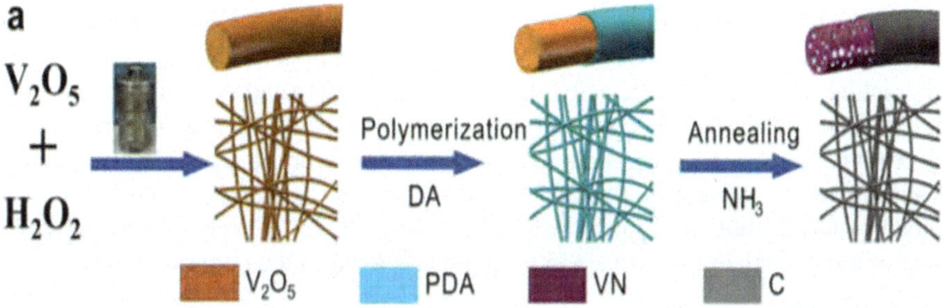

Fig. 4 Schematic diagram of the synthesis of nitrogen-doped carbon encapsulated mesoporous Vanadium nitride (MVN@NC) nanowires

Fig. 5 Schematic illustration of the synthesis of coaxial PANI/TiN/PANI nanotube arrays

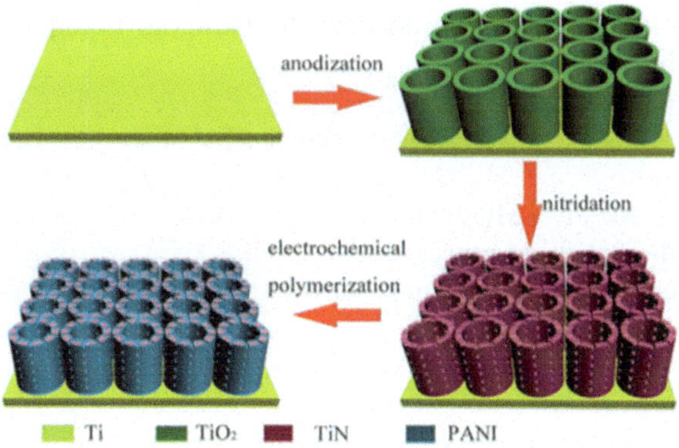

4 Applications of TMNs in Energy Storage Devices

4.1 Transition Metal Nitrides for Lithium-Ion Batteries

Electrochemical studies recently have made it easy to understand the electrochemistry of energy storage devices, especially Lithium-ion battery systems. Recently published articles show the substantial success of Li-ion batteries in Science and Technology in the last decade. The availability of LIBs in all shapes and sizes, renders them to be used for power needs irrespective of the size of the system. From a tiny LIB that can power a smartwatch to a massive one for charging an electronic vehicle; LIBs are used in various devices such as cell phones, laptops, and all other commonly used consumer electronic goods and digital cameras. Besides, LIBs provide power solutions beyond the spectrum from energy storage to portable energy solutions. The advantages of LIBs include high energy density, wide temperature windows, high operating voltage, low self-discharge and low to minimal maintenance. The power density of LIBs is 5–6 times and the energy density is also 2–3 times higher than other batteries. These batteries mainly consist of 3 components: (1) an anode (mostly of graphite with a specific capacity of 372 mA h g^{-1}), (2) a cathode (normally lithium metal oxide such as $LiCoO_2$; since inside a battery Li itself is highly unstable), (3) an electrolyte which is Li-ion conducting and (4) a separator for separating the cathode from the anode. Normally, the LIB storage technique is classified into 3 trials for anode materials that are (a) insertion, (b) alloying and (c) conversion reactions. The insertion mechanism takes place in graphite as shown in the equation below.

$$\mathrm{C} + x\mathrm{Li}^{+} + x\mathrm{e}^{-} \rightarrow \mathrm{Li}_{x}\mathrm{C} \tag{5}$$

Some elements undergo alloying and dealloying mechanisms; for example, Tin, Germanium, Silicon, Antimony and etc. This reaction is also reversible and the mechanism can be detailed by taking an example of Tin shown in the equation below.

$$\mathrm{Sn} + x\mathrm{Li}^{+} + x\mathrm{e}^{-} \rightarrow \mathrm{Li}_{x}\mathrm{Sn} \tag{6}$$

Mostly TMOs exhibit the last mechanism which is the conversion reaction where the reduction process yields lithium oxide (Li_2O) and bare transition metal in the end. This mechanism is shown through the general equation below.

$$\mathrm{M}_{x}\mathrm{O}_{y} + 3y\mathrm{Li}^{+} + 3\mathrm{e}^{-} \rightarrow x\mathrm{M}^{0} + y\mathrm{Li}_{2}\mathrm{O} \tag{7}$$

Although LIBs are abundant sources of energy storage, the above-mentioned mechanisms have limitations like both conversion and alloying and dealloying mechanisms face low conductivity, Solid Electrolyte Interphase (SEI) film configuration and volume expansion that can cause poor rate capability and cyclic stability. Another reason for low capacity is less availability of vacancies in the insertion mechanism for Li-ion reversible storage. TMNs are the appropriate ones to undergo conversion mechanism among all the other materials due to their stable performance. The first row of TMNs starting from Scandium to Zinc is suitable for alternate electrode materials owing to their easy preparation, availability and economic factors.

TMNs are promising electrode materials because of their high Li-ion diffusion and also, they have ordered and disordered phases with remarkable levels of Li vacancies (charge carriers in Li_3N) because these are the popular fast ion conductor with the highest Li^+ conductivity in the crystalline solid electrolyte. Li-ion diffusion consisting of inter- and intralayer diffusion can be enhanced by the formation of additional vacancies in the nitride phases. Sun and Fu [18] studied the electrochemical behavior of Vanadium mononitride (VN) with Lithium-ion by using in-situ spectro-electrochemical method and Selected Area Electron Diffraction (SAED) measurement and explained the formation of metal V and Li_3N confirming the decomposition of VN after the film is discharged to 0.01 V and while charged to 3.5 V, formation of VN supports the reversible conversion reactions just like the TMOs exhibit with Li^+. Zhao et al. [19] synthesized carbon-coated VN hollow spheres assembled from porous nanosheets by a facile template-assisted strategy as shown in Fig. 6. VN hollow spheres with high surface area (66 m^2 g^{-1}), graded microporous (250 nm), mesoporous (5–30 nm) structure and the uniform mesoporous shells with a thickness of 55 nm show superb Lithium storage applications and high catalytic activity in the Oxygen Reduction Reaction (ORR). The reported theoretical capacity of the VN HS electrode is 1043 mA h g^{-1}, while after 40 cycles, the reversible capacity of the electrode slightly increased to 720 mA h g^{-1} much higher than that of bulk VN which is about 280 mA h g^{-1} at a current rate of 50 mA g^{-1} with 100% capacity retention and a coulombic efficiency of $98.5 \pm 1.1\%$. When the same cell was cycled at higher current rates 2, 3 and 4 A g^{-1}, the VN hollow spheres can have a capacity of 462, 420 and 460 mA g^{-1} respectively after 100 cycles with almost no capacity fading. The stimulating electrochemical performance is offered by the porous structure of VN nanosheets fabricated by Peng et al. [20] through a simple hydrothermal method that provides active sites for the accommodation of Li-ions, change in volume due to hollow structures, short Li^+ diffusion path and faster charge transfer kinetics. Wang et al. [21] synthesized VN quantum dots dispersed over nitrogen-doped graphene homogeneously (VNQD-NG) by a simple hydrothermal procedure of ammonium metavanadate (NH_4VO_3) with graphene oxide (GO) followed by annealing with ammonia. The zero-dimensional VN quantum dots have a size of 3–6 nm and VNQD-NG nanosheets can be used to form a 3D porous network with maximized active sites for ORR and increased electron transfer. The electrochemical measurements show that the capacity of VN with graphene (715 mA h g^{-1} at 0.2C, where 1C = 372 mA h g^{-1}) is much higher than that of VN (365 mA h g^{-1}) and also nitrogen-doped graphene (433 mA h g^{-1}) (Fig. 7).

If TMNs are modified by an embodiment with carbonaceous materials then it will enhance the electrode performance to ease the volumetric change by promoting the electron flow and Li-ion transport. These carbonaceous materials will behave as shock absorbers and also prevent agglomeration and demolition of TMNs. Xiu

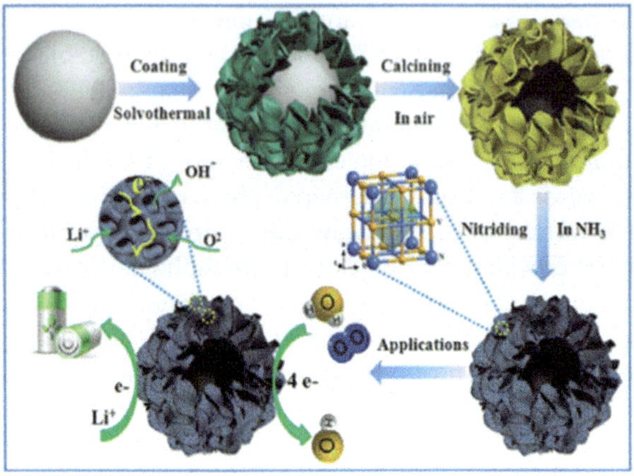

Fig. 6 Schematic presentation of the synthesis process of VN hollow spheres and applications in lithium storage and oxygen reduction reaction

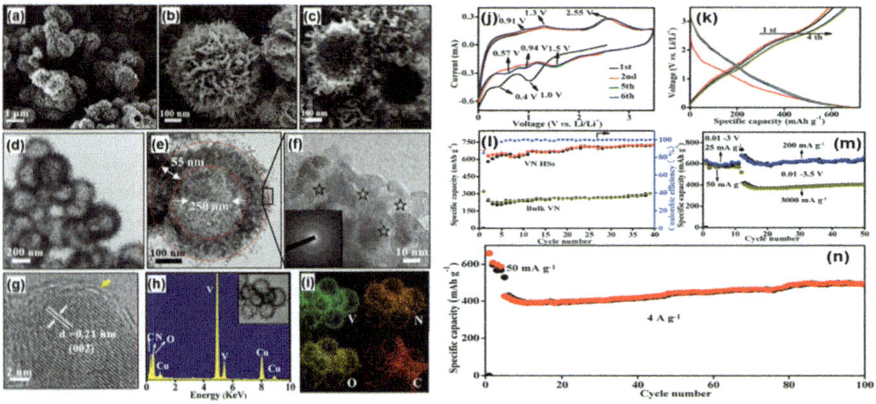

Fig. 7 Morphological studies of VN hollow spheres **a–c** FESEM images with different resolutions, **d–e** TEM images at different resolutions, **f** HRTEM image with the SAED pattern as the inset, **g** HRTEM image showing the clear lattice planes, **h** EDX spectrum, **i** images showing elemental mapping, **j** CV curves ranging from 0.01–3.5 V at a scan rate 0.5 mV s^{-1}, **k** galvanostatic charge/discharge profile at current density 50 mA g^{-1}, **l** cycle life of bulk VN and VN HS at current density 50 mA g^{-1}, **m** cycle life of VN HSs within different voltage ranges and current densities and **n** cycle life of VN HS electrode at 4 A g^{-1} for the first 3 cycles after activation at 0.05 mA g^{-1}

et al. [22] synthesized porous TiN nanoparticles embedded with N-doped carbon composite electrode for Li-ion battery by metal–organic framework method exhibited excellent electrochemical properties like high reversible lithium specific capacity (561 mA h g^{-1} at 50 mA g^{-1}), good rate capability (281 mA h g^{-1} at 2 A g^{-1}) and excellent cycle stability (310 mA h g^{-1} at 2 A g^{-1} for 400 cycles). Graphene-based materials are superior host materials for lithium-ion intercalation because of their two-dimensional structure, large surface area and high electrical conductivity. For example, Yousefi et al. [23] prepared TiN/graphene nanocomposite through the reaction of $TiCl_4$ and NaN_3 in a benzene medium followed by ammonia treating at 1000 °C for 10 h which delivered an excellent specific capacity of 115 mA h g^{-1} when cycled at 1.6C and good capacity retention at high charge/discharge rates. Many researchers have investigated Fe-based nitride materials for their use as LIB anode that showed an advantage of following a complete conversion reaction. Huang et al. [24] synthesized carbon-constrained Fe_3N@C nanoparticles by the DC arc discharge method with excellent capacity and rate performance. The Fe_3N@C electrode showed a high initial charge/discharge capacity of 676 mA h g^{-1} at a rate of 100 mA g^{-1} and a specific reversible capacity of 358 mA h g^{-1} maintained after 500 cycles. The carbon shell offered a dual capacitive interface and Fe_3N nanoparticles provided a high surface area for charge transportation. Researchers have tried to improve the capacity and rate capability of nickel nitrides to overcome the low theoretical capacity of 424.3 mA h g^{-1} for Ni_3N by using various carbon-based materials. Balogun et al. [25] fabricated Ni_3N nanosheets on carbon cloth by a simple hydrothermal and post-annealing process and this 3D flexible Ni_3N/carbon cloth composite electrode reached the initial capacity of 593 mA h g^{-1} at 1C and retained at 304 mA h g^{-1} at the current rate of 10C because of the thin and porous structure of nanosheets (Fig. 8).

Jiang et al. [26] successfully fabricated 3D ordered mesoporous CoN electrode that showed a wonderful long life cycle stability with a stable specific capacity of 300 mA h g^{-1} at the current rate of 2 A g^{-1} even after 2000 cycles which outperformed those of non-porous, disordered mesoporous and bulk CoN. Chromium delivers low lithium insertion potential which makes it interesting for the LIB application. Different concentrations of Nitrogen-doped Carbon Nanotubes (NCNTs) incorporated with mesoporous CrN were prepared by simple coprecipitation and further calcination and nitridation process that showed high charge transfer kinetics and reduced polarization [27]. The as-prepared CrN/0.08% NCNTs nanostructures offered a discharge capacity of 1172 mA h g^{-1} and the coulombic efficiency reached ~ 100% after 200 cycles. The electrochemical tests showed that the electrode offered a reversible capacity of 1042.9 mA h g^{-1} at a current rate of 20 C with 99.5% coulombic efficiency. This outstanding electrochemical performance of the aforementioned nanocomposites is because of their mesoporous channel and conductive CNTs framework. Figure 9 gives an overall idea about capacities versus current densities of already prepared TMNs for Li^+ storage [28].

4.2 *Transition Metal Nitrides in Sodium-Ion Batteries*

As all of us know that sodium is a cheap and uniformly distributed source, researchers hunting for long life cycle batteries found another EES system in the name of sodium-ion batteries (SIBs). The intercalation of Na^+ in a layer form, which is the same as of Li^+ can be used for the development of EES devices and was first observed in

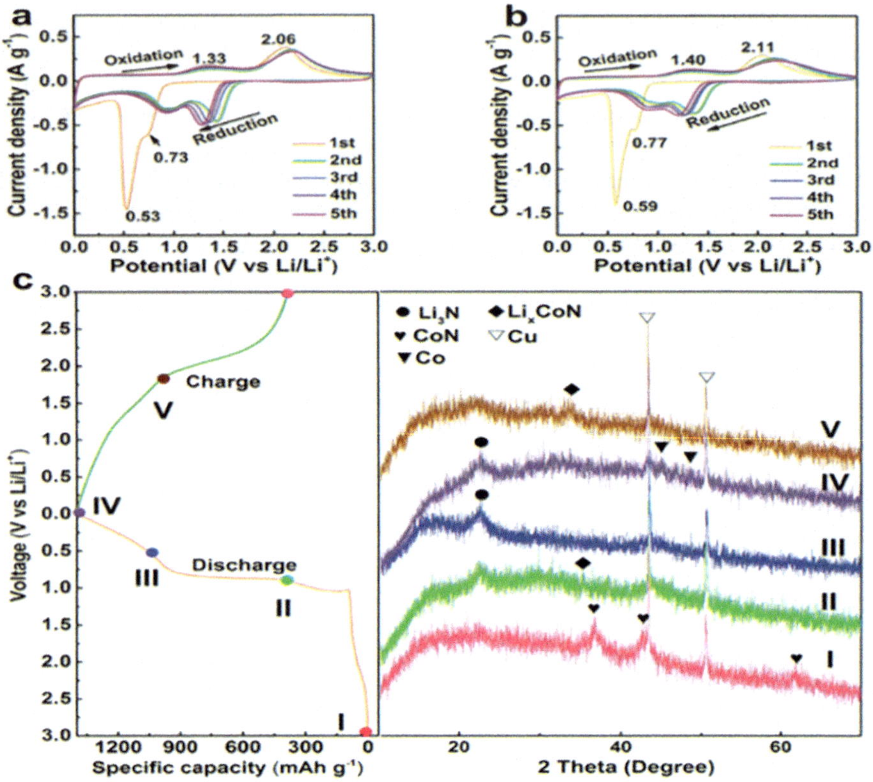

Fig. 8 **a** Cyclic voltammetry curves of om-CoN, **b** dm-CoN at 0.1 mV s^{-1} and **c** ex-situ XRD patterns of CoN electrodes at different charge/discharge states as indicated in the corresponding voltage profile

Fig. 9 A list of capacities v/s current densities attained from various literature cited TMNs for LIBs

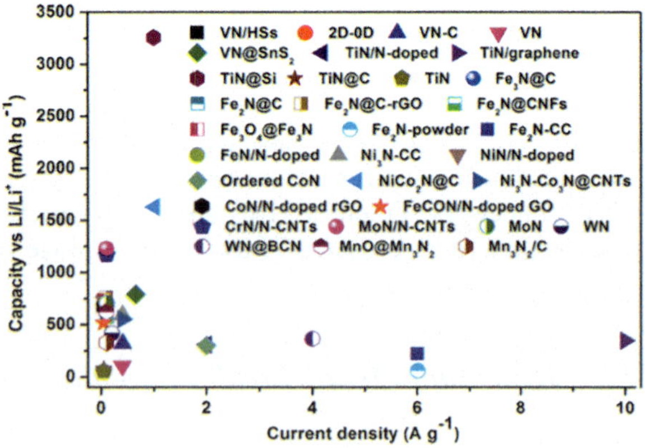

1971. But because of passive kinetics and the large ionic radius (1.02 A°) of Na^+, the electrode materials that were developed for LIBs could not perform efficiently in the SIBs system. Therefore, further research is required for the development of electrode systems in SIBs. Well, researchers successfully developed the cathode material for SIBs such as transition metal oxides [29], poly-anionic compounds like hexagonal BN/blue phosphorene heterostructure [30] and Prussian Blue Analogues (PBAs) [31]; but testing on anode materials is not fruitful yet. The property of anode material determines the characteristics of the battery like cycle life, rate capability and charge/discharge, etc. However, the previously synthesized anode materials for SIBs are Ti-based compounds (TiO_2/C nanofibers) [32], metal alloys (Sn@CNT) [33] and hard carbons [34]. Still, their application in anode material is restricted because of large polarization and change in volume, low potential and poor conductivity.

Nowadays researchers have explored some other anode materials like TMNs and carbides to increase the cycle life and capacity of SIBs. Liu et al. [35] reported a new anode material that is porous Mo_2N nanobelts with a single crystalline cubic structure synthesized by a facile hydrothermal method of MoO_3 followed by a nitridation process. The first discharge capacity of as-synthesized por–Mo_2N-NB was 437 mA h g^{-1} with a capacity retention rate of 85% over 200 cycles at the current rate of 0.1 A g^{-1}. Synthesis of MnN-decorated reduced graphene oxide (MnN@rGO) nanostructures were done using a simple microwave nitridation technique by Sridhar and Park [36]. The MnN@rGO electrode showed an exceptional sodium storage capacity of 716 mA h g^{-1} even after 180 cycles due to harmony between reduced graphene oxide and MnN nanoparticles (Fig. 10).

Generally, the iron-based nitride electrodes in SIBs show good electrochemical performance. One such example is, Liu et al. [37] prepared coral-like carbon-coated Fe_2N nanoparticles and applied them in SIBs for the first time. The as-synthesized nanoparticles exhibited a reversible capacity of 60 mA h g^{-1} after 1000 cycles at a current rate of 0.5 A g^{-1} with a retention of about 80%. The active redox behavior of Vanadium-based materials draws attention to energy storage applications. A facile electrospinning method was used to fabricate a carbon nanofiber-based VN composite followed by immunization to be used as anode material in SIBs by Xu et al. [38]. The as-synthesized VN/CNFs delivered a large capacity of 403 mA h g^{-1} after 100 cycles and 237 mA h g^{-1} after 4000 cycles at current rates of 0.1 A g^{-1} and 2 A g^{-1} respectively with very less capacity fading which was one of the best performances of vanadium-based anode material for SIB (Fig. 11).

4.3 Transition Metal Nitrides in Supercapacitor

Recently, TMNs-based supercapacitors have gained noteworthy recognition as high-capacity electrochemical energy storage materials for their use from the portable device to electric vehicles. SCs and batteries are similar in configurations with cathode, anode, electrolyte and separator; but, SCs are preferred because of their long life cycle which is almost 100 times that of battery life and high power density.

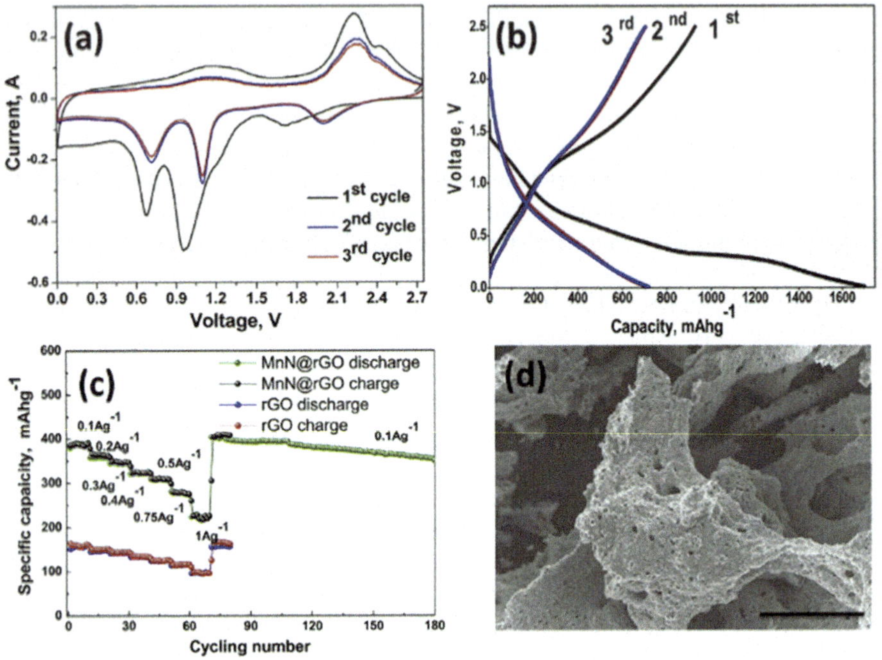

Fig. 10 **a** Cyclic-voltammetry curve, **b** galvanostatic discharge–charge curve, **c** specific capacity v/s number of cycles and **d** SEM morphology of MnN@ rGO after 180 cycles with 3 μm scale bar

Fig. 11 A record of the capacities obtained for different TMNs in SIBs at specific current densities

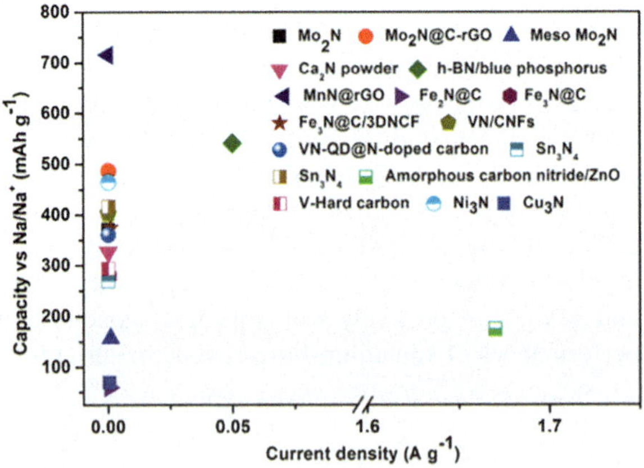

Based on charge/ discharge mechanism, SCs are classified into 2 types: (i) Electric Double-Layer Capacitors (EDLC) and (ii) Pseudo-capacitors. EDLCs use carbon electrodes or derivatives and charge separation is achieved in a Helmholtz double layer at the interface between the surface of a conductive electrode and an electrolyte. EDLCs can be charged and discharged up to 10^6 times (~1000–2000 m^2/g) with a

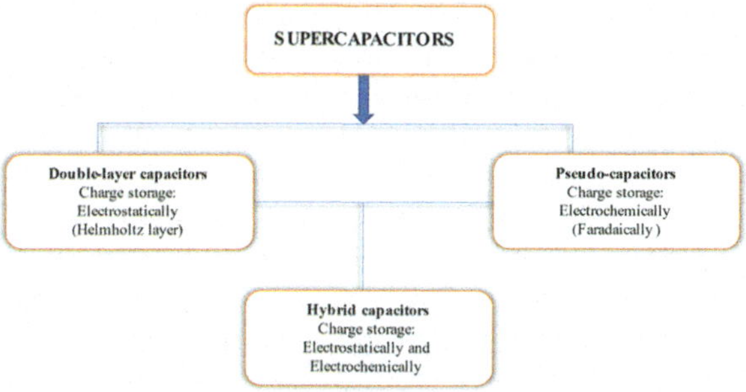

Fig. 12 Hierarchical classification of supercapacitor

high specific power density without any power deficiency. EDLCs are more advantageous than batteries due to long life cycles, high power densities and fast charge–discharge kinetics. Whereas, electrochemical pseudo-capacitors use metal oxide or conducting polymer electrodes and pseudo-capacitance is achieved by Faradaic electron charge transfer with redox reactions, intercalation, or electrosorption. Pseudo-capacitors have large load and discharge rates and can store much larger power than a supercapacitor (Fig. 12).

TMNs have gained a lot of interest in all EES devices, especially in SCs owing to their excellent property combinations such as large electrical conductivity, mechanical stability, high reaction selectivity, long durability and good physical and electrical properties. The electrical conductivity of TMNs in SCs is about 4000–5500 S cm^{-1} which is much higher than most metal oxides. TMNs can produce large power density in SCs due to their excellent capacitance; much higher than those of metal oxides and carbon materials. Among TMNs, VN is proven to be a noteworthy electrode material for SCs. VN thin film electrode with 25 nm thickness prepared by DC magnetron sputtering shows the specific capacitance of 422 F g^{-1} in an aqueous 1 M KOH electrolyte. VN films with less than 100 nm thickness provide a volumetric power of 125 W cm^{-3} for electrolytic capacitors with a volumetric energy density of 0.01 W h cm^{-3} [39]. Bi et al. [8] fabricated meso-crystal nanosheets of VN-based flexible supercapacitors offering an electrical conductivity of 1.44×10^5 S m^{-1} at room temperature with a volumetric capacitance of 1937 mF cm^{-3}. Mesoporous TiN-VN fibers prepared by coaxial electrospinning and subsequent annealing in NH_3 exhibited a specific capacitance of 247.5 F g^{-1} at a scan rate of 2 mV s^{-1} (mainly due to VN) and a rate capability of 160.8 F g^{-1} at 50 mV s^{-1} rate (due to TiN) and this TiN-VN hybrid is a promising candidate for high-performance supercapacitor [40]. Individually, TiN fibers electrode has low capacitance but better rate capability at higher scan rates; while VN fiber electrode has a large capacitance and worst retention power (Fig. 13).

Molybdenum nitrides (Mo_xN_y) group when used as electrodes in SCs also provide high electrochemical performance and stability against electrolyte reactions. For example, single-crystalline mesoporous Mo_3N_2 NWs were synthesized by topotactic reaction process and nitridation of MoO_3 NWs which delivered a specific capacitance of 220 F g^{-1} at a scan rate of 50 mV s^{-1} much larger than commercial Mo_3N_2 (66

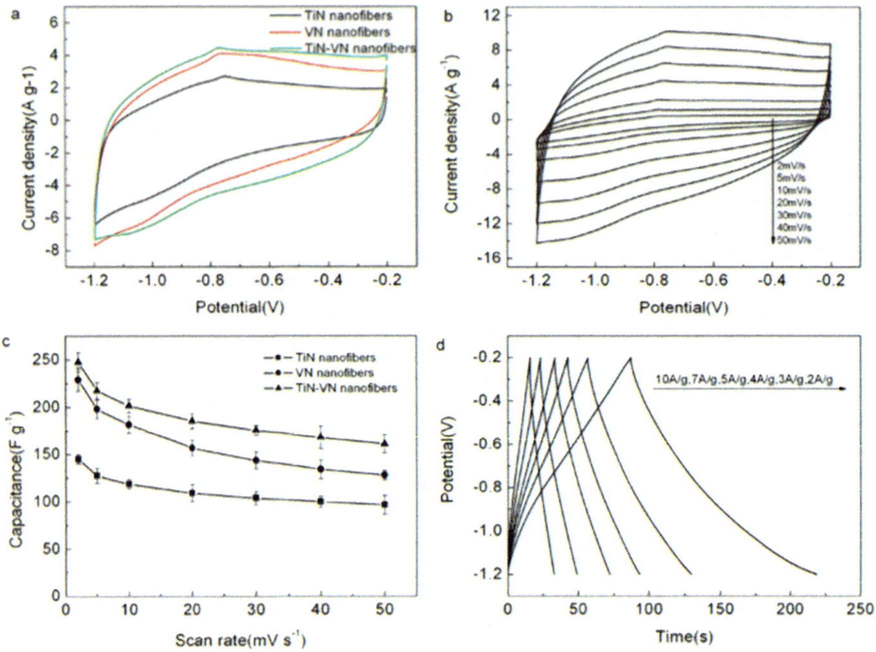

Fig. 13 **a** Cyclic-voltammetry curves of TiN, VN, TiN-VN nanofibers at a scan rate of 20 mV s^{-1}, **b** CV curves of TiN-VN fibers at different potential sweep rates from 2 to 50 mV s^{-1}, **c** Specific capacitances of the electrodes with TiN, VN, TiN-VN fibers respectively at scan rates from 2 to 50 mV s^{-1} and **d** The charge/discharge curves of TiN-VN fibers at rates varied from 2 to 10 A g^{-1}

F g^{-1}). Also, with the increase in scanning rate from 25 to 200 mV s^{-1}, the specific capacitance of meso-Mo_3N_2 NWs reduced by 4.5% while comm-Mo_3N_2 reduced by a huge 48.1% [41] (Fig. 14).

Fig. 14 A list of capacities obtained for different TMNs in SCs at specific current densities

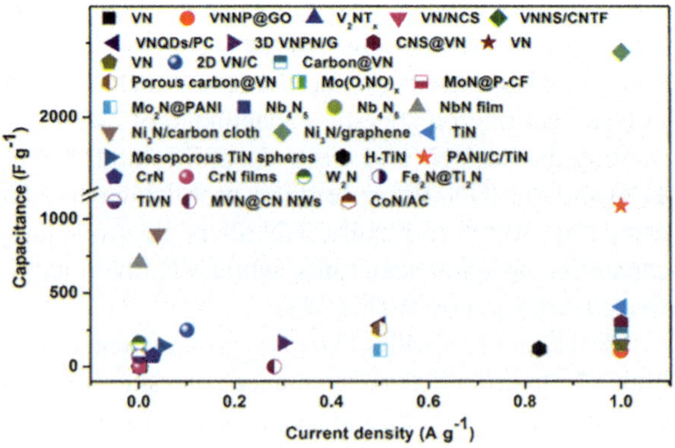

5 Conclusion

Electrical energy storage leads to a clean, flawless and dirt-free economy validating the societal changes and developments in the form of portable electronic devices like smartphones, computers and tablets, electric vehicles and renewable energy grid storage. Electrode material determines the electrochemical capacitive performance playing a key role in EES devices. Various metal oxides, conducting polymers and carbonaceous materials as efficient electrode materials have been discussed earlier. But recently, TMNs have become prominent because of their morphology, good conductivity, better cyclic characteristics, mechanical stability, large volumetric energy density and better catalytic activity. Nanostructured TMNs offer better electrochemical performances than their bulk counterparts due to smaller particle size, larger surface area and thus providing sufficient contact between the active materials and electrolytes and enabling shorter diffusion paths for the motion of electrons and ions. This chapter discusses the progress in TMNs as promising electrode material in LIBs, SIBs and SCs. Nitrides derived from transition metals like Fe, Ni, Mo, V, Nb, Co, Ti, etc. deliver better electrochemical performance than their oxide counterparts while this comparison does not sit very well in the case of SC. Some of the TMNs show weakness and instability for a long life cycle that has been improved largely by making composites with various carbonaceous materials and other TMNs as supporting materials. It has been concluded that the electrochemical performance of TMNs depends upon the structure, morphology, composition and conductivity of TMNs.

References

1. Larcher D, Tarascon J-M (2014) Towards greener and more sustainable batteries for electrical energy storage. Nat Chem 7(1):19–29. https://doi.org/10.1038/nchem.2085
2. Balogun M-S, Qiu W, Wang W, Fang P, Lu X, Tong Y (2015) Recent advances in metal nitrides as high-performance electrode materials for energy storage devices. J Mater Chem A 3:1364–1387. https://doi.org/10.1039/C4TA05565A
3. Gao B, Li X, Ding K, Huang C, Li Q, Chuc PK, Huo K (2019) Recent progress in nanostructured transition metal nitrides for advanced electrochemical energy storage. J Mater Chem A 7:14. https://doi.org/10.1039/c8ta05760e
4. Kubel F, Lengauer W, Yvon K, Junod A, Knorr K (1988) Structural phase transition at 205 K in stoichiometric vanadium nitride. Phys Rev B 38:12908. https://doi.org/10.1103/PhysRevB.38.12908
5. Christensen AN, Lebech B (1978) A reinvestigation of the structure of 8-tantalum nitride. Acta Cryst B34:261–263. https://doi.org/10.1107/s0567740878002733
6. Browne JD, Liddell PR, Street R, Mills T (1970) An investigation of the antiferromagnetic transition of CrN. Phys Stat Sol A 1:715–723. https://doi.org/10.1002/pssa.19700010411
7. Lengauer W (2000) Transition metal carbides, nitrides, and carbonitrides. In: Handbook of ceramic hard materials, pp 202–252. https://doi.org/10.1002/9783527618217.ch7
8. Bi W, Hu Z, Li X, Wu C, Wu J, Wu Y, Xie Y (2015) Metallic mesocrystal nanosheets of vanadium nitride for high-performance all-solid-state pseudocapacitors. Nano Res 8:193–200. https://doi.org/10.1007/s12274-014-0612-y

9. Gao B, Li X, Guo X, Zhang X, Peng X, Wang L, Fu J, Chu PK, Huo K (2015) Nitrogen-doped carbon encapsulated mesoporous vanadium nitride nanowires as self-supported electrodes for flexible all-solid-state supercapacitors. Adv Mater Interf 2:1500211. https://doi.org/10.1002/admi.201500211
10. Xiong Z, Yang J, Gao Z, Yang Q, Shi D (2020) Orthorhombic Mo_3N_2 nanobelts with improved electrochemical properties as electrode material for supercapacitors. Results Phys 16:102941. https://doi.org/10.1016/j.rinp.2020.102941
11. Xiao X, Yu H, Jin H, Wu M, Fang Y, Sun J, Hu Z, Li T, Wu J, Huang L, Gogotsi Y, Zhou J (2017) Salt-templated synthesis of 2D metallic MoN and other nitrides. ACS Nano 11(2):2180–2186. https://doi.org/10.1021/acsnano.6b08534
12. Xiao X, Peng X, Jin H, Li T, Zhang C, Gao B, Hu B, Huo K, Zhou J (2013) Freestanding mesoporous VN/CNT hybrid electrodes for flexible all-solid-state supercapacitors. Adv Mater 25:5091–5097. https://doi.org/10.1002/adma.201301465
13. Lu X, Liu T, Zhai T, Wang G, Yu M, Xie S, Ling Y, Liang C, Tong Y, Li Y (2014) Improving the cycling stability of metal-nitride supercapacitor electrodes with a thin carbon shell. Adv Energy Mater 4(4):1300994. https://doi.org/10.1002/aenm.201300994
14. Zhang L, Holt CMB, Luber EJ, Olsen BC, Wang H, Danaie M, Cui X, Tan XH, Lui VW, Kalisvaart WP, Mitlin D (2011) High rate electrochemical capacitors from three-dimensional arrays of vanadium nitride functionalized carbon nanotubes. J Phys Chem C 115(49):24381–24393. https://doi.org/10.1021/jp205052f
15. Yang P, Chao D, Zhu C, Xia X, Zhang Y, Wang X, Peng S, Tay BK, Shen ZX, Mai W, Fan HJ (2016) Ultrafast-charging supercapacitors based on corn-like titanium nitride nanostructures. Adv Sci 3:1500299. https://doi.org/10.1002/advs.201500299
16. Haldorai Y, Arreaga-Salas D, Rak CS, Huh YS, Han YK, Voit W (2016) Platinized titanium nitride/graphene ternary hybrids for direct methanol fuel cells and titanium nitride/graphene composites for high performance supercapacitors. Electrochim Acta 220:465–474
17. Peng X, Huo K, Fu J, Zhang X, Gao B, Chu PK (2013) Coaxial PANI/TiN/PANI nanotube arrays for high-performance supercapacitor electrodes. Chem Commun 49(86):10172–10174. https://doi.org/10.1039/c3cc45997g
18. Sun Q, Fu ZW (2008) Vanadium nitride as a novel thin film anode material for rechargeable lithium batteries. Electrochim Acta 54(2):403–409. https://doi.org/10.1016/j.electacta.2008.07.057
19. Zhao D, Cui Z, Wang S, Qin J, Cao M (2016) VN hollow spheres assembled from porous nanosheets for high-performance lithium storage and the oxygen reduction reaction. J Mater Chem A 4(20):7914–7923. https://doi.org/10.1039/c6ta01707j
20. Peng X, Li W, Wang L, Hu L, Jin W, Gao A, Zhang X, Huo K, Chu PK (2016) Lithiation kinetics in high-performance porous vanadium nitride nanosheet anode. Electrochim Acta 214:201–207. https://doi.org/10.1016/j.electacta.2016.08.023
21. Wang J, Wang L, Yang S (2018) VN quantum dots anchored uniformly onto nitrogen-doped graphene as efficient electrocatalysts for oxygen reduction reaction. NANO: Brief Rep Rev 13(4):1850041. https://doi.org/10.1142/S1793292018500418
22. Xiu Z, Kim D, Alfaruqi MH, Gim J, Song J, Kim S, Duong PT, Baboo JP, Mathew V, Kim J (2016) Porous TiN nanoparticles embedded in a N-doped carbon composite derived from metal–organic frameworks as a superior anode in lithium-ion batteries. J Mater Chem A 4:4706–4710. https://doi.org/10.1039/c5ta10342h
23. Yousefi E, Ghorbani M, Dolati A, Yashiro H (2016) Facile synthesis of titanium nitride-graphene nanocomposite and its improved rate-dependent electroactivity with respect to lithium storage. Mater Res Bull 84:388–396. https://doi.org/10.1016/j.materresbull.2016.08.040
24. Huang H, Gao S, Wu AM, Cheng K, Li XN, Gao XX, Zhao JJ, Dong XL, Cao GZ (2017) Fe_3N constrained inside C nanocages as an anode for Li-ion batteries through post-synthesis nitridation. Nano Energy 31:74–83. https://doi.org/10.1016/j.nanoen.2016.10.059
25. Balogun MS, Zeng Y, Qiu W, Luo Y, Onasanya A, Olaniyib TK, Tong Y (2016) Three-dimensional nickel nitride (Ni_3N) nanosheets: free standing and flexible electrodes for lithium ion batteries and supercapacitors. J Mater Chem A 4:9844–9849. https://doi.org/10.1039/c6ta02492k

26. Jiang G, Han H, Zhuang W, Xu X, Kaskel S, Xu F, Wang H (2019) Three-dimensional ordered mesoporous cobalt nitride for fast-kinetics and stable-cycling lithium storage. J Mater Chem A 7:17561–17569. https://doi.org/10.1039/C9TA03391B
27. Idrees M, Abbas SM, Rehman AU, Ahmad N, Mushtaq MW, Naqvi RA, Nam KW, Muhammad B, Iqbal Z (2017) Mechanistic insights into high lithium storage performance of mesoporous chromium nitride anchored on nitrogen-doped carbon nanotubes. Chem Eng J 327:361–370. https://doi.org/10.1016/j.cej.2017.06.095
28. Idrees M, Mukhtar A, A-U-Rehman, Abbas SM, Zhang Q, Li X (2021) Transition metal nitride electrodes as future energy storage devices: a review. Mater Today Commun 27:102363. https://doi.org/10.1016/j.mtcomm.2021.102363
29. Wang Y, Xiao R, Hu YS, Avdeev M, Chen L (2015) P_2-$Na_{0.6}[Cr_{0.6}Ti_{0.4}]O_2$ cation-disordered electrode for high-rate symmetric rechargeable sodium-ion batteries. Nat Commun 6:6954. https://doi.org/10.1038/ncomms7954
30. Bao J, Zhu L, Wang H, Han S, Jin Y, Zhao G, Zhu Y, Guo X, Hou J, Yin H, Tian J (2018) Hexagonal boron nitride/blue phosphorene heterostructure as a promising anode material for Li/Na-Ion batteries. Phys Chem C 122(41):23329–23335. https://doi.org/10.1021/acs.jpcc.8b07062
31. Mukherjee S, Bates A, Schuppert N, Son B, Kim JG, Choi JS, Choi MJ, Lee D-H, Kwon O, Jasinski J, Park S (2015) A study of a novel Na ion battery and its anodic degradation using sodium rich prussian blue cathode coupled with different titanium based oxide anodes. J Power Sources 286:276–289. https://doi.org/10.1016/j.jpowsour.2015.03.167
32. Xiong Y, Qian J, Cao Y, Ai X, Yang H (2016) Electrospun TiO_2/C nanofibers as a high-capacity and cycle-stable anode for sodium-ion batteries. ACS Appl Mater Interfaces 8:16684–16689. https://doi.org/10.1021/acsami.6b03757
33. Xie X, Kretschmer K, Zhang J, Sun B, Su D, Wang G (2015) Sn@CNT nanopillars grown perpendicularly on carbon paper: a novel free-standing anode for sodium ion batteries. Nano Energy 13:208–217. https://doi.org/10.1016/j.nanoen.2015.02.022
34. Bai Y, Wang Z, Wu C, Xu R, Wu F, Liu Y, Li H, Li Y, Lu J, Amine K (2015) Hard carbon originated from polyvinyl chloride nanofibers as high-performance anode material for Na-ion battery. ACS Appl Mater Interfaces 7:5598–5604. https://doi.org/10.1021/acsami.5b00861
35. Liu S-L, Huang J, Liu J, Lei M, Min J, Li S, Liu G (2016) Porous Mo_2N nanobelts as a new anode material for sodium-ion batteries. Mater Lett 172:56–59. https://doi.org/10.1016/j.matlet.2016.02.121
36. Sridhar V, Park H (2019) Manganese nitride stabilized on reduced graphene oxide substrate for high performance sodium ion batteries, super-capacitors and EMI shielding. J Alloys Compd 808:151748. https://doi.org/10.1016/j.jallcom.2019.151748
37. Liu S, Liu J, Wang W, Yang L, Zhu K, Wang H (2016) Synthesis of coral-like Fe_2N@C nanoparticles and application in sodium ion batteries as a novel anode electrode material. RSC Adv 6:86131–86136. https://doi.org/10.1039/C6RA17251B
38. Xu L, Xiong P, Zeng L, Liu R, Liu J, Luo F, Li X, Chen Q, Wei M, Qian Q (2020) Facile fabrication of vanadium nitride/carbon fibers composite for half/full sodium-ion and potassium-ion batteries with long-term cycling performance. Nanoscale 12:10693–10702. https://doi.org/10.1039/C9NR10211F
39. Lucio-Porto R, Bouhtiyya S, Pierson JF, Morel A, Capon F, Boulet P, Brousse T (2014) VN thin films as electrode materials for electrochemical capacitors. Electrochim Acta 141:203–211. https://doi.org/10.1016/j.electacta.2014.07.056
40. Zhou X, Shang C, Gu L, Dong S, Chen X, Han P, Li L, Yao J, Liu Z, Xu H, Zhu Y, Cui G (2011) Mesoporous coaxial titanium nitride-vanadium nitride fibers of core-shell structures for high-performance supercapacitors. ACS Appl Mater Interfaces 3:3058–3063. https://doi.org/10.1021/am200564b
41. Lee K-H, Lee Y-W, Ko A-R, Cao G, Park K-W (2013) Single-crystalline mesoporous molybdenum nitride nanowires with improved electrochemical properties. J Am Ceram Soc 96:37–39. https://doi.org/10.1111/jace.12096

Chapter 3
Electrode Materials in Lithium-Ion Batteries

R. Dash, P. Kommu, and A. S. Bhattacharyya

1 Introduction

Electrochemical storage batteries are used in fuel cells, liquid/fuel generation, and even electrochemical flow reactors. Vanadium Redox flow batteries are utilized for CO_2 conversion to fuel, where renewable energy is stored in an electrolyte and used to charge EVs, and telecom towers, and act as a replacement for diesel generators, providing business back up and rural electrification. LIBs (lithium-ion batteries) are one of the most significant achievements of modern electrochemistry. They have a high energy density and are easily cyclable. LIBs have dominated the portable electronics industry for several decades. LIBs are increasingly being used in high-performance Electric Vehicles (EVs) due to their dependability, overall high performance at a wide range of temperatures, cycle behavior, stability, and safety [1–3]. Due to events such as the battery failures in the Boeing 787 Dreamliner, which delayed the aircraft's introduction and prompted an extensive investigation, as well as numerous battery problems in both Tesla and Chevrolet's electric vehicles, the emphasis on Li-ion battery design as a system has grown significantly. In today's world, Battery Manufacturing Systems (BMS) must take climatic conditions into account. The world is separated into two temperature zones: tropical and subtropical. Because India is in a tropical zone, it has hurdles in any technology, and thus EV standards in India differ from those in other advanced countries, as there are challenges in terms of road conditions and heat management systems. Cathode, anode, separator, electrolyte, can, and lid are the major components of LiBs. For Li storage,

R. Dash · P. Kommu · A. S. Bhattacharyya (✉)
Department of Metallurgical and Materials Engineering, Central University of Jharkhand, Ranchi 835 205, India
e-mail: arnab.bhattacharya@cuj.ac.in

A. S. Bhattacharyya
Centre of Excellence in Green and Efficient Energy Technology, Central University of Jharkhand, Ranchi 835 205, India

B. P. Swain (ed.), *Energy Materials*, Materials Horizons: From Nature to Nanomaterials,
https://doi.org/10.1007/978-981-99-3866-7_3

cylindrical- and pouch-shaped batteries are utilized. In many systems, the cathode is an aluminum foil coated with the active cathode material.

Lithium-ion batteries most frequently use the following cathode chemistry blends: LFP (Li Fe phosphate), NMC (Li Ni Mn Co), LCO (Li Co oxide), NCA (Li Ni-Co Al), and LMO (Li Mn oxide) [4]. These five basic chemistries and their combinations are used in a variety of ways to reach varied performance results like high-power capabilities, low cost, and safety. Modification of electrodes by lattice doping and coatings may play a critical role in improving their electrochemical properties, cycle life, and thermal behavior. This chapter discusses the different material combinations for use as a cathode material.

2 Lithium Iron Phosphate (LFP)

LFP is one of the most popular chemical formulations used in automotive applications. It can swiftly deliver an accelerating discharge and accept a regenerative brake charge. Compared to the other chemistries, LFP has a lower energy density, which means it would require less energy to fail. However, LFP is more resilient to harmful circumstances like overcharging the cell and extreme heat. Moving phase boundaries result from spinodal breakdown or nucleation for small currents. The spinodal disappears and particles fill uniformly over a certain current density (in the Tafel regime), which could account for the higher rate capability and prolonged cycle life of nano-$LiFePO_4$ cathodes [5]. The EV lifespan has to be improved using graphite in LFP [6]. LFP nanocrystals have been used for light-assisted charging by a combination with dye-sensitized solar cells. Dye sensitization generates electron–hole pairs causing delithiation [7].

LFP is a widely utilized chemical formulation in automotive applications. It is capable of rapidly delivering an accelerating discharge and accepting a regenerative braking charge. LFP has a lower energy density than the other chemistries, which implies it would take less energy to fail. LFP, on the other hand, is more resistant to hazardous conditions such as overcharging the cell and severe heat. For tiny currents, moving phase boundaries emerge from spinodal breakdown or nucleation. In the Tafel regime, the spinodal disappears and particles fill uniformly over a specific current density, which could explain the increased rate capability and longer cycle life of nano-$LiFePO_4$ cathodes [5]. A thermally modulated LFP battery has been constructed to run in any ambient environment at a working temperature of roughly 60 °C, which is promising for EVs. The limited working period at high temperatures allows for the use of graphite with reduced surface areas, potentially extending the EV lifespan [6]. LFP nanocrystals have been used for light-assisted charging by a combination with dye-sensitized solar cells. Dye sensitization generates electron–hole pairs causing delithiation [7]. The charge–discharge profile of NCA/$LiPF_6$/Cu can be seen in Fig. 1. An initial improvement in the values was suggestive of the fact that some materials were still inactive while the decay in the higher cycles was due to dead Li found in anode-free LiB [8].

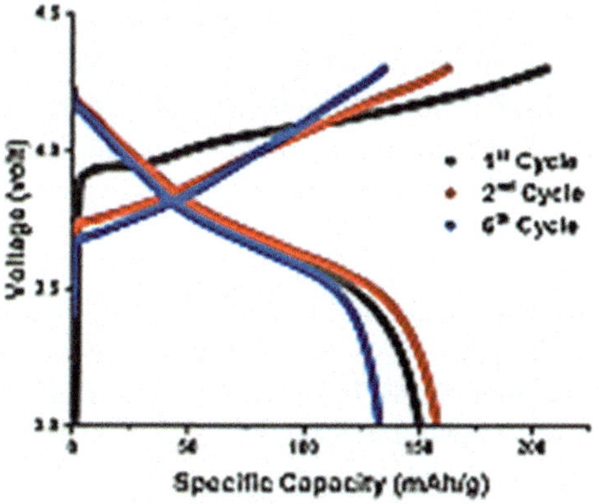

Fig. 1 The charge–discharge profile of NCA/$LiPF_6$/Cu (reproduced under CC-BY Licence, Open Access) [8]

3 Nickel Manganese Cobalt (NMC/NCM) Oxides

With the introduction of manufacturing EVs, lithium NMC chemistries have begun to take off due to their superior energy density and voltage. If the chemistry has a higher amount of cobalt than manganese, it may be referred to as nickel cobalt manganese (NCM) depending on the material combination. NMC has one of the highest energy densities in a manufacturing cell today, ranging between 140 and 180 Wh/kg in production applications, with certain chemistries exceeding 200 Wh/kg, and a relatively high nominal voltage of around 3.6–3.8 V per cell [9]. The performance of Ni-rich NMC ($LiNi_xMn_yCo_{1-x-y}O_2, x \geqslant 0.5$) cathode material can be done by dopants, gradient layers, surface coatings, carbon matrixes and utilizing some alterations in synthesis methods. They usually possess enhanced specific capacity and thermal stability [10].

The NMC-based cathodes, when compared to $LiMO_2$ (M = Ni, Mn, or Co) cathodes, also have the combined advantages of the three transition metals, with nickel being able to offer high specific capacities and Co and Mn being able to provide layered structures and improved structural integrity. NMC-based cathodes can also perform better than lithium iron phosphate (LFP) cathodes in many other ways, especially when it comes to operational voltage, as LFP-based LIBs can only produce voltages below 3.4 V and have high self-discharge rates. Li_2MnO_3 and lithium titanate, for example, have low industrial potential as cathodes due to their short lifetimes and high costs [11].

4 Drawbacks Due to Excessive Ions

While manganese is used sparingly as a structural stabilizer, high levels of Ni^{4+} on cathode surface layers/regions might generate side reactions, whereas Ni^{2+} can cause cation mixing. As a result, with these Ni-rich cathode materials, increased mass-specific capacity comes at the expense of rate capability and structural stability, resulting in severe capacity fading. When compared to low Ni content NMC-based cathodes, they experience much higher capacity loss and impedance growth. This is due to the fact that significant concentrations of Ni^{4+} on cathode surface regions might generate side reactions, while Ni^{2+} can cause cation mixing. Furthermore, excessive Li is commonly used in the fabrication of NMC-based cathodes, where it tends to build on the surface of particles and react with electrolytes to form lithium carbonate and hydroxide. Additionally, the continual charging and discharging of these cathodes might produce secondary particle cracking. One typical solution to this problem is to combine LMO and NMC, which can improve capacity retention [12, 13]. Due to low octahedral site stability energy, NMC-based cathodes possess a tendency to form spinel-structured $LiMn_2O_4$ on surfaces [14] and Mn^{4+} possesses a similar tendency to migrate to the Li-layer [15] and decrease capacity. Nevertheless, diffused Mn^{4+} on particle surfaces can stabilize structures during long-term cycling [16]. NCM811 and NCM622 are known to form microcracks along the phase boundaries during cycling, leading to an increase in surface areas for side reactions [17].

5 Structural Stability

NMC-based cathodes with higher Mn content have higher cycle stability [18, 19], but NMC-based cathodes with higher Ni content are thought to be incapable of reaching high cut-off voltages due to a lack of Mn^{4+} as a structural stabilizer [20]. Aside from Ni^{2+}, structural instability also applies to Mn^{3+}, albeit Mn migration is not considered important due to the high Ni content. Active material dissolving, oxygen release, and primary particle intergranular cracking are among the factors that contribute to cathode degradation. Dopants, gradient layers, coatings (both Li-free and Li-contained), carbon matrices, and enhanced synthesis techniques were used to improve the performance of Ni-rich NMC-based cathodes.

6 Effect of Doping

Doping of Li $[Ni_xCo_yMn_z]O_2$ Ni-rich cathode materials with various cations (such as Fe^{3+}, Cu^{2+}, Cr^{3+}, Mg_{2+}, Al^{3+}, and $Zr^{4+)}$ has been reported to enhance their electrochemical behavior [21–26]. Dopants have a variety of functions in the material. On the one hand, they can operate as pillars in the lithium layers, suppressing c-lattice contraction near the end of the charge and so improving the electrode's rate capability. Furthermore, by occupying lithium sites, the Li+/Ni_2+ mixing can be decreased, so suppressing the production of inactive rock salt on the surface, resulting in lower charge-transfer-resistance increase with cycling. The surface lithium may react with the dopants and surface coatings to generate lithium conductive layers [27, 28]. After 50 galvanostatic cycles, the Zr-doped material performs better at higher cycling rates and has a more stable capacity. The impedance behavior of the Zr-doped NCM, which exhibits orders of magnitude lower responses and reduced impedance growth during cycling, demonstrates the increased rate capacity. Li_2ZrO_3 is formed by Zr-doped NCM, showing the development of lithium-rich phases on the particles with excess lithium [21, 29, 30].

Preferential Aluminium (Al^{+3}) doping at Mn, Co, or Ni sites occurs due to the highest negative substitution energy of Al at the Ni sites and results in lower capacity fading of the electrodes. The reason being, Al-doped electrodes partially suppress the unavoidable formation of LiF, stabilizing the electrode/solution interface and, hence, leading to lower impedance and more stable charge/discharge behavior. The modified interface comprising the Li+ -ion conducting active centers such as $LiAlO_2$, AlF_3, and Li [AlF_4] promotes enhanced Li+ transport to the electrode bulk and facilitates the charge-transfer reactions [22, 26, 31].

7 Delithiation

The increase in the nickel content in NCMs leads to an increase in capacity from 155 to 195 $mAh{\cdot}g^{-1}$. The higher delithiation degree (~70%) is substantial and leads to multiple phase changes during charge and discharge [32, 33]. The oxygen layers become less protected as the lithium layer depletes because the number of lithium ions keeps on decreasing leading to an increase in repulsive force and c-lattice parameter. As Li is withdrawn further, the emptier lithium layer is likely to be partially filled with transition metals (e.g., Ni) to preserve the lattice. This results in strong contacts between high-oxidation-state transition metals between the lithium and oxygen layers, further lowering the c-parameter. The rapid contraction of the c-parameter induces a volume shift and stress on the primary particle phase boundaries[32].

8 Coating Cathode Materials

In addition to doping approaches, one prominent strategy for enhancing electrode cycle life and thermal stability is to cover cathode materials with thin surface layers. The purpose is to cover the surface area of the active material with an inert thin coating of organic or inorganic compounds. The coating will act as a barrier between the active material and the electrolyte at high voltages, decreasing parasitic side reactions. Furthermore, some coatings are known to be HF scavengers, which is particularly advantageous if the coating is not uniform or easily breaks during cycling [34]. The majority of the coating techniques discussed to use a wet chemical method to deposit the coating layer over the active component, which is then heat treated. Although this process is straightforward, the coatings usually shatter or are placed in flakes. Powder atomic layer deposition (p ALD) has recently been shown to be a viable alternative with numerous advantages [35]. Several coatings for various NCMs (e.g., AlF_3, Al_2O_3, ZrO_2, TiO_2, SiO_2) have been researched. All of the coatings were demonstrated to improve the cycle stability of the investigated active material when applied in modest amounts. Coatings are also used to improve thermal stability, particularly for Ni-rich NCMs [36–40].

9 Lithium Cobalt Oxide (LCO)

LCO ($LiCoO_2$) is typically found in portable electronics such as laptop computers, cameras, and cell phones. It is currently used in more than 31% of manufactured LIBs due to its well-ordered, –$NaFeO_2$ layered structure, which enables simple scalable production and quick and reversible lithium intercalation [41, 42]. LCO has a longer cycle life and a higher energy density in general than other chemistries, although it is less stable at higher temperatures and more reactive. The cell will enter the thermal runaway stage at temperatures above 130 °C, which is lower than that of other lithium-ion chemistries. As a result, while LCO is still used in small consumer electronics, it has not been extensively accepted in large applications, and several automakers have actively opposed it. LCO is a significantly more expensive chemical since it contains a significant proportion of cobalt, a rather uncommon substance. Using a simple and scalable hydrothermal-assisted hybrid surface treatment, lithium, aluminum, fluorine (LAF)-modified lithium cobalt oxide with a stable and conductive layer reduces active cobalt loss and forms a thin doping layer composed of a Lithium-Aluminium-cobalt-oxide-fluorine solid solution, which suppresses the phase transition of lithium cobalt oxide when operated at voltages greater than 4.55 V [43].

10 Nickel Cobalt Aluminium Oxides (NCA)

NCA is often utilized in portable power applications, but because of its high power, it is also being considered for automotive applications. However, NCA does have several disadvantages that are preventing it from gaining traction in the EV arena, including high cost and low safety [44, 45]. Nickel-rich layered, mixed lithium transition metal oxides have been examined as possible cathode materials for the next generation of lithium-ion batteries due to their high energy density and low cost. However, their usage in large-scale applications is limited due to poor cycling and rate performance produced by acute side reactions between Ni_4+ and carbonate electrolytes [46, 47].

Cathode materials based on nickel have a high specific capacity and discharge voltage. NMC ($LiNiMnCoO_2$) and NCA ($LiNiCoAlO_2$) batteries are commonly utilized in electric and plug-in hybrid automobiles. While cycling, NCA-based batteries degrade quickly, resulting in capacity and power loss. Its degradation is exacerbated by overcharging, over-discharging, high-current charging (or fast

Fig. 2 SEM of NCA synthesized by means of oxalate co-precipitation (reproduced under CC-BY Licence, Open Access) [8]

charging), and operating outside of the permissible temperature window. Electrochemical Impedance Spectroscopy (EIS), Incremental Capacity Analysis (ICA), and the galvanostatic intermittent titration technique are all methods for monitoring ongoing battery degradation. These processes are also important for establishing the State of Health (SoH) of batteries before recycling and reusing them [48].

Layered-type lithium nickel cobalt aluminum oxide (NCA) is regarded as one of the most promising and cutting-edge cathode materials for Li-ion batteries due to its favorable properties such as high columbic capacity, gravimetric energy density, and power density. Because nickel is less poisonous and less expensive than cobalt, NCA with a high nickel concentration is preferred. In addition, the Ni-rich cathode material has a higher capacity than high cobalt content $LiCoO_2$. As a result, NCA/ graphite batteries are a strong power. NCA cathode material was produced via the oxalate co-precipitation technique which is cheaper and simpler compared to other synthesis procedures. The particles were found to be loose, irregular, or quasi-cubical shaped with smooth surfaces as shown in Fig. 2. [48].

11 Lithium Manganese Oxide (LMO)

Long run periods are desirable in portable power applications, but not necessarily in automotive applications where LMO is utilized, which gives high energy and high power but has a reduced cycle life. When activated chemically with acid or electrochemically above 4.5 V versus Li+ /Li in lithium cells, Li_2MnO_3 has a theoretical capacity of 459 mAh/g, which corresponds to the extraction of 2 Li per formula unit [46–48].

A comparison of the four forms of LiB studied thus far, namely LFP, NMC, LMO, and NCA, revealed that LFP and NCA had a stronger hysteresis impact than the other two. The modeling of batteries is critical for the safe and effective operation of battery applications. The battery management system (BMS) typically uses the comparable circuit model to monitor and manage Li-ion batteries [49].

12 Cathode Materials for EVs

There are mainly two classes of cathode materials that are considered in LIBs, especially for EVs. The first one being Li- and Mn-rich $x\text{Li}_2\text{MnO}_3(1-x)\text{Li}[\text{Ni}_a\text{Co}_b\text{Mn}_c]\text{O}_2$ and Ni-rich ($y > 0.5$) $\text{LiNi}_y\text{Co}_x\text{Mn}_{1-y-x}\text{O}_2$. The second one is Lithiated transition metal (TM) oxides Lithium and manganese-rich layered composites from the $x\text{Li}_2\text{MnO}_3(1-x)\,\text{Li}[\text{Ni}_a\text{CobMn}_c]\text{O}_2$ family, which are typically defined as consisting of two layered structural phases [29, 50–52]. Li-cells made of the $x\text{Li}_2\text{MnO}_{3(1-x)}$ Li $[\text{Ni}_a\text{Co}_b\text{Mnc}]\text{O}_2$ materials have high discharge capabilities of more than 250 mAh/g [53, 54]. Lithiated Transition Metal (TM) oxides with the general formula $\text{LiNi}_y\text{Co}_x\text{Mn}_{1-y-x}\text{O}_2$ on the other hand have shown a high capacity of 200 mAh/g between 2.5 and 4.6 V, or 155 mAh/g if the anodic voltage is restricted to 4.3 V [29, 49, 55].

The recycling of EV batteries is being investigated these days due to rising costs and other concerns associated with disposal. Demand for Co-based batteries has decreased, while demand for Li- and Ni-based batteries has increased. Recently, a recycling system for extracting valuable metals from medium-sized lithium-ion batteries using a wet process was developed. However, new materials must be developed to replace the primary metal in LiBs, as well as cost-competitive new materials to replace pricey and highly volatile metals such as lithium and cobalt, as well as a secondary battery with improved performance, low cost, and high battery energy density by examining the NCM content ratio [56].

13 Li and Mn Layered Structures

To achieve high capacities, electrodes consisting of Li- and Mn-rich materials must be activated by charging to more than 4.5 V during the first cycle [58]. Both redox centers and Li+ intercalation sites play important roles in charging. When Li- and Mn-rich materials are charged, initial Ni $^{2+/4+}$ and Co $^{3+/4+}$ oxidation occurs in Li $[\text{Ni}_a\text{Co}_b\text{Mn}_c]\text{O}_2$ layered materials [59, 60]. In addition to creating resistance to Li+ re-intercalation, this migration of transition metals to the Li-layer results in significant voltage hysteresis between charge and discharge [61–64]. A Li/Mn-Rich layer $\text{Li}_{1.2}\text{Mn}_{0.54}\text{Ni}_{0.13}\text{Co}_{0.13}\text{O}_2$ which was again surface treated with graphene and carbon nanotubes layered cathode has been used for fast charging. Its semi-hollow microsphere structure as shown in Fig. 3 enabled high tap density and bidirectional ion diffusion pathways [65].

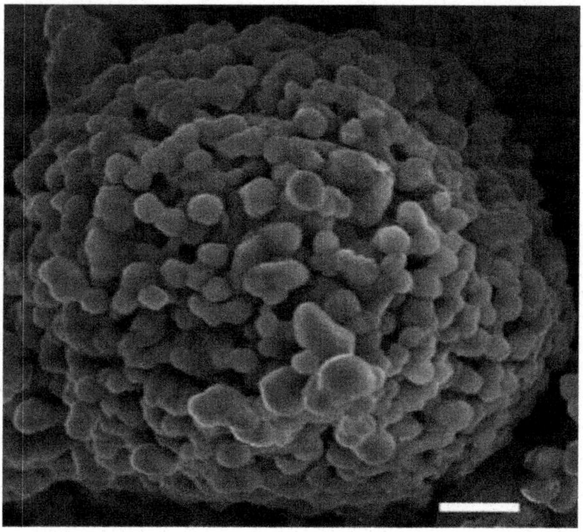

Fig. 3 Semi-microsphere structure of Li/Mn-Rich layer $Li_{1.2}Mn_{0.54}Ni_{0.13}Co_{0.13}O_2$ which was again surface treated with graphene and carbon nanotubes (Bar scale 500 nm) (reproduced under CC-BY Licence—Open Access) [65]

14 Summary

This chapter presents a review of recent advances and limitations to overcome in the development and use of inorganic compounds as cathode in lithium-ion batteries. Different cathode configuration like LFP, LCO, NMO, LCA, and LMO exists having their specific advantages and drawbacks. The Li- and Mn-rich layered composites are used which have two layered structures. Also employed are the Ni-rich layered structure materials which are lithiated transition metal oxides composed of Ni, Co, and Mn. The lattice doping Al^{3+}, and Zr^{4+} and the deposition of thin surface coatings have significantly improved the cycling behaviour of electrodes by reducing capacity fading and stabilizing average voltage, for example. Additionally, surface coatings help cathodes become more thermally stable in charged states and reduce their side reactions with solution species.

References

1. Goodenough JB, Park K-S (2013) The Li-ion rechargeable battery: a perspective. J Am Chem Soc 135:1167–1176
2. Manthiram A (2011) Materials challenges and opportunities of lithium-ion batteries. J Phys Chem Lett 2:176–184
3. Armand M, Tarascon J-M (2008) Building better batteries. Nature 451:652–657
4. Warner J (2015) The handbook of lithium-ion battery pack design. Elsevier. ISBN: 978-0-12-801456-1
5. Peng B, Cogswell DA, Bazant MZ. arXiv:1108.2326v1 [cond-mat.mtrl-sci]
6. Yang XG, Liu T, Wang CY (2021) Thermally modulated lithium iron phosphate batteries for mass-market electric vehicles. Nat Energy 6:176–185

7. Paolella A, Faure C, Bertoni G et al (2017) Light-assisted delithiation of lithium iron phosphate nanocrystals towards photo-rechargeable lithium-ion batteries. Nat Commun 8:14643
8. Yudha C, Hutama A, Rahmawati M, Arinawati M, Aliwarga H, Widiyandari H, Purwanto A (2022) Production of nickel-rich $LiNi_{0.89}Co_{0.08}Al_{0.03}O_2$ cathode material for high capacity NCA/graphite secondary battery fabrication. Open Eng 12(1):501–510
9. Li T, Yuan XZ, Zhang L et al (2020) Degradation mechanisms and mitigation strategies of nickel-rich NMC-based lithium-ion batteries. Electrochem Energy Rev 3:43–80
10. Iturrondobeitia A, Aguesse F, Genies S et al (2017) Post-mortem analysis of calendar-aged 16 Ah NMC/graphite pouch cells for EV application. J Phys Chem C 121:21865–21876
11. Thackeray MM, Johnson CS, Vaughey JT, Li N, Hackney SA (2005) Advances in manganese-oxide 'composite' electrodes for lithium-ion batteries. J Mater Chem 15:2257–2267
12. Yang J, Xia YY (2016) Suppressing the phase transition of the layered Ni-rich oxide cathode during high-voltage cycling by introducing low-content Li_2MnO_3. ACS Appl Mater Interfaces 8:1297–1308
13. Wu SH, Lee PH (2017) Storage fading of a commercial 18650 cell comprised with NMC/LMO cathode and graphite anode. J Power Sources 349:27–36
14. Erickson EM, Schipper F, Tian R, Shin J-Y, Erk C, Chesneau FF, Lampert JK, Markovsky B, Aurbach D (2017) Enhanced capacity and lower mean charge voltage of Li-rich cathodes for lithium-ion batteries resulting from low-temperature electrochemical activation. RSC Adv 7:7116–7121
15. Rozier P, Tarascon JM (2015) Review-Li-rich layered oxide cathodes for next-generation Li-ion batteries: chances and challenges. J Electrochem Soc 162:A2490–A2499
16. Lee H, Kim Y, Hong Y-S, Kim Y, Kim MG, Shin N-S, Cho J (2006) Structural characterization of the surface-modified $Li_xNi_{0.9}Co_{0.1}O_2$ cathode materials by MPO4 coating (M = Al, Ce, Sr, and Fe) for Li-ion cells. J Electrochem Soc 153:A781–A786
17. Schipper F, Erickson EM, Erk C, Shin J-Y, Chesneau FF, Aurbach D (2017) Review—recent advances and remaining challenges for lithium-ion battery cathodes: I. Nickel-rich, $LiNi_xCo_yMn_zO_2$. J Electrochem Soc 164:A6220–A6228
18. Shukla AK, Ramasse QM, Ophus C, Duncan H, Hage F, Chen G (2015) Unravelling structural ambiguities in lithium- and manganese-rich transition metal oxides. Nat Commun 6:8711
19. Nayak PK, Grinblat J, Levi E, Levi M, Markovsky B, Aurbach D (2017) Understanding the influence of Mg doping for the stabilization of capacity and higher discharge voltage of Li- and Mn-rich cathodes for Li-ion batteries. Phys Chem Chem Phys 19:6142–6152
20. Qiao Q-Q, Qin L, Li G-R, Wang Y-L, Gao X-P (2015) Sn-stabilized Li-rich layered $Li(Li_{0.17}Ni_{0.25}Mn_{0.58})O_2$ oxide as a cathode for advanced lithium-ion batteries. J Mater Chem A 3:17627–17634
21. Schipper F, Dixit M, Kovacheva D, Talianker M, Haik O, Grinblat J, Erickson EM, Ghanty C, Major DT, Markovsky B et al (2016) Stabilizing nickel-rich layered cathode materials by a high-charge cation doping strategy: Zirconium-doped $LiNi_{0.6}Co_{0.2}Mn_{0.2}O_2$. J Mater Chem A 4:16073–16084
22. Aurbach D, Srur-Lavi O, Ghanty C, Dixit M, Haik O, Talianker M, Grinblat Y, Leifer N, Lavi R, Major DT et al (2015) Studies of aluminum-doped $LiNi_{0.5}Co_{0.2}Mn_{0.3}O_2$: electrochemical behavior, aging, structural transformations, and thermal characteristics. J Electrochem Soc 162:A1014–A1027
23. Jin X, Xu Q, Liu H, Yuan X, Xia Y (2014) Excellent rate capability of Mg doped $Li[Li_{0.2}Ni_{0.13}Co_{0.13}Mn_{0.54}]O_2$ cathode material for lithium-ion battery. Electrochim Acta 136:19–26
24. Liu L, Sun KN, Zhang NQ, Yang TY (2009) Improvement of high-voltage cycling behavior of $Li(Ni_{1/3}Co_{1/3}Mn_{1/3})O_2$ cathodes by Mg, Cr, and Al substitution. J Solid State Electrochem 13:1381–1386
25. Liu DT, Wang ZX, Chen LQ (2006) Comparison of structure and electrochemistry of Al- and Fe-doped $LiNi_{1/3}Co_{1/3}Mn_{1/3}O_2$. Electrochim Acta 51:4199–4203
26. Chen CH, Liu J, Stoll ME, Henriksen G, Vissers DR, Amine K (2004) Aluminum-doped lithium nickel cobalt oxide electrodes for high-power lithium-ion batteries. J Power Sources 128:278–285

27. Jung S-K, Gwon H, Hong J, Park K-Y, Seo D-H, Kim H, Hyun J, Yang W, Kang K (2014) Understanding the degradation mechanisms of $LiNi_{0.5}Co_{0.2}Mn_{0.3}O_2$ cathode material in lithium ion batteries. Adv Energy Mater 4:1300787
28. Shin DW, Bridges CA, Huq A, Paranthaman MP, Manthiram A (2012) Role of cation ordering and surface segregation in high-voltage spinel $LiMn_{1.5}Ni_{0.5-x}M_xO4$ (M = Cr, Fe, and Ga) cathodes for lithium-ion batteries. Chem Mater 24:3720–3731
29. Kalyani P, Chitra S, Mohan T, Gopukumar S (1999) Lithium metal rechargeable cells using Li_2MnO_3 as the positive electrode. J Power Sources 80:103–106
30. Nayak PK, Grinblat J, Levi E, Markovsky B, Aurbach D (2016) Effect of cycling conditions on the electrochemical performance of high capacity Li and Mn-rich cathodes for Li-ion batteries. J Power Sources 318:9–17
31. Park BC, Kim HB, Myung ST, Amine K, Belharouak I, Lee SM, Sun YK (2008) Improvement of structural and electrochemical properties of AlF_3-coated $Li[Ni_{1/3}Co_{1/3}Mn_{1/3}]O_2$ cathode materials on high voltage region. J Power Sources 178:826–831
32. Li J, Shunmugasundaram R, Doig R, Dahn JR (2016) In situ x-ray diffraction study of layered Li–Ni–Mn–Co oxides: effect of particle size and structural stability of core–shell materials. Chem Mater 28:162–171
33. Ghanty C, Markovsky B, Erickson EM, Talianker M, Haik O, Tal-Yossef Y, Mor A, Aurbach D, Lampert J, Volkov A et al (2015) Li+-ion extraction/insertion of Ni-rich Li1+x(NiyCozMnz)wO2 (0.005 <x <0.03; y:z=8:1, w $\approx$ 1) electrodes: insitu XRD and raman spectroscopy study. ChemElectroChem 2:1479–1486
34. Amalraj SF, Sharon D, Talianker M, Julien CM, Burlaka L, Lavi R, Zhecheva E, Markovsky B, Zinigrad E, Kovacheva D (2013) Study of the nanosized Li2MnO3: Electrochemical behavior, structure, magnetic properties, and vibrational modes. Electrochim Acta 97:259–270
35. Li X, Liu J, Meng X, Tang Y, Banis MN, Yang J, Hu Y, Li R, Cai M, Sun X (2014) Significant impact on cathode performance of lithium-ion batteries by precisely controlled metal oxide nanocoatings via atomic layer deposition. J Power Sources 247:57–69
36. Ghanty C, Dahiya PP, Basu RN, Chang J-K, Majumder SB (2015) Improvement of the electrochemical characteristics of lithium and manganese rich layered cathode materials: effect of surface coating. J Electrochem Soc 162:A1957–A1965
37. Chen Y, Zhang Y, Chen B, Wang Z, Lu C (2014) An approach to application for LiNi0.6Co0.2Mn0.2O2 cathode material at high cutoff voltage by TiO2 coating. J Power Sources 256: 20–27. [Google Scholar]
38. Qin C, Cao J, Chen J, Dai G, Wu T, Chen Y, Tang Y, Li A, Chen Y (2016) Improvement of electrochemical performance of nickel rich $LiNi_{0.6}Co_{0.2}Mn_{0.2}O_2$ cathode active material by ultrathin TiO2 coating. Dalton Trans 45:9669–9675
39. Cho W, Kim S-M, Song JH, Yim T, Woo S-G, Lee K-W, Kim J-S, Kim Y-J (2015) Improved electrochemical and thermal properties of nickel rich $LiNi_{0.6}Co_{0.2}Mn_{0.2}O_2$ cathode materials by SiO_2 coating. J Power Sources 282:45–50 [Google Scholar]
40. Myung S-T, Izumi K, Komaba S, Sun Y-K, Yashiro H, Kumagai N (2005) Role of alumina coating on Li–Ni–Co–Mn–O particles as positive electrode material for lithium-ion batteries. Chem Mater 17:3695–3704
41. Goodenough JB, Kim Y (2010) Challenges for rechargeable li batteries. Chem Mater 22:587–603
42. Schmid R, Pillot C (2014) Introduction to energy storage with market analysis and outlook. AIP Conf Proc 1597:3–13
43. Qian J, Liu L, Yang J et al (2018) Electrochemical surface passivation of $LiCoO_2$ particles at ultrahigh voltage and its applications in lithium-based batteries. Nat Commun 9:4918
44. Blomgren GE (2016) The development and future of lithium ion batteries. J Electrochem Soc 164:A5019–A5025
45. Diaz F, Wang Y, Moorthy T, Friedrich B (2018) Degradation mechanism of nickel-cobalt-aluminum (NCA) cathode material from spent lithium-ion batteries in microwave-assisted pyrolysis. Metals 8:565

46. Vadivel S, Phattharasupakun N, Wutthiprom J, Duangdangchote S, Sawangphruk M (2019) High-performance Li-ion batteries using nickel-rich lithium nickel cobalt aluminium oxide–nanocarbon core–shell cathode: in operando X-ray diffraction. ACS Appl Mater Interfaces 11(34):30719–30727
47. Li Q, Yao Z, Lee E et al (2019) Dynamic imaging of crystalline defects in lithium-manganese oxide electrodes during electrochemical activation to high voltage. Nat Commun 10:1692
48. Kemeny M, Ondrejka P, Mikolasek M (2023) Comprehensive degradation analysis of NCA Li-Ion batteries via methods of electrochemical characterisation for various stress-inducing scenarios. Batteries 9:33
49. Rossouw M, Thackeray M (1991) Lithium manganese oxides from Li_2MnO_3 for rechargeable lithium battery applications. Mater Res Bull 26:463–473
50. Tran M-K, DaCosta A, Mevawalla A, Panchal S, Fowler M (2021) Comparative study of equivalent circuit models performance in four common lithium-ion batteries: LFP, NMC, LMO, NCA. Batteries 7:51
51. Amalraj F, Talianker M, Markovsky B, Burlaka L, Leifer N, Goobes G, Erickson EM, Haik O, Grinblat J, Zinigrad E et al (2013) Studies of Li and Mn-rich $Li_x[MnNiCo]O_2$ electrodes: electrochemical performance, structure, and the effect of the aluminum fluoride coating. J Electrochem Soc 160:A2220–A2233
52. Erickson EM, Ghanty C, Aurbach D (2014) New horizons for conventional lithium-ion battery technology. J Phys Chem Lett 5:3313–3324
53. Berg EJ, Villevieille C, Streich D, Trabesinger S, Novak P (2015) Rechargeable batteries: grasping for the limits of chemistry. J Electrochem Soc 162:A2468–A2475
54. Yan JH, Liu XB, Li BY (2014) Recent progress in Li-rich layered oxides as cathode materials for Li-ion batteries. RSC Adv 4:63268–63284
55. Amalraj F, Talianker M, Markovsky B, Sharon D, Burlaka L, Shafir G, Zinigrad E, Haik O, Aurbach D, Lampert J et al (2012) Study of the lithium-rich integrated compound $xLi_2MnO_3\cdot(1-x)LiMO_2$ (x around 0.5; M = Mn, Ni, Co; 2:2:1) and its electrochemical activity as positive electrode in lithium cells. J Electrochem Soc 160:A324–A337
56. Lu ZH, MacNeil DD, Dahn JR (2001) Layered cathode materials $Li(Ni_xLi_{(1/3-2x/3)}Mn_{(2/3-x/3)})O_2$ for lithium-ion batteries. Electrochem Solid State Lett 4:A191–A194
57. Mo JY, Jeon W (2018) The impact of electric vehicle demand and battery recycling on price dynamics of lithium-ion battery cathode materials: a vector error correction model (VECM) analysis. Sustainability 10(8):2870
58. Kang SH, Johnson CS, Vaughey JT, Amine K, Thackeray MM (2006) The effects of acid treatment on the electrochemical properties of $0.5\ Li_2MnO_3\cdot 0.5\ LiNi_{0.44}Co_{0.25}Mn_{0.31}O_2$ electrodes in lithium cells. J Electrochem Soc 153:A1186–A1192
59. Abe M, Matsumoto F, Saito M, Yamamura H, Kobayashi G, Ito A, Sanada T, Hatano M, Ohsawa Y, Sato Y (2012) Activation of a Li-rich solid-solution layered $LiNi_{0.18}Li_{0.20}Co_{0.03}Mn_{0.58}O_2$ cathode and retention of high capacities via an electrochemical pretreatment with a low discharge voltage limit. Chem Lett 41:418–419
60. Zheng JM, Xu PH, Gu M, Xiao J, Browning ND, Yan PF, Wang CM, Zhang JG (2015) Structural and chemical evolution of Li- and Mn-rich layered cathode material. Chem Mater 27:1381–1390
61. Rana J, Kloepsch R, Li J, Stan M, Schumacher G, Winter M, Banhart J (2016) Structural changes in a Li-rich $0.5Li_2MnO_3\cdot 0.5LiMn_{0.4}Ni_{0.4}Co_{0.2}O_2$ cathode material for Li-ion batteries: a local perspective. J Electrochem Soc 163:A811–A820
62. Armstrong AR, Dupre N, Paterson AJ, Grey CP, Bruce PG (2004) Combined neutron diffraction, NMR, and electrochemical investigation of the layered-to-spinel transformation in LiMnO2. Chem Mater 16:3106–3118 [Google Scholar] [CrossRef]
63. Gallagher KG, Croy JR, Balasubramanian M, Bettge M, Abraham DP, Burrell AK, Thackeray MM (2013) Correlating hysteresis and voltage fade in lithium- and manganese-rich layered transition-metal oxide electrodes. Electrochem Commun 33:96–98 [Google Scholar] [CrossRef]

64. Croy JR, Gallagher KG, Balasubramanian M, Long BR, Thackeray MM (2014) Quantifying hysteresis and voltage fade in xLi2MnO3·(1 − x)$LiMn_{0.5}Ni_{0.5}O_2$ electrodes as a function of Li2MnO3 content. J Electrochem Soc 161:A318–A325
65. Zhao Z, Sun M, Wu T, Zhang J, Wang P, Zhang L, Yang C, Peng C, Lu H (2021) A bifunctional-modulated conformal Li/Mn-rich layered cathode for fast-charging, high volumetric density and durable Li-ion full cells. Nano-Micro Lett 13:118

Chapter 4
State-of-the-Art of Dye-Sensitized Solar Cells

Rahul Singh and Ragini Raj Singh

1 Introduction

Fossil fuel diminution increases the concept of energy sources which are renewable and environment friendly. In exploring sustainable energy, natural resources like wind, sun, and water anticipate humankind for energy harvesting devices. Sun, an abundant, clear and renewable energy source, has endless energy that can be consumed using Photovoltaic (PV) devices. PV devices work in the photovoltaic process, where sunlight is converted into electrical energy. Advancement in photovoltaic devices has developed three generations of PV devices till now, classified as first-, second-, and third-generation solar cells. The first generation includes silicon-based solar cells with mono-crystalline and poly-crystalline PV cells. These PV cells are commercialized and are ruling the global market because of the highest efficiency attained, i.e. ~ 27%. The benefits of these PV cells include abundant and safe environmental material, high efficiency under full light illumination and stable performance under all climate conditions, but also disadvantages like low efficiency at indoor applications, high production cost, and wastage of silicon wafers. Thin films solar cells are considered second-generation solar cells, including amorphous silicon, Copper Indium Gallium Di-Selenide (CIGS)-based, and Cadmium Telluride (CdTe)-based solar cells. Although these PV cells were light-weighted and used powder-based materials efficiency of these cells was comparatively low, and materials were also limited. Therefore, third-generation solar cells were discovered in the search for economic PV cells [1, 2].

Third-generation solar cells include dye-sensitized solar cells (DSSC), Quantum Dot-Sensitized Solar Cells (QDSSC) and perovskite solar cells, which have been developed using low-cost materials for economical solar cells. Amongst them, DSSC

R. Singh · R. R. Singh (✉)
Nanotechnology Laboratory, Department of Physics and Materials Science, Jaypee University of Information Technology, Waknaghat, Solan, Himachal Pradesh 173234, India
e-mail: raginirajsingh@gmail.com; ragini.rajsingh@juit.ac.in

B. P. Swain (ed.), *Energy Materials*, Materials Horizons: From Nature to Nanomaterials,
https://doi.org/10.1007/978-981-99-3866-7_4

has many advantages and first evolved as an alternative to p-n junction solar cells. These are fabricated using low-cost materials and by a simple and cost-effective process. They are lightweight, easy to fabricate, give reasonable efficiency, and work in low-sunlight conditions, i.e. suitable for indoor applications [3–5]. The evolution of DSSC as an alternative to p-n junction got a hope in the late 1960s with the research on sensitization of wide gap semiconductors with absorbing material, and the discovery that illuminated dyes can generate electricity in electrochemical cells [6]. Chlorophyll, later used as a sensitizer, was extracted from spinach at Berkeley at the University of California. Thus for the first time in 1972, Zinc Oxide (ZnO) was sensitized with Chlorophyll to make a Chlorophyll-sensitized electrode. The behaviour of excited Chlorophyll dye molecules was studied and resulted in the generation of electric current from the photons with the electron injection from the excited levels of Chlorophyll to the conduction band of the ZnO electrode [7].

On the contrary, the use of TiO_2 as an electrode was also first published in 1972 [8]. However, due to very low efficiency, i.e. less than 1%, further attempts were made to increase efficiency using a multi-dye layer but were unsuccessful. Therefore, the roughness of wide band gap semiconductors was increased to enhance the efficiency of devices, mainly including an increment in the porosity of the wide band gap semiconductors. This will improve the sensitization of absorbers/dyes on the wide band gap electrodes, further enhancing the absorption of more photons and increasing efficiency. Considering all the points, Brian O'Regan and Michael Grätzel created a milestone by inventing the DSSC, also known as Grätzel cells, in 1988. Finally, they succeeded in producing DSSC in 1991 at the Ecole Polytechnique Federale de Lausanne (EPFL) [9]. To construct the device, they used an optically transparent, 10 μm thick, high surface area film of titanium dioxide nanoparticles, which was sensitized with a monolayer of Ru-based N749 dye. Based on the dye and TiO_2 characteristics, 7.1–7.9% efficiency was achieved in simulated solar light, and 12% efficiency was achieved in diffuse daylight. The device absorbed 46% of incident solar light and converted 80% of incident photons into current [9]. DSSC efficiency was enhanced to 9.6% in 1993, then in 1997 at National Renewable Energy Laboratory (NREL) was reported at 10% by the Grätzel group [3]. NREL achieved a maximum efficiency of 12.3% [10], whereas the maximum efficiency of DSSC recorded till now is 14.3% [11].

Theoretical efficiency, i.e. Power Conversion Efficiency (PCE) of DSSC, was calculated to be ~ 20%, but we are still lagging in the efficiency value and to implement DSSC in the commercialized market. Various reasons behind this are the challenges in different parts of DSSC that include less abundant materials, leakage problems due to hazardous organic solvents and short-term stability, low efficiency in environment-friendly DSSCs, and a lifetime of DSSC. Nevertheless, research has been going on to overcome these hurdles. Therefore in this book chapter, we will briefly discuss the current scenario, i.e. state-of-the-art for DSSC and its respective components. Along with this, future scopes that could enhance DSSC efficiency will also be discussed.

2 Insight into DSSC (Efficiency Determining Parameters, DSSC Parts, and Working Mechanism)

2.1 *Efficiency Determining Parameters*

The performance of the DSSC is dependent and determined using various essential parameters, which include open circuit voltage (V_{oc}), short circuit current (I_{sc}), fill factor (FF), overall efficiency (η), and external quantum efficiency [1, 3, 4]. Each parameter has significance in the DSSC when exposed to a constant light source, thus deciding whether the cell's capability is commercialized. The detailed discussion of these parameters is as follows:

2.1.1 Open Circuit Voltage (V_{oc})

It is defined as the voltage measured across the open terminals of the cell, which means that the positive and negative terminals of the cells are kept under open circuit conditions and voltage is measured at zero milli-ampere (mA) current (no current is flowing in the circuit and resistance is infinite between both the electrodes). This also represents the potential difference between the electrolyte's redox potential and the conduction energy level of the dye (absorbance/semiconducting) material. Also, V_{oc} decreases with the increase in the temperature inside the cell. Therefore, the V_{oc} of a cell (measured in volts) can be expressed as:

$$V_{oc} = V_t \ln\left[\frac{I_{SC}}{I_o} + 1\right] \tag{1}$$

where V_t is the terminal voltage of the cell.

2.1.2 Short Circuit Current (I_{sc})

It is defined as the current measured at the short-circuited terminals, i.e. when the positive and negative electrodes of the cells are kept at short circuit conditions and the current is measured at zero mili-volt (mV) voltage (means the voltage between both the electrodes and load resistance, both are zero). Also, I_{sc} increases with the increase in the temperature of the cell. Therefore, the I_{sc} of a cell (measured in mA cm^{-2}) can be expressed as:

$$I_{sc} = I + I_o\left[\exp\frac{v}{v_t} - 1\right] \tag{2}$$

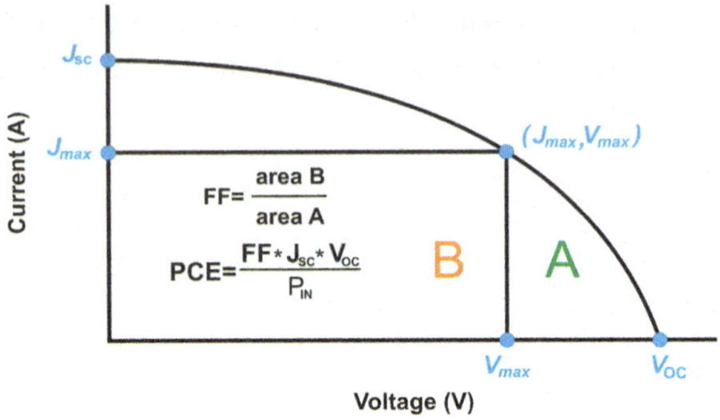

Fig. 1 *I-V* curve for efficiency determining parameters [3]

2.1.3 Fill Factor (FF)

It is the ratio of maximum output power, *i.e.* the product of maximum current (I_m) and maximum voltage (V_m), to the power at circuit conditions, i.e. product of current at short circuit condition (I_{sc}), and voltage at open circuit condition (V_{oc}). FF of a solar cell can be expressed as:

$$\text{FF} = \frac{I_m * V_m}{I_{sc} * V_{oc}} \tag{3}$$

2.1.4 Overall Efficiency (η)

It is also known as power conversion efficiency (PCE) and expressed in percentage (%). Cell can convert solar power into electrical power. It is also defined as the ratio of electrical power to the incident solar power and can be expressed as (shown in Fig. 1).

$$\eta = \frac{\text{FF} * V_{oc} * I_{sc}}{P_{in}} \tag{4}$$

The efficiency value (η) increases with the increase in the value of V_{oc} and FF, and with the decrease in the value of I_{sc}.

2.1.5 External Quantum Efficiency

It is also known as an incident Photon to Current Conversion Efficiency (IPCE), which is defined as the generation of several electrons flowing in the external circuit concerning the number of photons incident on the solar cell at a particular wavelength and can be expressed as:

$$\text{IPCE}(\%) = 1240 * \frac{I_{\text{sc}}}{P_{\text{in}} * \lambda} \tag{5}$$

The above-stated parameters depend on the parts of the DSSC and the nature of the material used in the parts, which will help in the efficient working of the DSSC and give excellent parameter values which can emphasize the commercialization of DSSC.

2.2 DSSC Parts and Working Mechanism

The working of DSSC depends on the constitutional parts (shown in Fig. 2), which include the working electrode (consists of substrate and oxide semiconductor material), dye or photosensitizer, electrolyte, and Counter Electrode (CE). The working electrode is also known as a photo-electrode or anode. Substrates are used for the deposition of semiconductor material and sensitizers which should be conductive and transparent for the better conductivity of electrons excited by the absorption process of photons by the sensitizers. Semiconductor materials are used for sensitizing the absorbing materials (dye in case of DSSC), which helps in electron injection and transportation of charge carriers. Photosensitizers are the main parts of DSSC since it helps in the absorption of photons/solar energy. Different photosensitizers absorb different regions of the solar spectrum, thus helping determine the determination efficiency of DSSC.

On the contrary, an electrolyte helps provide the holes for the sensitizer, thus also known as a Hole Transport Material (HTM). The counter electrode (also known as the cathode) helps stabilize HTM by providing the electrons from the external

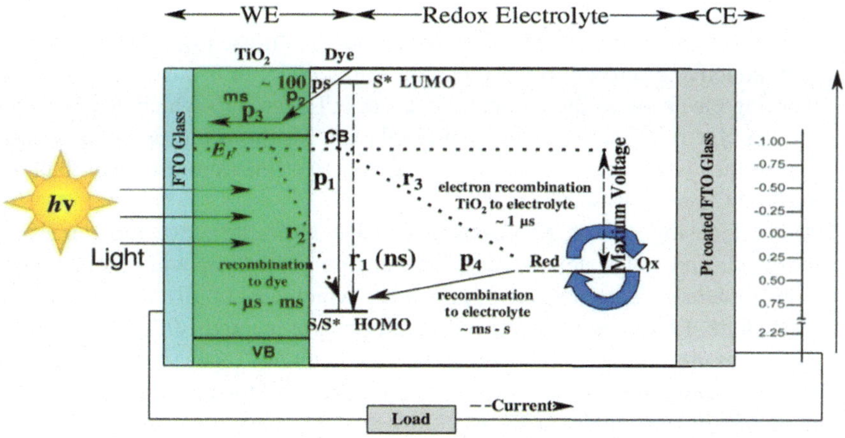

Fig. 2 Schematic diagram explaining the working of DSSC, generation, and recombination of electrons [12], along with time scale kinetics [13]

circuit and collecting the holes from HTM. Along with this, CE also completes the circuit by connecting through the external circuit to the working electrode. A detailed description of the required properties of the different parts of DSSC is mentioned in the next section, 'State-of-the-Art of DSSC'. Therefore, these other parts of the DSSC help in the working since the working principle depends on these essential parts' characteristics.

The working principle can be best described in four steps: absorption of solar energy (photons) and excitation of electrons, injection of excited electrons, transportation of charge carriers, and collection of electrical power in the form of current. These basic steps formulate the working principle, which in detail are following as:

(i) Illuminated light falls on the working electrode of the DSSC, which passes through the transparent substrate and Semiconducting Metal Oxide (SMO) material having a large band gap. Since SMO used to have a huge band gap, they absorb only a fraction of photons, thus allowing maximum absorption of light using dye. Therefore, photosensitizers usually dye molecules, i.e. S(dye), which are adsorbed on the SMO, absorb the photons which cause the excitation of electrons from the HOMO level of dye to the LUMO level of the dye, resulting in the excited state of dye (S^*), shown in Eq. (6).

$$\mathrm{S(dye)} + h\nu \rightarrow \mathrm{S^*(dye)} \tag{6}$$

(ii) Electrons from the excited state or LUMO level of dye are then injected to the LUMO level of SMO which is just below the LUMO level of dye, thus inducing the charge separation and then oxidizing the dye (means dye have lost electrons), shown in Eq. (7).

$$\mathrm{S^*(dye)} \rightarrow \mathrm{S^+(dye)} + \mathrm{e^-_{cb}(TiO_2)} \tag{7}$$

(iii) Injected electrons in the SMO get transferred and diffuse towards the substrate or conductive electrode. From there, an external circuit is connected from the working electrode to the counter electrode (CE) through which electrons are used as electrical energy to the connected load. Electrons through this external circuit then reach the CE, which has a distinctive function to counter balance the DSSC's overall functioning.

(iv) Again focusing on the photosensitizer (dye molecule), dye molecules are oxidized, and holes generated by the charge separation with excited electrons remain vacant on the dye molecules, which are stabilized by the HTM/electrolyte. Electrons from the red-ox electrolyte mediator filled the holes of dye molecules, thus oxidizing the red-ox mediator and regenerating the ground state or HOMO level of dye by activating it again for the absorption process in photosensitizers, shown in Eq. (8).

$$\mathrm{S^+(dye)} + 3\mathrm{I^-(electrolyte)} \rightarrow \mathrm{S(dye)} + \mathrm{I_3^-(electrolyte)} \tag{8}$$

(v) This reduction process of the red-ox mediator changes its concentration by generating a concentration gradient across the electrolyte, which tends to get into the equilibrium by diffusing towards the CE and accepting the electrons from CE to neutralize itself for the proper functioning of DSSC (shown in Eq. (9)). Electrolyte functions as a charge transporter between the working electrode and the counter electrode. This process completes the circuit and repeats the whole cycle for the continued working of DSSC.

$$I_3^-(\text{electrolyte}) + e^-(\text{CE}) \rightarrow 3I^-(\text{electrolyte}) \tag{9}$$

Processes in step (iv) and step (v) work simultaneously with the process occurring in step (i), step (ii), and step (iii).

However, the recombination of excited electrons in the above-stated process hinders the proper functioning of DSSC from being efficient and occurs as a competitive process (shown in Fig. 2). In the electron injection process from LUMO of dye molecule to LUMO of SMO, both recombination processes (radiative and non-radiative) of dye compete with it. They terminate the injection process, i.e. recombining excited electrons directly from the LUMO to the HOMO dye level also contributes to electron loss. Still, it happens only in fractions which can be neglected since the electron injection rate from LUMO of dye to LUMO of SMO is 1000 times faster than the direct recombination process from the LUMO to HOMO level of dye [14, 15]. Excited electrons in the LUMO level of SMO can also recombine directly with the oxidized dye or with the oxidized red-ox mediator, thus acting as a barrier to the better performance of DSSC. Therefore, for an efficient DSSC, transportation time and transportation rate from dye's LUMO to SMO's LUMO should be faster, i.e. in the order of pico- to femto-seconds and 1000 times, respectively, as compared to the direct decay process of excited electrons from LUMO to HOMO level of dye i.e. nano-seconds. Additionally, the LUMO level of the dye molecules/sensitizers should be more negative when compared with the LUMO level of SMO, and the HOMO level of the sensitizer should be more positive when compared to the red-ox potential. Along with this, charge transfer to the SMO from the dye must take place with better quantum yield [16]. Also, ideally, stabilizing oxidized dye by the red-ox mediator of the electrolyte should be 100 times faster than the direct recombination process of dye's excited electron from LUMO to HOMO level [17]. Also, restoration of the red-ox mediator utilizing CE should be very fast to stop the recombination losses of photo-excited electrons which help in the kinetics of transportation of electrons, thus enhancing the performance of DSSC. Therefore to enhance the performance of DSSC and to make it efficient, we have to look at the nature, working, and properties of its constitutional parts along with the above-stated key points. Also, choosing and applying different materials to get better photovoltaic properties and to relate them with the future prospective, state-of-the-art constitutional parts of DSSC is essential, which is thoroughly discussed in the next section.

3 State-of-the-Art of DSSC Parts

Each part of the DSSC has an important role, and all the parts depend on each other to produce a better DSSC photovoltaic performance. Since substrates are used as a base for SMO and CE, sensitizers are used for absorption and electrolyte as a mediator to connect the working electrode with the counter electrode, improving each part to enrich the DSSC properties and approaches towards commercialization.

3.1 Substrate

The basic construction of DSSC starts with using two substrate sheets of transparent and conducting materials, which are therefore used for depositing semiconducting metal oxide material on one sheet and catalyst on the other. Connecting both sheets with an external circuit will allow electrons to transport and help convert them into electrical energy. These substrate sheets, collectively with SMO material and with catalyst, are known as the working electrode and counter electrode, respectively, which act as a current collector [18]. Substrate sheets should have two main features/properties to be utilized in DSSC, which are as follows:

(i) Substrate sheets should have transparency more significant than 80% to allow optimal sunlight to fall on the productive area of the DSSC, i.e. transmittance value should be very high.
(ii) Substrate sheets should also have a high electrical conductivity value to ease the charge transfer from SMO to CE through the external circuit and to reduce the charge and energy losses in the cell.

Thus to obtain these two main characteristics to form the substrate sheets, a layer of conductive transparent oxide is coated on the glass. Generally, Indium Tin Oxide (ITO), i.e. In_2O_3:Sn and Fluorine Tin Oxide (FTO), i.e. SnO_2:F are used as transparent conductive oxide layers on soda lime glass. In terms of transmittance, ITO has a transmittance value of 80%, whereas FTO has a transmittance value of 75%. However, FTO has more conductivity value since the sheet resistance value is only ~ 8.5 Ω/cm^2, which is less than the sheet resistance value of ITO, i.e. ~ 18 Ω/cm^2. However, these are the typical choices of the researchers but still are not suitable due to their rigidity issues [19]. Sima et al. have compared the performance of DSSC based on ITO and FTO. They have reported FTO to be better as a substrate because its sheet resistance was found to be constant upon sintering, whereas the sheet resistance of ITO was found to be increased from 18 to 52 Ω/cm^2. Also, overall efficiency (η) was better for FTO-based DSSC, i.e. 9.4%, whereas it was observed to have a value of 2.4% for ITO [20]. Currently, Aluminium-doped Zinc Oxide (AZO) is an alternative to the ITO and FTO. Patni et al. have also compared the ITO and FTO with the AZO and demonstrated AZO and FTO to be better options than the ITO [19, 21]. Other alternatives to the ITO are metal sheets such as stainless steel, titanium and tungsten, and plastic foils. However, metal substrates have issues like transparency, corrosion due to electrolytes and high cost, whereas plastic substrates have stability issues. Thus, both substrates affect the performance of the DSSC. These problems can be solved using polymers because of their low cost and high flexibility, but temperature also restricts their use in DSSC [18].

3.2 Semiconductor Metal Oxide Materials Layer

Semiconductor oxide materials play an essential role in providing surface area for dye adsorption, transparency for light illumination on dye, acceptance of excited electrons and electron mobility. Therefore, these above-stated characteristics should be in SMO to enhance the DSSC photovoltaic performance, which is thoroughly discussed below [12, 18]:

(i) SMO should be porous material with a very high surface area so dye loading would be maximum and dye could be adsorbed properly on the SMO. This will increase the possibility of full light illumination on the dye.
(ii) SMO should also be capable of binding the photosensitizer at the SMO surface so that photo-generated electrons can be efficiently collected at the interface of SMO-dye when the light is illuminated on the DSSC.
(iii) SMO material should have a high band gap so that it cannot absorb photons (especially from the visible and infrared region) and follow the transparency for the absorption of photons by the dye/photosensitizers on the light illumination on the DSSC so that efficient light harvesting could be possible.
(iv) Conduction band energy levels of the SMO and dye molecules should be aligned appropriately for the better injection of the excited electrons, i.e. LUMO level of the SMO should be more negative and must lie just below the CB of the photo-sensitizer/dye. It will provide efficient electron injection form dye to SMO. This improved faster electron injection will increase the transit of electrons.
(v) SMO should have high electron mobility so efficient transmission of the electrons through the SMO material network would be possible. It will provide a better path for the transportation of the electrons with a minimum recombination rate. High electron mobility will improve the transit time for the electrons by emphasizing the minimum electron recombination.
(vi) SMO should be inactive for the SMO-electrolyte interface for charge recombination.

Overall to attain the high performance of the DSSC, high surface area, faster electron injection, faster transit time, high electron mobility, and better electron collection of the SMO materials could enhance the photovoltaic performance of the DSSC. The electron transport rate mainly depends on the surface area of SMO, its crystallinity, and morphology. Thus nature of the SMO material will be helpful in its selection for the DSSC. Along with this, the porosity, structures, particle size, and technique used in coating SMO layers on the substrate (preferably FTO) are also significant for the performance of DSSC parameters. Therefore in the current section, we will discuss the SMO materials used by researchers, the present scenario of SMO materials for the DSSC, and SMO preparation methods on the substrate glass.

ZnO and TiO_2 were first used as an electrode in 1972, where ZnO was sensitized with Chlorophyll [7, 8], and the first DSSC formed in 1991 used TiO_2 as SMO with Ru-based N749 dye [9]. Other metal oxides such as SnO_2, In_2O_3, WO_3, Al_2O_3, SiO_2, and Nb_2O_3 were also used as SMO in DSSC. However, lower efficiency was observed in DSSC when compared with the TiO_2 [22–26]. However, Fukai et al. have shown that SnO_2 as an SMO in DSSC shows a higher value of $J_{sc,}$ i.e. 15.6 mA/cm^2 and a comparable value of I_{sc} under one sun condition. In addition, it gives a high charge collection value resulting from both SMO materials' comparable electron diffusion lengths [27].

TiO_2 is the researcher's primary choice in the DSSC because of its easy availability, non-toxic nature, and low cost. It exists in two forms: anatase and rutile. The anatase form has a band gap value of 3.2 eV, whereas the rutile form has a band gap value of 3 eV. Since the high band gap value makes the crystalline form more stable, the anatase form is used mainly. Also, due to slow electron transport and less surface area, which results in less dye adsorption on the surface, rutile form leads to the less efficient DSSC [28]. Also, less efficiency was counted due to the recombination of the electrons injected from the dye to SMO with the electrolyte, which leads to the formation of a dark current. This electron recombination can be decreased again by several methods like changing morphologies, surface treatment of TiO_2 and coating of the insulating layer on the SMO layer.

Methods to reduce recombination losses are discussed above using different ways, such as insulating layers, and different morphologies are discussed in detail below. The introduction of an insulating layer (Al_2O_3, SiO_2, ZnO, and Nb_2O_3) on the SMO will reduce the interaction of the injected electrons of the SMO with the electrolyte [24, 26, 29–31]. Otherwise, the resulting current from recombining the electrons with the electrolyte (also known as dark current) will diminish the efficiency of DSSC. Another way to reduce the recombination is the surface treatment of SMO (TiO_2) with the $TiCl_{4,}$ which increases the electron transfer resistance at the SMO-electrolyte interface [32]. Different morphologies also produce different efficiency values in the DSSC. Govindaraj et al. synthesized spherical TiO_2 nanoparticles using the sol-gel technique, where a hierarchically highly crystalline anatase structure was formed. The band gap of mesoporous TiO_2 nanoparticles was found to be 3.2 eV, where the average size of bi-modal-like pores was 7.1 nm and 9.3 nm. Because of this mesoporous nature, a surface area formed for the dye loading was calculated as 31.76 m^2 g^{-1}. The calculated cell parameters were 13.206 mA cm^{-2} for I_{sc}, 0.607 V for V_{oc}, and 42.56% for fill factor. The overall efficiency recorded was 3.415% [33]. Since TiO_2 nanorods as SMO have a higher electron transfer rate but it lowers the efficiency due to low surface area for dye loading. Therefore, Wan et al., using a hydrothermal method, produced single crystalline long nanorods of TiO_2 of the rutile phase. The low surface area problem was solved using the etching of synthesized TiO_2 nanorods at a different time using hydrochloric acid again by hydrothermal method. PCE for

DSSC without etching was found to be 1.34% which shows poor dye loading due to less surface area available. However, the best results for DSSC with 8 h of etching of TiO_2 nanorods (shown in Table 1) recorded a PCE value of 11.14%, thus showing a high loading surface area, longer electron lifetime, and higher efficiency value. I_{sc} and V_{oc} values calculated for the etched TiO_2 nanorods were 21.59 mA cm^{-2} and 0.79 V respectively [34]. Rice-like morphological nanoparticles of TiO_2 were formed of a highly crystalline nature using a one-pot hydrothermal method. This morphology is the result of the mixture of ammonia solution, ethanol, and titanium isopropoxide with the decoupling process used in hydrolysis and condensation. BET characteristic and average particle size of rice-like TiO_2 NPs was found to be quite close with the TiO_2 available commercially i.e. Degussa P25, however, the PCE value was found to be 20% higher than the Degussa P25 TiO_2 under 1 sun illumination [35]. Li et al., used the solvothermal method to produce TiO_2 Microspheres (MS) using acetone as a solvent which often gives efficient PV performances because of excellent light scattering effect, high dye loading, and unique microstructure. Due to its highly interconnected nanocrystal array, the dye loading surface area was observed to be 138.47 m^2/g. The prepared TiO_2 microspheres gives a higher PCE value when compared with 18 nm-sized TiO_2 NPs paste of Dyesol, the result of higher dye loading area availability, and longer electron lifetime. Post-treatment of TiO_2 with $TiCl_4$ resulted in a better efficiency value of 10.32% [36].

Similarly, DSSC based on SnO_2 photoanode having nano-flower morphology gives enhancement in the performance of the DSSC. A hydrothermal method without surfactant was used for the synthesis of SnO_2 nano-flowers, which resulted in the pure rutile phase. The band gap of prepared samples was found to be higher than bulk SnO_2, where with the increase in the reaction time of the hydrothermal method, it varied from 3.64 to 3.81 eV. Reaction time also changes the morphology i.e. from nanoparticles to manipulated to nano-flowers when the reaction time changed from 2 to 24 h. SnO_2 nano-flowers were sized from 300 to 700 nm, where 0.43 V value of V_{oc} and 4.36 mA cm^{-2} value of I_{sc} with 1.11% PCE value were recorded. This

Table 1 Calculated parameters (Photovoltaic, resistance (R_s, Rc_t), BET, dye loading, etc.) of the DSSC w.r.t. the etching time of TiO_2 nanorods [34]

Parameters	0 h	6 h	8 h	10 h
R_s (Ω)	22.51	19.12	13.64	24.84
Rc_t (Ω)	561.3	27.77	12.48	35.27
τ_n (ms)	1.46	17.57	30.80	14.74
Specific area (m^2g^{-1})	3.65	15.98	21.59	21.45
Dye adsorption (nanomolar cm^{-2})	38.30	131.92	257.62	234.04
Jsc (mA/cm^2)	3.65	15.98	21.59	21.45
Voc (V)	0.7 6	0.78	0.79	0.77
FF	47.89	63.82	65.2	58.32
PCE (%)	1.34	7.95	11.02	9.64

efficiency enhancement may be due to the nano-flower-like structure which provides the charge transport path [37]. On the contrary, micro-flowers shaped ZnO were deposited on the ZnO NPs to form the bilayer structure. The advantage of deposition of ZnO micro-flower was reflected in terms of the scattering layer which resulted in an increment in PCE value from 2.31 to 3.20% due to the high absorption efficiency, higher generation of electrons, smaller resistance of charge, and longer lifetime of electrons when compared with the single-layered ZnO photoanode [38].

Other than the morphology, SMO's structure also helps enhance DSSC. ZnO Nanorods (NR)-Nanosheets (NS)-based architecture allows the 90% increment in the PCE value of DSSC when compared with the ZnO nanorods-based DSSC. ZnO NR-NS structure has long nanorods and thin nanosheets prepared using the hydrothermal method, which helps increase the value of short circuit current, leading to higher dye loading, and increased light scattering [39]. The double-layered structure of TiO_2, one with larger nano spindles and another layer with smaller nano spindles thus enhances light scattering and dye loading, respectively. This double-layered structure enhances the PCE value from 4.3 to 8.32%, with the I_{sc} value increasing from 8.4 to 16.4 mA/cm^2 and V_{oc} value from 0.755 to 0.760 V compared with the larger TiO_2 nano spindles-based DSSC [40]. Dye loading on SMO surface area also affects the efficiency of DSSC. Satapathi et al. used graphene-TiO_2 nanocomposite in the DSSC with the different-sized graphene sheets (184 nm, 292 nm, 444 nm, 1.2 μm) and compared the performance parameters with the standard DSSC, i.e. without graphene sheet. Photovoltaic properties were calculated and observed that the smallest graphene sheet (184 nm) had recorded the highest I_{sc} (14.66 mA/cm^2), V_{oc} (0.68 V), FF (66%), and PCE (6.62%) amongst all designed DSSC. Also, it gives a 49% improvement in the DSSC compared to the standard manufactured DSSC. It shows the efficiency improvement by providing smaller graphene sheets which effectively enhance dye loading on the surface of graphene [41]. Similarly, Carbon Nanotubes (CNT) were incorporated into the TiO2 SMO, forming the hierarchically porous photoanode structure. The roughness of the surface was 15.4 nm, whereas the size of the micropores was 15 nm and the size of the macropores was 280 nm. The incorporation of CNT has increased the surface area, leading to more dye loading on the photoanode, faster charge injection, faster charge transport, and increased electron lifetime, thus overall resulted in increased Isc and PCE. The DSSC performance parameters recorded were 10.12 mA/cm^2 (I_{sc}), 0.68 V (V_{oc}), 56% (FF), and the PCE value was 3.58% [42]. Different synthesis techniques also resulted in better photovoltaic parameters. A wet chemical method was used to synthesize the core-shell structure of $ZnO–Zn_2SnO_4$, where ZnO was in the form of nanorods and Zn_2SnO_4 was in microcubes. DSSC parameters were calculated for DSSC with only ZnO NRs and DSSC with core-shell structure. PCE value was enhanced from 0.78 to 3.36% which was due to the agglomeration of zinc ions and reduction in the energy barrier across the SMO-dye interface on the introduction of Zn_2SnO_4 [43]. For the comparison of the performance of TiO_2 photoanode-based DSSC, TiO_2 photoanode was prepared using two methods: solvothermal and hydrothermal, using the same precursor. It was observed that maximum efficiency was monitored by the hydrothermal method, i.e. 7.44%, whereas the efficiency calculated for a solvothermal method is 5.98%. It is also worth noting that the dye loading area using the hydrothermal method is also greater ($9.34 * 10^{-6}$ mol/cm^2) than the solvothermal method, i.e. $8.43 * 10^{-6}$ mol/cm^2. Detailed calculations of the surface area of SMO, pore size and volume, charge transfer and sheet resistance, and electron lifetime were also done to analyze all the parameters affecting PCE value [44]. Mohammadi et al. have also used the solvothermal method to prepare g-C_3N_4 and ZnO-modified TiO_2 nanotube (TNTs) arrays. This structure has increased the charge collection rate, increased the electron lifetime, and reduced the recombination rate which resulted in the 135% enhancement in the PCE value compared with the pure TNTs [45].

Therefore, photoanode is an essential part of the DSSC, which collects the excited electrons from the dye and transports it for the further working of DSSC. To enhance the DSSC parameters, surface loading, morphology, structure, porosity, band gap, and SMO thickness is very important.

3.3 *Photosensitizer (Dye)*

Sensitizer or dye is the main part of the DSSC since it plays that role from where the process of transport of charge carriers starts. Since sensitizers are adsorbed on the photoanode surface, as discussed in the above section, the purpose of dye is to absorb the incident light/photons so that electrons get excited to the conduction band, leaving a hole behind. Then excited electrons get transported from the conduction band of dye to the LUMO level of the photoanode, whereas a hole gets transferred to the electrolyte simultaneously. Since charge transfer across the interface is only possible when the charge separation at the interface of photoanode-dye-electrolyte has occurred well, which also depends on the nature of the dye, thus can affect the performance of DSSC. Therefore, the dye should have some unique characteristics which are discussed below, so that photon absorption, charge separation, and electron injection can happen efficiently [3, 4, 12, 18, 46].

(i) Dye should be much more efficient in absorbing the solar spectrum from UV-Visible to the near-infrared region and be luminescent.
(ii) Dye should have a LUMO level above the conduction band of SMO (usually TiO_2) material so that the electron injection is fast and efficient. In contrast, the HOMO level of the dye should be lower than the redox potential of the electrolyte so that the hole can be transported to the electrolyte, and oxidized dye should be regenerated.
(iii) Dye should also be stable in its oxidized state to get reduced by the electrolyte.
(iv) Strong binding and interaction between the adsorbed dye and the SMO layer material should be present to transfer electrons efficiently. For this purpose, reactive, co-absorbent, and anchoring groups should be added between the SMO layer and dye. This process also avoided the aggregation of the dye molecules on the SMO material layer and resulted in decreased recombination.
(v) Dye should have a high extinction coefficient.

(vi) Dye should have long stability and be able to operate for 20 years, i.e. it should have ~ 108 turnovers. Concerning this, the dye should be hydrophobic to minimize the interface contact of SMO/electrolyte. Otherwise, a dye can be distorted from the SMO materials, thus reducing the stability. Also, the dye must exhibit both chemical and heat stability.

Above-stated features can be achieved by changing the Opto-physical, electrochemical properties, arrangement, and molecular structure of dyes on the SMO material's surface, which can enhance the PV parameters and performance of DSSC. Therefore, detailed information on the dyes and their current state is discussed in this section.

Dyes are basically of two types: natural and synthetic. Synthetic dyes can be further classified into organic dyes (metal-free dyes) and organometallic dyes (metal-complex dyes). Organic dyes contain organic chromophores, whereas organometallic dyes contain transition metals in their structure.

3.3.1 Natural Photosensitizer

Natural sensitizers were first used as sensitizing material to make sensitized electrodes. For example, Chlorophyll is the first natural dye-sensitized to ZnO to make Chlorophyll-sensitized ZnO electrode [7]. The advantages of natural dyes are that they are economical, environmentally safe with non-toxicity, easy preparation techniques, and have significant absorption in the visible region. Natural dyes can be extracted as pigments from natural sources such as vegetables, flowers, bacteria, and fruits exhibiting a specific colour range [47]. To date, natural dyes employed in the DSSC are discussed in detail below (where some natural pigments are shown in Table 2). Also, the chemical structure of some natural dyes is shown in Fig. 3.

Chlorophylls: It is a green-coloured pigment that can be extracted from algae, plants, cyanobacteria, dinoflagellates, and diatoms. Based on the extraction source, Chlorophyll, a chelate compound, can be divided into four types i.e. Chlorophyll (a, b, c, and d), which contains magnesium, carbon, hydrogen, oxygen, and nitrogen. Chlorophyll 'a' can be extracted from algae, cyanobacteria, and higher plants; Chlorophyll 'b' can be extracted from green algae and plants; Chlorophyll 'c' can be extracted from dinoflagellates, green algae, and diatoms; whereas Chlorophyll 'd' can be extracted only from red algae. Chlorophyll 'a' has a composition of $C_{55}H_{72}O_5N_4Mg$, whereas Chlorophyll 'b' only changes by the composition of 'H' and 'O', i.e. $C_{55}H_{70}O_6N_4Mg$. The most efficient Chlorophyll in DSSC is methyl trans-32-carboxy-pyropheophorbide which is a derivative of the Chlorophyll 'a' [4].

Plant Pigments: These can be classified into three groups: *flavonoids, anthocyanin, and carotenoids* (shown in Fig. 3) [18]. Flavonoids are the group of biological pigments present in gymnosperms, ferns, bryophytes, and angiosperms, having an organic structure with 15 carbon atoms and two phenyl rings connected by a carbon framework of C_6–C_3–C_6 [3]. Flavonoids can further be divided into *anthocyanins,*

Table 2 Natural dyes pigment, their categories, and sources [4]

Pigment	Category	Source
Flavonoids	Flavones	Bryophytes
	Flavonols	Gymnosperms
	Flavanones	Angiosperms
	Flavanonol	Ferns
	Flavanols or Catechins	
	Anthocyanins	
	Chalcones	
Chlorophyll	Chlorophyll 'a'	Al photosynthetic plants
	Chlorophyll 'b'	Cyanobacteria
	Chlorophyll 'c'	Algae
	Chlorophyll 'd'	
Carotenoids	Carotenes	Crustaceans
	Xanthophyll	Bacteria
		Photosynthetic plants
Tannins	Hydrolysable tannins	Gymnosperms
	Condensed tannins	Angiosperms
Betalains	Betacyanin	Caryophyllates
	Betaxanthin	Fungi

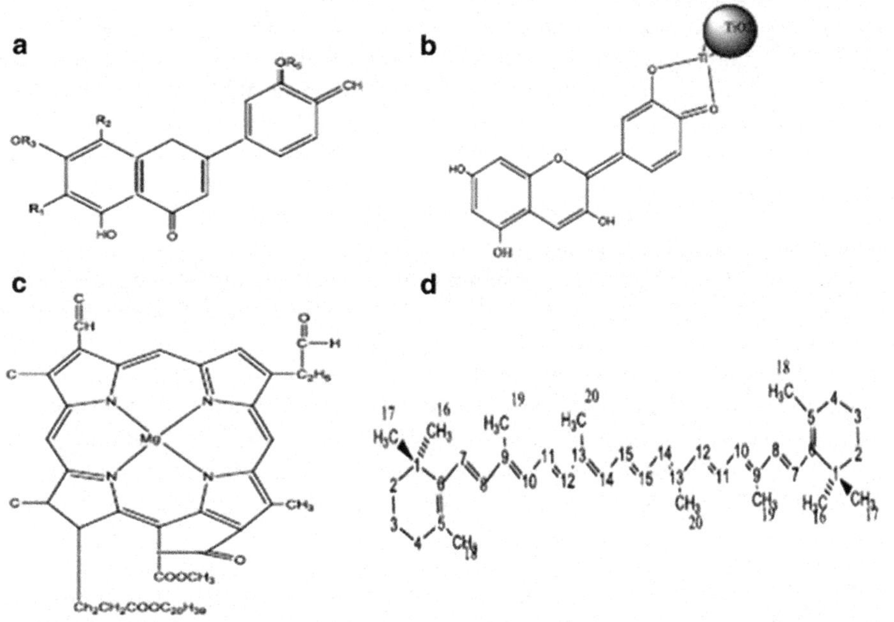

Fig. 3 Chemical structure of natural photosensitizer: **a** anthocyanin, **b** flavonoid, **c** β, β-carotene, and **d** Chlorophyll [3]

flavonols, and *proanthocyanidins.* Amongst these three, only anthocyanins exhibit colours from pink to dark blue through red and violet of most fruits, leaves, and flowers. These groups contain hydroxyl and carbonyl groups in their structure which bind the dye on the SMO layer and in the electron transfer of an excited electron from dye to photoanode. The structure of anthocyanins is based on C_6–C_3–C_6 skeleton and is glycoside salts of phenyl-2-benzopyrilium. Anthocyanin transfers the absorbed energy to the anthocynanin pair of the photosystem using resonance energy transfer. Its maximum absorption ranges from 510 to 548 nm [48]. Carotenoids, also known as tetraterpenoids, are the family of isoprenoids having the eight isoprenoids unit, which the presence of a C40 hydrocarbon backbone can identify. These can be found in plants, bacteria, and crustaceans [4]. In carotenoids, its molecules 'p' delocalized electrons are transformed by light, thus achieving light absorption. During this, carotenoid electrons are transferred to Chlorophyll molecules, where the singlet state of Chlorophyll is formed with a slightly higher energy level. Natural dye tannins are also used as a sensitizer in DSSC. These are the pigments of biological plants present in angiosperms and gymnosperms. These can also be classified in non-hydrolysable and hydrolysable. Another natural dye known as betalains which are the pigments, are also classified as betacynin extracted from the caryophyllates, and betaxanthin extracted from the fungi. These are light absorbing dyes and have the properties of antioxidant.

Sawhney et al. used four dyes extracted from natural fruits like plum, berries, Indian Jamun, and black currant and used them as sensitizers in DSSC. Anthocyanin pigment was extracted using acidified ethanol since these naturally occurring sources are inexpensive and rich in anthocyanin, making them appropriate candidates for DSSC. The highest achieved efficiencies in used sensitizers are from berries and black currant sensitizers, which are observed to be 0.53 and 0.55% respectively [49]. Sreeja and Pesala have studied the effect of co-sensitization on the formed DSSC using Chlorophyll- 'a' and betanin. It was observed that efficiency was increased because of co-sensitization. The PCE value for Chlorophyll-based DSSC, betanin-based DSSC, and both Chlorophyll-betanin-based co-sensitized DSSC, was 0.047, 0.562, and 0.601%. This enhancement was identified using the computational method where it was found that required optimal band levels alignment conditions were satisfied, i.e. HUMO and LUMO levels of the dye are appropriately aligned with the electrolyte's redox potential and the LUMO level of photoanode, respectively [50]. They have again tried to enhance the efficiency value by incorporating the Ag NPs in the betanin-based and co-sensitized (Chlorophyll and betanin) DSSC. They have used the concept of Localized Surface Plasmon Resonance (LSPR) exhibited by the Ag NPs. The recorded PCE value for betanin solar cells, with and without Ag NPs showing LSPR, was 0.538 and 0.658%. However, more enhancement in the DSSC was recorded when the co-sensitized solar cell was used. The efficiency value was recorded at 0.612 and 0.793% for co-sensitized and plasmon-enhanced co-sensitized DSSC [51]. Muhammed et al. extracted anthocyanin dyes from the Bottlebrush and Bougainvillea flowers and fabricated the DSSC, and the corresponding efficiency was recorded at 0.23 and 0.31%, respectively. This could be due to the interaction of the hydroxyl and carbonyl groups on the photoanode (TiO_2) surface and the dye extracts and to the absorption region of both dyes between 400 and 600 nm [52]. To enhance the DSSC performance, a dye extracted from the flower Malvaviscus penduliflorus in the ethanol medium has sensitized on the TiO_2 and MnO_2–TiO_2 incorporated DSSC. The PV parameters were: $V_{oc} = 0.24$ V, $J_{sc} = 2.50$ mA/cm^2, FF = 52.20%, and PCE was 0.31% for TiO_2 incorporated DSSC, whereas $V_{oc} = 0.23$ V, $J_{sc} = 7.00$ mA/cm^2, FF = 54.70%, and PCE was 0.88% for MnO_2–TiO_2 incorporated DSSC [53]. Munawaroh et al. studied colour stability on anthocyanin using the mangosteen pericarp extract co-pigmentation method. Ascorbic acid and Malic acid were also used as co-pigment on the anthocyanin, and it was observed that anthocyanin-malic acid has the highest colour retention ratio of 1:5. In contrast, anthocyanin-ascorbic acid has the colour retention of 1:3 representing the ability to protect from degradation process of the colour [54].

3.3.2 Synthetic Photosensitizer

Organometallic photosensitizers: These are also known as metal-complex sensitizers and are composed of high transition metals (M) such as ruthenium (Ru), Iridium (Ir), and Osmium (Os) along with both ligands, i.e. anchoring (ACL) and ancillary (ALL) ligands, and water substituent groups or halides, thiocyanate, thiocarbamate etc., thus have the general formula ML_2X_2. ACLs function for the sensitizer's adhesion on the SMO, whereas ALLs are used in the tunability ability for the sensitizer's properties. Polypyridine complexes of d^6 metal ions show a high absorption coefficient from visible to NIR due to the high Metal-to-Ligand Charge Transfer (MLCT) along with the thermal and chemical stability [3, 4, 18]. Amongst all the meat complex ions, Ru-based polypyridine complexes are the primary choice of researchers due to the following reasons:

(i) The addition of ligand with the metal ions becomes smooth due to its octahedral structure.
(ii) Oxidation states from I to IV are affordable and stable.
(iii) It is stable in many solvents due to its ML_2X_2 structure.
(iv) Desired properties can be obtained by tuning the electrochemical, photochemical/physical properties.

The first DSSC fabricated by the Grätzel group used the ruthenium dye, which they again used to enhance the DSSC efficiency to 10% [3, 9]. A new complex using dual anchored Ruthenium(II) dye and 4-Nitro-phenylenediamine Schiff base ligand was used for the DSSC, and its PV parameters were: PCE value recorded was 3.42%, J_{sc} value was 7.12 mA/cm^2, and V_{oc} value was 0.79 V. Dye behaves as the better anchoring unit due to the electron without a group of pyridine and nitro group [55]. Naik et al. synthesized four DSSCs using the co-sensitization of Ru(II)-based N3 dye with D-A configured organic dyes i.e. $A_{1\text{-}4}$. These organic dyes have the N, N-dimethylaniline ring with the anchoring units. The best PV efficiency was obtained with A_3 dye (thiobarbituric acid), where V_{oc} value was 0.671 V, J_{sc} value was 15.27 mA/cm^2, FF was 68.47%, and PCE value was 7.02% [56]. N719 dye has almost the same structure as N3 but has a slow adsorption speed (24 h) compared to N3 (3 h). Venkatraman et al. created a database collected for DSSC, and ~ 250 articles were found for N719 dye. PCE value ranged from 2 to 11% from 329 results obtained from these 250 articles of N719 dye [57]. $Zn_{1-x}Sn_xO$ nanocomposite was sensitized by N719 dye for the fabrication of solid state DSSC with the spiro-OMeTAD was used as HTM. The highest efficiency was recorded for the composition $Zn_{0.95}Sn_{0.05}O$ with value of 4.3%. The other PV parameters were $V_{oc} = 0.74$ V, $J_{sc} = 12.45$ mA/cm^2, and FF = 46.70% [58].

The use of π-system extension in the dye is also in the current scenario due to the low absorption coefficients of Ru complexes. Therefore, ligands with the conjugated π- system are used in place of N3 dyes so that the absorption coefficient can be enhanced by introducing hydrophobic side chains to repel the tri-diode and water from the photoanode. In this respect, Rawashdeh et al. recorded the 0.45% PCE value using the N749 dye-sensitized in the ethanolic solution with the graphene-based electrode [59].

Organic photosensitizers: These are also known as metal-free sensitizers and are used to replace expensive metal-based sensitizers. However, these dyes give low efficiency as compared to the organometallic photosensitizers but have many advantages like easily tuned absorption regions i.e. broadening the absorption region by using organic dye as both sensitizers and co-sensitizers, low cost, non-toxic in nature, different structures, easy synthesis process, and high molar extinction coefficient. At higher temperatures, these dyes also get unstable. However, in a few cases, the efficiency of organic dyes-based DSSC has reached near to the organo-metallic ones. These dyes form the D-π-A configuration (shown in Fig. 4), where D is an electron donor that transfers the electrons from the π-conjugation bridge and is accepted by the acceptor level (A) [60]. Generally, electron-rich halves are selected for D, such as indoline, phenylamine, carbazoles, polyenes, thiophenes, and benzothiadiazole as π-conjugated groups, and pyridines, rhodamines, and cyanoacrylic as acceptors (A) [3, 4]. Different organic dyes can be used in DSSC, such as coumarin dyes, indole dyes, porphyrins, triarylamine dyes, phenothiazine (PTZ) dyes, carbazole dyes, and phenoxazine dyes, each one having a different molecular structure [3]. Indoline dyes were used with the quasi-solid gel, where highly reactive nitroxide molecules were used as a redox mediator. This interaction at the dye-electrolyte interface induced the charge recombination process, which is suppressed by using a long alkyl chain. The obtained PCE value under one sun illumination was 10.1% [61]. A novel indole-based push–pull metal-free ion dye named thieno[2, 3-b] was synthesized for DSSC and recorded efficiency values for four dyes: IK-3, IK-4, IK5, and IK-6. The maximum efficiency was obtained at 6.3% for IK-3 under the illumination of AM 1.5G radiations. The lower efficiency value for other dyes is due to the aggregation and intermolecular π-stacking [62]. Recently, Krishna et al. designed a novel sensitizer, i.e. porphyrin, based on D-π-A configuration. This novel porphyrin has a strong electron donating ethynyl carboxy phenyl group and methyl phenanthroimidazole ring, which attacked the hexyl phenyl chains to reduce the loss of

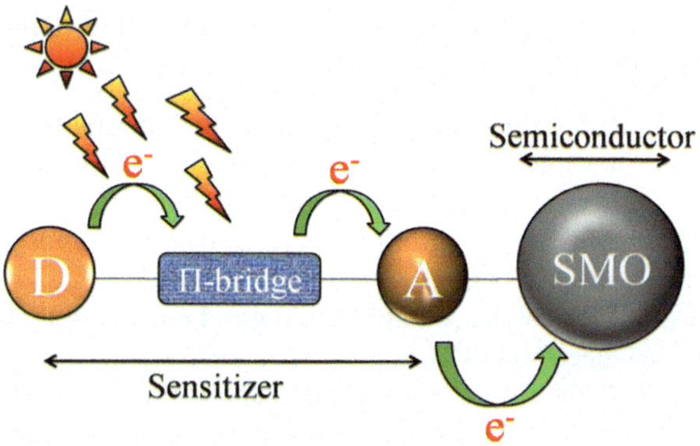

Fig. 4 D-π-A (donor- π bridge-acceptor) configuration to enhance the PCE value in DSSC

open circuit voltage, thus enhancing the V_{oc} to 650 mV [63]. Another dye, i.e. triarylamines derivatives, reduces the charge combination, degrades the dye, and increases the metal-free dyes' molar extinction coefficientdyes using π-π stacking. Structural modification in the triarylamines derivatives can enhance the PV parameters of the DSSC. Therefore, recently, three new dyes were synthesized named as T2, T3, and T4 (T1 is used as reference dye) using triphenylamine (TPA) as donor, where different substitutions of TPA in terthiophene act as a π-conjugated linker. Amongst all, at AM 1.5 light irradiation, T2 dye exhibited the highest efficiency value of 8.08%, which was 41% higher than the efficiency calculated for the reference cell (5.42%). Also, when this dye T2 was used with the co-adsorbent CDCA, the PCE value exceeded 9.02%, representing that dye modification using a π-conjugated linker in the D-π-A configuration enhances the efficiency value [64]. The effect of PTZ dye with TEG, which becomes a hydrophilic sensitizer, has been studied with choline-chlorine-based DES. Glucuronic acid, which was added as a co-adsorbent, interacted with the hydrophilic sensitizer and participated in the interactions of hydrogen binding. The reported efficiency value was 0.50% at the equal ratio of dye and co-absorbant [65]. Quantum dots, specially II–VI and III–V semiconductors QDs, are also used as co-sensitizers with the dyes because of their tunable band gap, high absorption coefficient, and Multiple Exciton Generation (MEG). Saad et al. prepared six QDs samples of varying wt % of ZnS and fixed wt% of CdSe and then immersed them in the N719 dye. It has resulted that 40 wt% of ZnS is suitable for the DSSC photoanode, and 50 wt% is the best composition to enhance the absorbance intensity [66].

3.4 *Electrolyte (HTM)*

In DSSC, the electrolyte reactivates the oxidized photosensitizers in their ground state by providing electrons from the CE. Since the life span and stability of DSSC mainly depends on the electrolyte, it should have the following properties [1, 3, 4, 12, 18, 46]:

(i) It should be electrochemically and thermally stable with a long-term span.
(ii) It should be non-corrosive with other DSSC parts and have the excellent ability to regenerate oxidized sensitizer into its initial state.
(iii) Absorption region of the electrolyte should not match the sensitizer's absorption spectra.
(iv) Electrolytes should have a good contact at the interfaces: electrolyte-dye and electrolyte-CE.
(v) It should have negligible vapour pressure, low diffusion, and viscosity, high dielectric constant, high boiling point, and excellent electrical conductivity to transfer the charge carriers.

Electrolytes are divided into three categories: liquid electrolytes, solid-state electrolytes, and quasi-solid-state electrolytes, which in detail are discussed below:

3.4.1 Liquid Electrolytes

These electrolytes are further classified into two types: Room-Temperature Ionic Liquid electrolytes (RTIL) based on the solvent used and organic solvent-based electrolytes.

Organic electrolytes: Different components of the organic electrolytes, such as additive (co-absorbents), solvent, and redox couple, affect the PCE value of DSSC, amongst which red-ox couple helps optimize the potential for the efficient transfer of the charge carriers. Different red-ox couples have been used till now, such as $SeCN^-$/$(SeCN)_2$, Br^-/Br_3, I_3^-/I^-, substituted bipyridyl cobalt (III/II) and $SCN^-/(SCN)_2$. I_3^- (triiodide)/I^- (iodine) is the most used electrolyte by researchers, since it enhances the efficiency of DSSC because of aligned red-ox potential, good solubility, low absorption coefficient, fast dye regeneration, and low recombination kinetics between the excited electron in SMO and electrolyte. But creates a hindrance by limited stability and life span of DSSC. Another part of electrolyte, which is an organic solvent, helps in the diffusion of triiodide/iodide ions, including ethylenecarbonate (EC), N-methyl pyrrolidone (NMP), acrylonitrile (AcN), 3-methoxypropionitrile (MePN), propylene carbonate (PC), and γ-butyrolactone (GBL) as organic solvents. Organic solvents based on stability can be arranged in decreasing order as alkyl pyridinium, picolinium, and imidazolium. Solvents have many characteristics like the dielectric number, donor number, and dielectric constant, amongst which donor number affects the photovoltaic parameters of DSSC, i.e. V_{oc} and I_{sc}. Improvement in the PV parameters also noticed in electric additives like 4-tert-butylpyridine (TBP), guanidinium thiocyanate (GuSCN), and N-methylbenzimidazole (NMBI) [67]. The third part of the electrolyte, i.e. co-absorbent, plays its role in DSSC performance by reducing the charge recombinations of injected electrons from photoanode to electrolyte. Long-term stability and reduction in dye aggregation on SMO materials surface can also be achieved by co-absorbent, which tunes the energy levels position of CB of photoanode, thus resulting in higher PV parameters. Optimized concentration of the red-ox couple (I_3^-/I^-) is also an essential factor in the PCE value because if the concentration is deficient, it will not lead to the required conductivity. If the concentration is relatively high, it will also lead to the absorption of the photons in the visible region.

Room temperature ionic liquid electrolytes (*RTIL*): Organic electrolytes hinder the performance of DSSC by limiting the long-term stability through its leakage and high evaporation rates. Therefore, ionic liquid electrolytes reduce these two problems successfully. Ionic electrolytes are organic salts having anions from pseudohalide and halide groups, whereas cations such as imidazolium and pyridinium. Ionic electrolytes act as both: a solvent and a red-ox couple. They have numerous advantages like high room temperature electrical conductivity, negligible vapour pressure, wide electrochemical window, and low flammability. However, due to its higher viscosity, it limits the transportation of charge carriers which reduces the conductivity of the electrolyte and a decrease in PCE values will be observed.

The long stability of DSSC was analyzed by mixing the gel electrolyte with the liquid electrolyte, where the solvent effect was also investigated. Type I used distilled water, whereas type II used acetonitrile (ACN) with a concentration of KI and iodine. The efficiency value was obtained better for type II (0.0556%) when compared to the kind I mixture (0.0456) [68]. The stability of the current characteristics of the DSSC was recently analyzed from the second quadrant of the *I–V* characteristics. It explains that in the second quadrant of *I–V* characteristics, the forward current continues flowing and gradually shifts towards the more negative value ascribed to the operating voltage. The stated reason behind this is the increase in the ratio of I^- (iodide) to I_3^- (triiodide) [69]. The efficiency of DSSC was also improved with the doping of liquid crystals (LC) in the liquid electrolyte. It was observed that the addition of the small amount of dopant (LC) in the electrolyte had deceased the electrochemical reaction for the iodine reduction, therefore decreasing the I_{sc} value of the DSSC and also decreasing the degradation rate, which is responsible for the declination of evaporation rate [70].

3.4.2 Solid-State Electrolytes (SSE)

Because of the short-term stability of DSSC due to the leakage problem of the liquid electrolytes, solid-state electrolytes came into existence. SSE is further divided into two categories: hole transfer material (HTM), a p-type semiconductor, and SSE, including transport medium of iodide/triiodide redox couple. HTMs are used because of their low chemically aggressive nature. Inorganic HTMs, which are copper-based HTMs used till now, are CuBr, CuI, and CuSCN, whereas organic HTM has advantages over inorganic ones, like easy deposition process and low cost [18]. Organic HTM is introduced with 2,2',7,7'-tetrakis (N,N-di-p-methoxyphenylamine) 9,9'-spirobifluorene (OMeTAD), giving a PCE value of 0.74% only [71]. The addition of silver ions to increase the dye adsorption in the dye solution has increased its efficiency up to 3.2% [72]. Copper-based electrolytes (CuI, CuBr, CuSCN) have increased the PVE value of DSSC because of good conductivity, amongst which CuSCN is the best HTM. However, it also resulted in poor efficiency below 4% due to a high crystallization rate, increased series resistance value, and poor contact between dye and CuSCN [73]. In their review, Yun et al. discussed the state-of-the-art integrated solar cell devices for DSSC and Perovskite Solar Cells (PSC) and mentioned many HTMs suitable for DSSC [74]. SSE, including the transport medium of iodide/triiodide redox couple, has more application than the HTMs because of the excellent interfacial and electronic contact properties between dye and electrolyte. PCE value of 4.2% was recorded for DSSC under the illumination of AM 1.5 using the poly (ethylene oxide) (PEO) matrix with the titania polymer adsorbed on TiO_2 bases photoanode [75].

3.4.3 Quasi-Solid State Electrolytes (QSSE)

Since the SSE solves the leakage problem, but the interfacial contact between the SMO and HTM is still weak, SSEs do not penetrate the pores of SMO. Therefore, this problem is solved by QSSEs, the hybrid structure consisting of liquid electrolytes and polymers. Because of this, QSSEs have properties of both states: liquid and solid, i.e. cohesive and diffusion transport, respectively, along with long-term stability. Also, the diffusion property or conductivity of the QSSE depends on several factors like morphology, working temperature, and molecular weight [76]. Most common polymers and HTMs used in QSSEs are bis(imidazolium) iodides, polymer gel-like PEO, 1-propargyl-3- methylimidazolium iodide and 1-ethyl-1-methylpyrrolidinium, and polyvinyl acetate and poly(vinylidinefluoride) containing redox couples [77, 78]. Sun et al. used Polyethylene terephthalate (PET) as a matrix in the electrolyte for DSSC fabrication which resulted in the enhancement of the absorption of an electrolyte by the membrane and avoiding short circuits because of excellent interfacial contact between the SMO and the electrolyte. Therefore, this optimized membrane gives the PCE value of 10.248% for the QSSE-based DSSC at 100 mW cm^{-2}. A longer lifetime of 15% was also recorded for QSSEC-based DSSC as compared to the liquid electrolyte-based DSSC because of the plasma-treated membrane with the oxygen and argon separately [79].

3.5 Counter Electrode (CE)

The counter electrode (Pt, Ag and Au) acts as a cathode in DSSC, and its function is to transfer electrons from the external circuit to the oxidized red-ox potential. Therefore, CE should have unique characteristics like:

(i) High specific area and low charge transfer resistance at the electrolyte-CE interface.
(ii) It should have a high exchange current density for efficient electrolyte generation.

Platinum (Pt) coated on the FTO glass is mainly used by researchers because it enhances the efficiency of the cell through its properties like transparency, high exchange current density, and good catalytic activity but it has a significant disadvantage like high cost. Therefore, the best alternatives with low-cost materials showing the above-mentioned unique properties are in search mode. In the last few years, the research is focused on amorphous carbon and carbon nanotubes (CNTs), graphene oxide, graphite and graphene nanoparticles, carbonylsulfide, and alloys like FeSe as CE for DSSC [1, 3, 12].

Thus, the efficiency of DSSC depends on its different parts and their combinations. Table 3 shows the photovoltaic parameters of DSSC using other working and counter electrodes.

Kokkonen et al. reviewed the advanced research trends in the various parts of the DSSC along with the sealing techniques used. Electrolytes includes copper (Cu), $[Co(phen)_3]^{2+/3+}$ and cobalt (Co) redox couple-based electrolytes, and PEDOT catalyst; sensitizers include inkjet-printed dyes; counter electrodes include SWCNT-based CEs; and UV-curable glue for advanced sealing techniques. This cutting-edge research can be the future of the DSSC which can pave the way for efficiency enhancement and commercialization of DSSCs [2].

4 Conclusions

DSSC, a third-generation solar cell, increases its opportunities for commercialization day by day by enhancing the overall efficiency value along with the economic nature, high and more extended stability, low weight, simple synthesis process, and working nature in low-light conditions. However, to date, the synthesized DSSC does not show their prophecy and is ultimately a hurdle for the commercialization process. DSSC consists of four parts: photoanode (transparent conductive oxide glass and semiconductor layers), sensitizers, electrolyte, and counter electrode. Therefore, this book chapter illustrates the state-of-the-art of different parts of DSSC, which shows the various reasons behind this unsuccessful nature. Till now, fluorine-based transparent oxide glass is superior amongst all in the TCO class, and TiO_2 in the semiconductor metal oxide (SMO)-layered materials. Counter electrode still has the

Table 3 Values of efficiency determining parameters of DSSC using different working electrodes and counter electrodes [3]

WE/CE	V_{OC} (mV)	J_{SC} (mA cm^{-2})	FF (%)	η (%)
WE: TiO_2 doped with tungsten	730	15.10	67	7.42
WE: TiO_2 doped with scandium	752	19.10	68	9.60
WE: TiO_2 doped with indium	716	16.97	61	7.48
WE: TiO_2 doped with boron	660	7.85	66	3.44
WE: TiO_2 doped with fluorine	754	11	76	6.31
WE: TiO_2 doped with carbon	730	20.38	57	8.55
WE: ONT/FTO	700	10.65	70	5.32
WE: G-TiO_2 NPs/TiO_2 NTs	690	16.59	56	6.29
WE: TiO_2 doped with Cu	591	6.84	56	2.28
WE: 7.5% SnO_2 doped TiO_2	790	14.53	58	6.7
WE: TiO_2:$Y_{1.86}Eu_{0.14}WO_6$	757	12.3	43	3.9
WE: Nb2O5	738	6.23	68.3	3.15
WE: Nanographite-TiO_2	720	1.69	35	0.44
CE: PtCo	717	16.96	66	7.64
CE:Pt+SLGO	670	7.9	65	3.4
CE: PtMo	697	15.48	62	6.75
CE: $PtCr_{0.05}$	739	13.07	71	6.88
CE: $CoNi_{0.25}$	706	18.02	66	8.39
CE: Ni-PANI-G	719	11.56	64	5.32
CE: PANI nanoribbons	720	17.92	56	7.23
CE: $Pd_{17}Se_{15}$	700	16.32	65	7.45
CE: PtCuNi	758	18.30	69	9.66
CE: g-C_3N_4/G	723	14.91	66	7.13
CE: FeN/N-doped graphene	740	18.83	78	10.86
CE: MoS_2 nanofilm	740	16.96	66	8.28
CE: $Ni_{0.33}Co_{0.67}Se$ microsphere	789	17.29	67	9.01
CE: Tubular orthorhombic $CoSe_2$	771	17.35	70	9.34
CE: $CoSe_2$	809	17.65	71	10.17
CE: $In_{2.77}S_4$@CC	750	17.34	67	8.71
CE: Electrochemically deposited Pt	750	17.16	60	7.72
CE: Fe_3O_4@RGO-NMCC	760	17.00	70	9.04
CE: CB-NPs/s-PT	764	17.21	69	9.02
CE: AC/MWCNTs	753	16.07	83	10.05
CE: $NiCo_2S_4$	148.6	2.98	55.8	0.24
CE: SS:Graphene	524	1.46	26	1.98

(continued)

Table 3 (continued)

WE/CE	V_{OC} (mV)	J_{SC} (mA cm^{-2})	FF (%)	η (%)
CE: Ru0.33Se	722	17.86	67.9	8.76
CE:5% Ag-doped SnS_2	740	16.7	70	8.70
CE: Cu_2O	680	11.35	47	3.62

best option of platinum for achieving high performances, where it functions with the electrolyte. Solid-state electrolyte is the choice of electrolyte nowadays. To avoid the challenges, photoanode still has to be improved for the porosity and specific surface area of the SMO, shape, morphology, and thickness of the SMO layer, and by using the heterojunction architecture for SMO-sensitized materials to reduce the charge recombination rates and electron lifetime, and increase the injection and transportation process. Since the dye is the core of the DSSC, its materials should be low cost, non-toxic, have a long life span, have a high absorption coefficient and well-aligned level with SMO material and red-ox potential of electrolyte. However, natural and organic dyes, low-cost and non-toxic sensitizers, provide low efficiency.

In contrast, metal-complex-based sensitizers provide high efficiency but are toxic and cost highly. Electrolytes also have problems like stability, leakage, and life span problems with the degradation of the photosensitizers and CE. The counter electrode is also dealing with high-cost materials and trying to replace them with other possible materials like carbon, graphene etc.

Various techniques are also in use nowadays to attain high efficiency, including co-sensitization and light scattering, transparent electrodes with compact layers of SMO, hydrogenation and protonation, and plasma and post-treatment. Various other possible scenarios can improve efficiency through state-of-the-art of parts in the future. Using recycled materials, low-cost production and non-toxic materials can be used and recycled, thus can increase the span and stability of the DSSC device. Perovskite sensitizers are also the new options because they can act as both hole and electron transport material. In contrast, their films can even deal efficiently with the charge generation, injection, and transportation process. Graphene-based, conductive, and transparent electrodes are also becoming the economic options for TCO like FTO and ITO. Cocktail dye sensitizers which can be a combination of natural dyes might increase the absorption values.

In contrast, water-based electrolytes, polymer-based electrolytes, and gel electrolytes can be the electrolyte options used in DSSC. Furthermore, advanced research trends (copper (Cu), $[Co(phen)_3]^{2+/3+}$ and cobalt (Co) redox couple-based electrolytes; inkjet-printed dyes; PEDOT catalyst; SWCNT-based CEs; advanced sealing techniques like UV-curable glue) for various parts of DSSC will also help in the better performance of DSSC. Thus, the enhanced DSSC efficiency with the optimized parameters and desirable economical, durable and sustainable status can be fulfilled in future, making it a commercialized solar cell.

References

1. Richhariya G, Kumar A, Tekasakul P, Gupta B (2017) Natural dyes for dye sensitized solar cell: a review. Renew Sustain Energy Rev 69:705–718. https://doi.org/10.1016/j.rser.2016.11.198
2. Kokkonen M, Talebi P, Zhou J, Asgari S, Soomro SA, Elsehrawy F et al (2021) Advanced research trends in dye-sensitized solar cells. J Mater Chem A 9:10527–10545. https://doi.org/10.1039/D1TA00690H
3. Sharma K, Sharma V, Sharma SS (2018) Dye-sensitized solar cells: fundamentals and current status. Nanoscale Res Lett 13:381. https://doi.org/10.1186/s11671-018-2760-6
4. Bera S, Sengupta D, Roy S, Mukherjee K (2021) Research into dye-sensitized solar cells: a review highlighting progress in India. J Phys Energy 3:032013. https://doi.org/10.1088/2515-7655/abff6c
5. Gong J, Sumathy K, Qiao Q, Zhou Z (2017) Review on dye-sensitized solar cells (DSSCs): advanced techniques and research trends. Renew Sustain Energy Rev 68:234–246. https://doi.org/10.1016/j.rser.2016.09.097
6. Kallmann H, Pope M (1959) Photovoltaic effect in organic crystals. J Chem Phys 30:585–586. https://doi.org/10.1063/1.1729992
7. Tributsch H, Calvin M (1971) Electrochemistry of excited molecules: photo-electrochemical reactions of chlorophylls*. Photochem Photobiol 14:95–112. https://doi.org/10.1111/j.1751-1097.1971.tb06156.x
8. Fujishima A, Honda K (1972) Electrochemical photolysis of water at a semiconductor electrode. Nature 238:37–38. https://doi.org/10.1038/238037a0
9. O'Regan B, Grätzel M (1991) A low-cost, high-efficiency solar cell based on dye-sensitized colloidal TiO_2 films. Nature 353:737–740. https://doi.org/10.1038/353737a0
10. Tomar N, Agrawal A, Dhaka VS, Surolia PK (2020) Ruthenium complexes based dye sensitized solar cells: fundamentals and research trends. Sol Energy 207:59–76. https://doi.org/10.1016/j.solener.2020.06.060
11. Kakiage K, Aoyama Y, Yano T, Oya K, Fujisawa J, Hanaya M (2015) Highly-efficient dye-sensitized solar cells with collaborative sensitization by silyl-anchor and carboxy-anchor dyes. Chem Commun 51:15894–15897. https://doi.org/10.1039/C5CC06759F
12. Kumar D (2021) A short review on the advancement in the development of TiO_2 and ZnO based photo-anodes for the application of dye-sensitized solar cells (DSSCs). Eng Res Express 3:042004. https://doi.org/10.1088/2631-8695/ac3b29
13. Baxter JB (2012) Commercialization of dye sensitized solar cells: present status and future research needs to improve efficiency, stability, and manufacturing. J Vac Sci Technol A 30:020801. https://doi.org/10.1116/1.3676433
14. Nazeeruddin MK, Kay A, Rodicio I, Humphry-Baker R, Mueller E, Liska P et al (1993) Conversion of light to electricity by cis-X2bis(2,2'-bipyridyl-4,4'-dicarboxylate)ruthenium(II) charge-transfer sensitizers (X = Cl-, Br-, I-, CN-, and SCN-) on nanocrystalline titanium dioxide electrodes. J Am Chem Soc 115:6382–6390. https://doi.org/10.1021/ja00067a063
15. Boschloo G (2019) Improving the performance of dye-sensitized solar cells. Front Chem vol 7. Available: https://doi.org/10.3389/fchem.2019.00077
16. Grätzel M (2004) Conversion of sunlight to electric power by nanocrystalline dye-sensitized solar cells. J Photochem Photobiol A 164:3–14. https://doi.org/10.1016/j.jphotochem.2004.02.023
17. Hagfeldt A, Graetzel M (1995) Light-induced redox reactions in nanocrystalline systems. Chem Rev 95:49–68. https://doi.org/10.1021/cr00033a003
18. Mehmood U, Rahman S, Harrabi K, Hussein IA, Reddy BVS (2014) Recent advances in dye sensitized solar cells. Adv Mater Sci Eng 2014:e974782. https://doi.org/10.1155/2014/974782
19. Wu S, Yuan S, Shi L, Zhao Y, Fang J (2010) Preparation, characterization and electrical properties of fluorine-doped tin dioxide nanocrystals. J Colloid Interface Sci 346:12–16. https://doi.org/10.1016/j.jcis.2010.02.031

20. Sima C, Grigoriu C, Antohe S (2010) Comparison of the dye-sensitized solar cells performances based on transparent conductive ITO and FTO. Thin Solid Films 519:595–597. https://doi.org/10.1016/j.tsf.2010.07.002
21. Patni N, Sharma P, Parikh M, Joshi P, Pillai SG (2018) Cost effective approach of using substrates for electrodes of enhanced efficient dye sensitized solar cell. Mater Res Express 5:095509. https://doi.org/10.1088/2053-1591/aad952
22. Green ANM, Palomares E, Haque SA, Kroon JM, Durrant JR (2005) Charge transport versus recombination in dye-sensitized solar cells employing nanocrystalline TiO_2 and SnO_2 films. J Phys Chem B 109:12525–12533. https://doi.org/10.1021/jp050145y
23. Kay A, Grätzel M (2002) Dye-sensitized core−shell nanocrystals: improved efficiency of mesoporous tin oxide electrodes coated with a thin layer of an insulating oxide. Chem Mater 14:2930–2935. https://doi.org/10.1021/cm0115968
24. Kumara GRRA, Tennakone K, Perera VPS, Konno A, Kaneko S, Okuya M (2001) Suppression of recombinations in a dye-sensitized photoelectrochemical cell made from a film of tin IV oxide crystallites coated with a thin layer of aluminium oxide. J Phys D: Appl Phys 34:868. https://doi.org/10.1088/0022-3727/34/6/306
25. Yoshida T, Minoura H (2000) Electrochemical self-assembly of dye-modified zinc oxide thin films. Adv Mater 12:1219–1222. https://doi.org/10.1002/1521-4095(200008)12:16%3c1219::AID-ADMA1219%3e3.0.CO;2-5
26. Nguyen T-V, Lee H-C, Alam Khan M, Yang O-B (2007) Electrodeposition of TiO_2/SiO_2 nanocomposite for dye-sensitized solar cell. Sol Energy 81:529–534. https://doi.org/10.1016/j.solener.2006.07.008
27. Fukai Y, Kondo Y, Mori S, Suzuki E (2007) Highly efficient dye-sensitized SnO_2 solar cells having sufficient electron diffusion length. Electrochem Commun 9:1439–1443. https://doi.org/10.1016/j.elecom.2007.01.054
28. Park N-G, van de Lagemaat J, Frank AJ (2000) Comparison of dye-sensitized rutile- and anatase-based TiO_2 solar cells. J Phys Chem B 104:8989–8994. https://doi.org/10.1021/jp9943651
29. Cho T-Y, Ko K-W, Yoon S-G, Sekhon SS, Kang MG, Hong Y-S et al (2013) Efficiency enhancement of flexible dye-sensitized solar cell with sol–gel formed Nb_2O_5 blocking layer. Curr Appl Phys 13:1391–1396. https://doi.org/10.1016/j.cap.2013.04.012
30. Chou C-S, Chou F-C, Kang J-Y (2012) Preparation of ZnO-coated TiO_2 electrodes using dip coating and their applications in dye-sensitized solar cells. Powder Technol 215–216:38–45. https://doi.org/10.1016/j.powtec.2011.09.003
31. Palomares E, Clifford JN, Haque SA, Lutz T, Durrant JR (2003) Control of charge recombination dynamics in dye sensitized solar cells by the use of conformally deposited metal oxide blocking layers. J Am Chem Soc 125:475–482. https://doi.org/10.1021/ja027945w
32. Choi H, Nahm C, Kim J, Moon J, Nam S, Jung D-R et al (2012) The effect of $TiCl_4$-treated TiO_2 compact layer on the performance of dye-sensitized solar cell. Curr Appl Phys 12:737–741. https://doi.org/10.1016/j.cap.2011.10.011
33. Govindaraj R, Pandian MS, Ramasamy P, Mukhopadhyay S (2015) Sol–gel synthesized mesoporous anatase titanium dioxide nanoparticles for dye sensitized solar cell (DSSC) applications. Bull Mater Sci 38:291–296. https://doi.org/10.1007/s12034-015-0874-3
34. Wan J, Tao L, Wang B, Zhang J, Wang H, Lund PD (2019) A facile method to produce TiO_2 nanorods for high-efficiency dye solar cells. J Power Sources 438:227012. https://doi.org/10.1016/j.jpowsour.2019.227012
35. Kumar D, Bist A, Dua P, Kuchhal P, Anand G, Parmar K (2019) Facile one-pot hydrothermal synthesis of nanorice-like TiO_2 for an efficient dye-sensitized solar cell (DSSC). Nanotechnol Perceptions 15:58–64. https://doi.org/10.4024/N03KU19A.ntp.15.01
36. Li Z-Q, Que Y-P, Mo L-E, Chen W-C, Ding Y, Ma Y-M et al (2015) One-pot synthesis of mesoporous TiO_2 micropheres and its application for high-efficiency dye-sensitized solar cells. ACS Appl Mater Interfaces 7:10928–10934. https://doi.org/10.1021/acsami.5b02195
37. Arote SA, Tabhane VA, Pathan HM (2018) Enhanced photovoltaic performance of dye sensitized solar cell using SnO_2 nanoflowers. Opt Mater 75:601–606. https://doi.org/10.1016/j.optmat.2017.11.023

38. Xu J, Fan K, Shi W, Li K, Peng T (2014) Application of ZnO micro-flowers as scattering layer for ZnO-based dye-sensitized solar cells with enhanced conversion efficiency. Sol Energy 101:150–159. https://doi.org/10.1016/j.solener.2013.12.039
39. Zhu S, Shan L, Tian X, Zheng X, Sun D, Liu X et al (2014) Hydrothermal synthesis of oriented ZnO nanorod–nanosheets hierarchical architecture on zinc foil as flexible photoanodes for dye-sensitized solar cells. Ceram Int 40:11663–11670. https://doi.org/10.1016/j.ceramint.2014.03.173
40. Qiu Y, Chen W, Yang S (2010) Double-layered photoanodes from variable-size anatase TiO_2 nanospindles: a candidate for high-efficiency dye-sensitized solar cells. Angew Chem Int Ed 49:3675–3679. https://doi.org/10.1002/anie.200906933
41. Satapathi S, Gill HS, Das S, Li L, Samuelson L, Green MJ et al (2014) Performance enhancement of dye-sensitized solar cells by incorporating graphene sheets of various sizes. Appl Surf Sci 314:638–641. https://doi.org/10.1016/j.apsusc.2014.07.003
42. Golobostanfard MR, Abdizadeh H (2014) Hierarchical porous titania/carbon nanotube nanocomposite photoanode synthesized by controlled phase separation for dye sensitized solar cell. Sol Energy Mater Sol Cells 120:295–302. https://doi.org/10.1016/j.solmat.2013.09.021
43. Siwatch S, Kundu VS, Kumar A, Kumar S, Chauhan N, Kumari M (2019) Effect of novel ZnO/Zn_2SnO_4 photoanode on the performance of dye sensitized solar cell. Optik 194:163117. https://doi.org/10.1016/j.ijleo.2019.163117
44. Madurai Ramakrishnan V, Pitchaiya S, Muthukumarasamy N, Kvamme K, Rajesh G, Agilan S et al (2020) Performance of TiO_2 nanoparticles synthesized by microwave and solvothermal methods as photoanode in dye-sensitized solar cells (DSSC). Int J Hydrogen Energy 45:27036–27046. https://doi.org/10.1016/j.ijhydene.2020.07.018
45. Mohammadi I, Zeraatpisheh F, Ashiri E, Abdi K (2020) Solvothermal synthesis of g-C_3N_4 and ZnO nanoparticles on TiO_2 nanotube as photoanode in DSSC. Int J Hydrogen Energy 45:18831–18839. https://doi.org/10.1016/j.ijhydene.2020.04.277
46. Francis OI, Ikenna A (2021) Review of dye-sensitized solar cell (DSSCs) development. Nat Sci 13:496–509. https://doi.org/10.4236/ns.2021.1312043
47. Semalti P, Sharma SN (2020) Dye sensitized solar cells (DSSCs) electrolytes and natural photosensitizers: a review. J Nanosci Nanotechnol 20:3647–3658. https://doi.org/10.1166/jnn.2020.17530
48. Ahliha AH, Nurosyid F, Supriyanto A, Kusumaningsih T (2018) Optical properties of anthocyanin dyes on TiO_2 as photosensitizers for application of dye-sensitized solar cell (DSSC). IOP Conf Ser: Mater Sci Eng 333:012018. https://doi.org/10.1088/1757-899X/333/1/012018
49. Sawhney N, Raghav A, Satapathi S (2017) Utilization of naturally occurring dyes as sensitizers in dye sensitized solar cells. IEEE J Photovolt 7:539–544. https://doi.org/10.1109/JPHOTOV.2016.2639343
50. Sreeja S, Pesala B (2018) Co-sensitization aided efficiency enhancement in betanin–chlorophyll solar cell. Mater Renew Sustain Energy. 7:25. https://doi.org/10.1007/s40243-018-0132-x
51. Sreeja S, Pesala B (2020) Efficiency enhancement of betanin–chlorophyll cosensitized natural pigment solar cells using plasmonic effect of silver nanoparticles. IEEE J Photovolt 10:124–134. https://doi.org/10.1109/JPHOTOV.2019.2953399
52. Muhammed T, Paulose P, George G, John J (2016) Dye-sensitized solar cell using natural dyes extracted from bougainvillea and bottle brush flower. Int J Eng Trends Technol 33:43–47. https://doi.org/10.14445/22315381/IJETT-V33P209
53. Datta S, Dey A, Singha NR, Roy S (2020) Enhanced performance of dye-sensitized solar cell with thermally stable natural dye-assisted TiO_2/MnO_2 bilayer-assembled photoanode. Mater Renew Sustain Energy 9:25. https://doi.org/10.1007/s40243-020-00185-3
54. Munawaroh H, adillah GF, Saputri LNMZ, Hanif QA, Hidayat R, Wahyuningsih S (2016) The co-pigmentation of anthocyanin isolated from mangosteen pericarp (Garcinia Mangostana L.) as natural dye for dye- sensitized solar cells (DSSC). IOP Conf Ser: Mater Sci Eng 107:012061. https://doi.org/10.1088/1757-899X/107/1/012061
55. Subramaniam K, Athanas AB, Kalaiyar S (2019) Dual anchored ruthenium(II) sensitizer containing 4-nitro-phenylenediamine schiff base ligand for dye sensitized solar cell application. Inorg Chem Commun 104:88–92. https://doi.org/10.1016/j.inoche.2019.03.043

56. Naik P, Abdellah IM, Abdel-Shakour M, Su R, Keremane KS, El-Shafei A et al (2018) Improvement in performance of N_3 sensitized DSSCs with structurally simple aniline based organic co-sensitizers. Sol Energy 174:999–1007. https://doi.org/10.1016/j.solener.2018.09.071
57. Venkatraman V, Raju R, Oikonomopoulos SP, Alsberg BK (2018) The dye-sensitized solar cell database. J Cheminform 10:18. https://doi.org/10.1186/s13321-018-0272-0
58. El-Shazly AN, Shalan AE, Rashad MM, Abdel-Aal EA, Ibrahim IA, El-Shahat MF (2018) Solid-state dye-sensitized solar cells based on $Zn_{1-x}Sn_xO$ nanocomposite photoanodes. RSC Adv 8:24059–24067. https://doi.org/10.1039/C8RA02852D
59. Al-Rawashdeh NAF, Albiss BA, Yousef MHI (2018) Graphene-based transparent electrodes for dye sensitized solar cells. IOP Conf Ser: Mater Sci Eng 305:012019. https://doi.org/10.1088/1757-899X/305/1/012019
60. Fuse S, Sugiyama S, Maitani MM, Wada Y, Ogomi Y, Hayase S et al (2014) Elucidating the structure–property relationships of donor–π-acceptor dyes for dye-sensitized solar cells (DSSCs) through rapid library synthesis by a one-pot procedure. Chem—A Eur J 20:10685–10694. https://doi.org/10.1002/chem.201402093
61. Suzuka M, Hayashi N, Sekiguchi T, Sumioka K, Takata M, Hayo N et al (2016) A quasi-solid state DSSC with 10.1% efficiency through molecular design of the charge-separation and -transport. Sci Rep 6:28022. https://doi.org/10.1038/srep28022
62. Irgashev R, Karmatsky A, Kim G, Sadovnikov A, Emets V, Grinberg V et al (2017) Novel push-pull thieno[2,3-b]indole-based dyes for efficient dye-sensitized solar cells (DSSCs). ARKIVOC: Arch Org Chem 2017:34–50. https://doi.org/10.3998/ark.5550190.p009.887
63. Vamsi Krishna N, Venkata Suman Krishna J, Singh SP, Giribabu L, Islam A, Bedja I (2017) Bulky nature phenanthroimidazole-based porphyrin sensitizers for dye-sensitized solar cell applications. J Phys Chem C 121:25691–25704. https://doi.org/10.1021/acs.jpcc.7b07692
64. Prachumrak N, Sudyoadsuk T, Thangthong A, Nalaoh P, Jungsuttiwong S, Daengngern R et al (2017) Improvement of D–π–A organic dye-based dye-sensitized solar cell performance by simple triphenylamine donor substitutions on the π-linker of the dye. Mater Chem Front 1:1059–1072. https://doi.org/10.1039/C6QM00271D
65. Boldrini CL, Manfredi N, Perna FM, Trifiletti V, Capriati V, Abbotto A (2017) Dye-sensitized solar cells that use an aqueous choline chloride-based deep eutectic solvent as effective electrolyte solution. Energ Technol 5:345–353. https://doi.org/10.1002/ente.201600420
66. Saad MA, Samsi NS, Hassan OH, Yahya MZA, Taib MFM, Ali AMM et al (2018) Studies of the absorbance peak on the N719 dye influence by combination between Cadmium Selenide (CdSe)QDs and Zinc Sulfide(ZnS)QDs. In: MATEC Web Conference, vol 154, p 01040. https://doi.org/10.1051/matecconf/201815401040
67. Stergiopoulos T, Rozi E, Karagianni C-S, Falaras P (2011) Influence of electrolyte co-additives on the performance of dye-sensitized solar cells. Nanoscale Res Lett 6:307. https://doi.org/10.1186/1556-276X-6-307
68. Puspitasari N, Amalia SSN, Yudoyono G, Endarko (2017) Effect of mixing dyes and solvent in electrolyte toward characterization of dye sensitized solar cell using natural dyes as the sensitizer. IOP Conf Ser: Mater Sci Eng 214:012022. https://doi.org/10.1088/1757-899X/214/1/012022
69. Iwata S, Shibakawa S, Imawaka N, Yoshino K (2018) Stability of the current characteristics of dye-sensitized solar cells in the second quadrant of the current–voltage characteristics. Energy Rep 4:8–12. https://doi.org/10.1016/j.egyr.2017.10.004
70. Huang C-Y, You C-F, Cheng C-E, Lei B-C, Jhang J-C, Yu F-C et al (2016) liquid crystal-doped liquid electrolytes for dye-sensitized solar cell applications. Opt Mater Express OME 6:1024–1031. https://doi.org/10.1364/OME.6.001024
71. Bach U, Lupo D, Comte P, Moser JE, Weissörtel F, Salbeck J et al (1998) Solid-state dye-sensitized mesoporous TiO_2 solar cells with high photon-to-electron conversion efficiencies. Nature 395:583–585. https://doi.org/10.1038/26936
72. Hao S, Wu J, Fan L, Huang Y, Lin J, Wei Y (2004) The influence of acid treatment of TiO_2 porous film electrode on photoelectric performance of dye-sensitized solar cell. Sol Energy 76:745–750. https://doi.org/10.1016/j.solener.2003.12.010

73. Hodes G, Cahen D (2012) All-solid-state, semiconductor-sensitized nanoporous solar cells. Acc Chem Res 45:705–713. https://doi.org/10.1021/ar200219h
74. Yun S, Qin Y, Uhl AR, Vlachopoulos N, Yin M, Li D et al (2018) New-generation integrated devices based on dye-sensitized and perovskite solar cells. Energy Environ Sci 11:476–526. https://doi.org/10.1039/C7EE03165C
75. Stergiopoulos T, Arabatzis IM, Katsaros G, Falaras P (2002) Binary polyethylene oxide/titania solid-state redox electrolyte for highly efficient nanocrystalline TiO_2 photoelectrochemical cells. Nano Lett 2:1259–1261. https://doi.org/10.1021/nl025798u
76. Wu J, Lan Z, Hao S, Li P, Lin J, Huang M et al (2008) Progress on the electrolytes for dye-sensitized solar cells. Pure Appl Chem 80:2241–2258. https://doi.org/10.1351/pac200880112241
77. Wang L, Zhang H, Wang C, Ma T (2013) Highly stable gel-state dye-sensitized solar cells based on high soluble polyvinyl acetate. ACS Sustain Chem Eng 1:205–208. https://doi.org/10.1021/sc300153b
78. Bai S, Bu C, Tai Q, Liang L, Liu Y, You S et al (2013) Effects of bis(imidazolium) molten salts with different substituents of imidazolium cations on the performance of efficient dye-sensitized solar cells. ACS Appl Mater Interfaces 5:3356–3361. https://doi.org/10.1021/am400337d
79. Sun KC, Sahito IA, Noh JW, Yeo SY, Im JN, Yi SC et al (2015) Highly efficient and durable dye-sensitized solar cells based on a wet-laid PET membrane electrolyte. J Mater Chem A 4:458–465. https://doi.org/10.1039/C5TA07720F

Chapter 5
Fabrication and Characterization of Silicon Nanowire Hybrid Solar Cells

Rabina Bhujel, Sadhna Rai, and Bibhu Prasad Swain

1 Introduction

A hybrid solar cell is a combination of both organic and inorganic semiconductors. In SiNWs hybrid solar cells, the inorganic part consists of 1-D SiNWs as the acceptor or the electron transport layer. SiNWs are amongst the most commonly used materials in solar cells due to their excellent charge career collection property and very efficient light-harvesting capability. The organic part of the hybrid photovoltaics consists of highly conjugated organic polymers that can absorb light in the form of a transport layer containing donor holes. Hybrid solar cells containing organic–inorganic heterojunctions are of great importance because of their significantly superior properties like low cost, high flexibility, low temperature, solution-based synthesis procedures, and comparatively lightweight. This book chapter includes various synthesis procedures utilized to synthesize SiNWs, graphene sheets and the fabrication of different types of SiNWs hybrid solar cells and the study of various properties using different characterization techniques.

1.1 Fundamentals of Solar Cell

See Fig. 1.

R. Bhujel (✉) · S. Rai
Sikkim Alpine University, Kamrang, Namchi, South Sikkim 737126, India
e-mail: rabina29@gmail.com

B. P. Swain
Department of Physics, National Institute of Technology Manipur, Langol, Imphal 795004, India

B. P. Swain (ed.), *Energy Materials*, Materials Horizons: From Nature to Nanomaterials,
https://doi.org/10.1007/978-981-99-3866-7_5

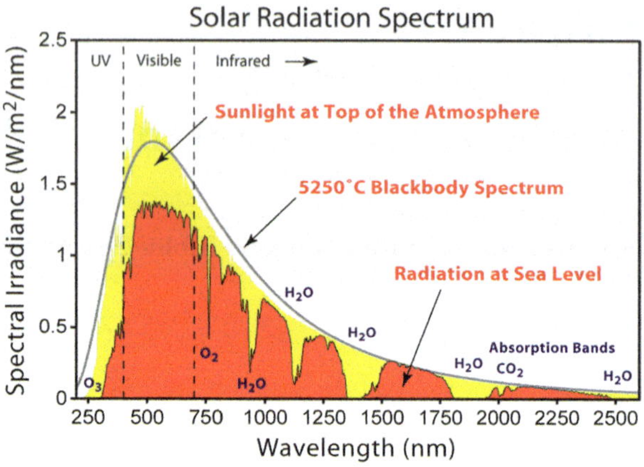

Fig. 1 Solar radiation spectrum [1]

1.2 Solar Spectrum

The Sun is believed to have a temperature of roughly 6000 K and therefore is considered a blackbody. According to Plank's radiation law [2], the maxima of the Sun's irradiation spectrum fall in the visible region. When the sunlight reaches the earth's surface, up to 45% of the solar energy is lost in the earth's atmosphere due to the solar energy interaction with the atmosphere and reflection of some of the solar radiations. The solar irradiation spectrum is shown in Fig. 1 [1]. The standard value of the solar spectrum used to obtain the solar cell performance is 1.5 AM with the visible solar spectrum in the range of 400–700 nm. The relationship between the Energy of photons (E_{ph}) and its wavelength λ in nm is shown by the following equation [3]:

$$E_{\text{ph}} = \frac{\text{hc}}{\lambda} = \frac{1240}{\lambda} \tag{1}$$

where c is considered as the speed of light and h corresponds to the Plank's constant. It is observed from the calculation that the visible solar range is in the range of 1.8–3.1 eV. The bandgap energy of Silicon (Si) is 1.12 eV, which is slightly beyond the range of the visible solar spectrum. This makes it an excellent candidate material for solar cells for achieving higher efficiency.

1.3 Limitation to the Theoretical Efficiency of Solar Cell

As explained by the law of thermodynamics, the maximum efficiency a solar cell can achieve is 85%. The theoretical consideration for the limitation of efficiency was obtained from the equation given below [4].

$$\eta = \left(1 - \frac{T_A^4}{T_s^4}\right)\left(1 - \frac{T_0}{T_A}\right) \tag{2}$$

Where, T_A gives the temperature of the ideal absorber for solar energy, the Sun's temperature $T_s = 5800$ K, and the environment temperature T_0 equal to 300 K. The amount of energy extracted from solar power is provided by the first term on the right-hand side of the Eq. (2). The energy obtained is specified as the difference between incident and emitted energy. The second term gives the amount of solar efficiency obtained using the extracted energy by the Carnot engine to generate electricity which is 85% at 2478 K.

The maximum efficiency that a single *p–n* junction solar cell can obtain is 31%, which is known as the Shockley–Queisser limit [5], or the detailed balance limit. This is the theoretical limit for obtaining efficiency by taking a single *p–n* junction system. In the past decade, researchers have worked extremely hard to overcome this Shockley–Queisser limit. The simple way is to fabricate a multi-junction solar cell that can easily overcome it due to the presence of more than one *p–n* junction. Another meaningful way is to use SiNWs instead of bulk Si. The one-dimensional structure of SiNWs significantly impacts light absorption, thereby increasing solar cell efficiency. The earlier studies show that the solar radiations having photon energy lower than the bandgap do not contribute to the performance of the solar cell and therefore, get reflected. The energy loss is called sub-bandgap loss. Moreover, the radiations with photon energy higher than the bandgap also lose part of their energy higher than the bandgap due to thermal relaxation. The bandgap of Si is 1.12 eV and hence can achieve a maximum efficiency of 29%. The bandgap of Si can be increased by converting its bulk counterpart to nanowires, such that it ranges between 1.12 and 2.6 eV [6]. The increment in the bandgap helps reduce the thermal loss hence increasing the efficiency and lifetime of the solar cell. This is one reason for choosing SiNWs as an essential material for the fabrication of hybrid solar cells.

1.4 The Basic Principle of SiNWs Hybrid Solar Cell

The essential components of SiNWs hybrid solar cells are *n*-type SiNWs, the inorganic component of the solar cell, and conducting polymers consisting of the organic part. When the sunlight of a particular wavelength is incident on the surface of the SiNWs hybrid solar cell, the excitation of electrons takes place from the Valence Band (VB) to the Conduction Band (CB) and becomes free. These highly energetic electrons then diffuse and some of them reach the junction where they get accelerated by an inbuilt Galvanic potential. This process generates electromotive force due to which the light energy gets converted into electrical energy. The electron that reaches the CB of SiNWs is being injected by the electrons present in the Lowest Unoccupied Molecular Orbital (LUMO) of organic conducting polymers. Thus, the energy level matching of the organic conducting polymers and the semiconductor is essential. It is considered that the LUMO of the organic conducting polymers has

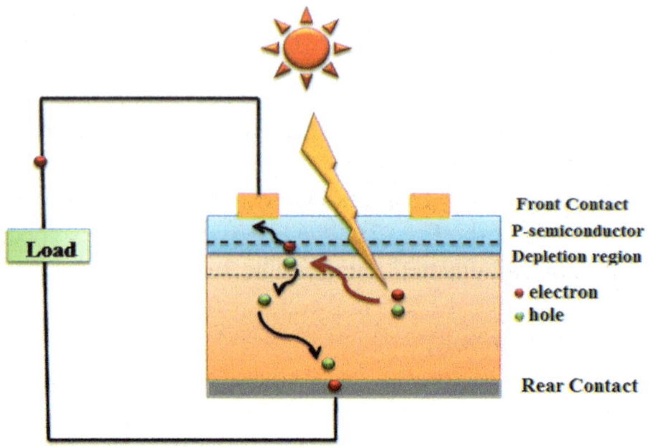

Fig. 2 Basic structure of *p–n* junction solar cell [4]

to be at least 0.16 eV higher than the CB of the *n*-type semiconductor to facilitate electron transfer from the LUMO to the CB.

The generation of photocurrent mainly includes four steps as illustrated in Fig. 2 (1) absorption of solar radiation, (2) electron–hole pair generation and its separation in a *p–n* junction, (3) diffusion of charge carriers (electrons and holes) in an *n*-type material and *p*-type material, and finally (4) circuit completion at rear contact. Amongst the above four steps, steps 1 and 2 are important where light absorption and electron–hole pair generation occur in the *p–n* junction [4].

The expression showing the dependence of current on applied voltage is given by [7]

$$J = J_{\text{ph}} - J_0\left[\exp\left\{\frac{q(V + \text{JR}_s)}{\text{nK}_B T}\right\} - 1\right] - \frac{V + \text{JR}_s}{R_{\text{sh}}} \tag{3}$$

where J_{ph} is photocurrent density which depends upon various factors such as properties of the materials used, device structure, the power of incident light, etc. J_0 stands for reverse saturation current density, q corresponds to the electronic charge, V is applied voltage, K_B stands for Boltzmann's constant, T is absolute temperature, R_s is series resistance, R_{sh} is the shunt resistance, and n is the ideality factor. The practically studied solar cell model shows J contains both series and shunt resistance as shown by the equivalent circuit model with resistance. From this circuit diagram, it is clear that the diode is connected in parallel with J_{ph} and R_{sh} and is connected in series with R_{sh}. The last term in the equation is called shunt current, which depends on the values of both resistances (Fig. 3).

The open-circuit voltage (V_{oc}) of the solar cell is described as the voltage without the current flow ($I = 0$). Again V_{oc} is obtained as given below [7]

$$V_{\text{oc}} = \frac{nK_B T}{q}\ln\left(\frac{J_{\text{ph}}}{J_0} + 1\right) \tag{4}$$

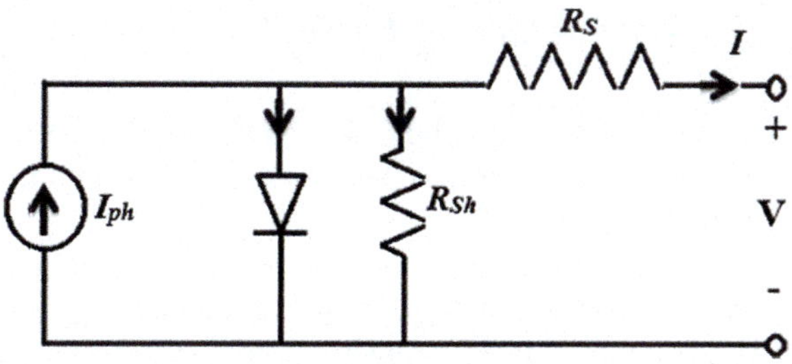

Fig. 3 Circuit diagram of a photovoltaic device

Finally, the Fill Factor (FF) of the solar cell is obtained as given below [8]

$$FF = \frac{P_{Max}}{V_{oc} \times J_{sc}} \times 100\% \tag{5}$$

It can be defined as the ratio of P_{Max} to the product of V_{max} and J_{max}. P_{Max} is the maximum power, and correspondingly V_{max} and J_{max} are the maximum voltage and current density generated by the solar cell. The FF is highly affected by the value of the resistance. FF increases proportionally with the increase in the value of R_{sh} whilst decreases with the rise in the value of R_s.

Finally, the power conversion efficiency of a solar cell can be obtained by the equation given below [8]

$$\text{Efficiency}(\eta)\% = \frac{V_{oc} \times J_{sc} \times FF}{P_{in}} \tag{6}$$

where J_{sc} is the short circuit current density and P_{in} is the power input. The schematic of the *J-V* characteristic curve of a solar cell is shown in Fig. 4. It has been observed from Eq. (6) that the efficiency of a solar cell is directly related to FF, V_{oc} and J_{sc}.

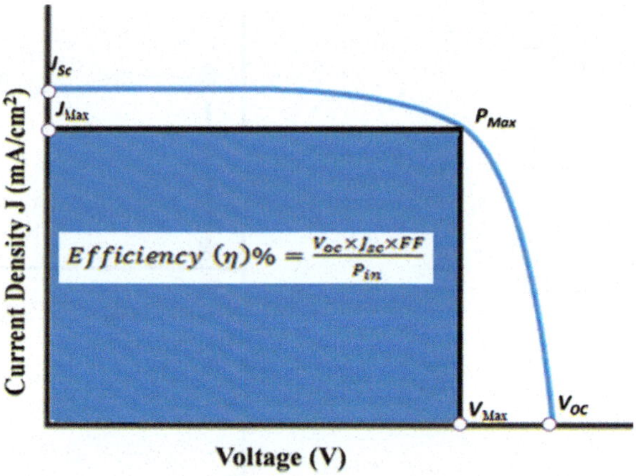

Fig. 4 *J-V* characteristic curve of a solar cell

1.5 Recombination Losses

The generated electron-hole pair remains separated and mobile for a very short period after which they return to the lowest energy state through the process called recombination. During the recombination process, the separated electrons and holes again recombine by emitting some amount of energy in terms of photons or phonons. The process of recombination is further divided into three main categories: (i) Radiative recombination, (ii) Auger recombination, and (iii) Shockley-Read-Hall (SRH) recombination. The schematic diagram of the recombination processes is shown in Fig. 5.

In the radiative recombination process band to band, recombination takes place. The electrons present in the conduction band and the holes present in the valence band combine and emit a photon. This type of recombination process takes place mostly in direct bandgap semiconductors.

Auger recombination process contains three carriers. In this process, the energy emitted by the recombination of an electron-hole pair is captured by the third carrier. This type of recombination is responsible for reducing the lifetime and efficiency of a solar cell device and only takes place in the impure and defective materials. The SRH recombination process takes place in two different steps. In the first step, the charge carriers (electrons or holes) are trapped in the forbidden energy level introduced due to the defects present in the crystal system. In the next step, the hole or electron gets excited to the same energy levels just before the electrons get thermally re-emitted into the conduction band and then the recombination occurs. All of the recombination processes explained above limit the lifetime of charge carriers. The generated electron-hole pair is lost due to these recombination processes and hence no power is generated by the solar cell device.

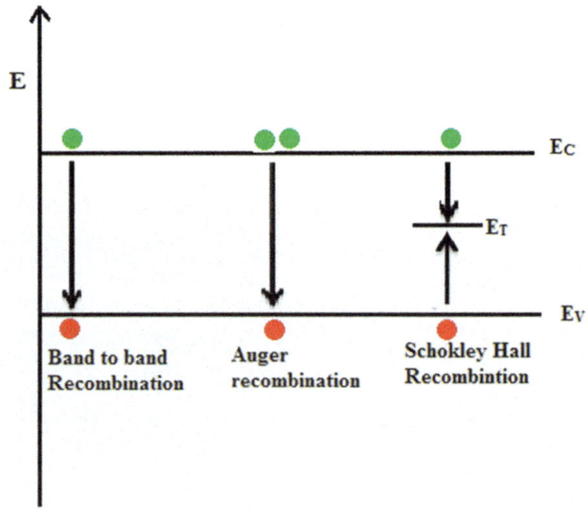

Fig. 5 Types of recombination that takes place in a solar cell

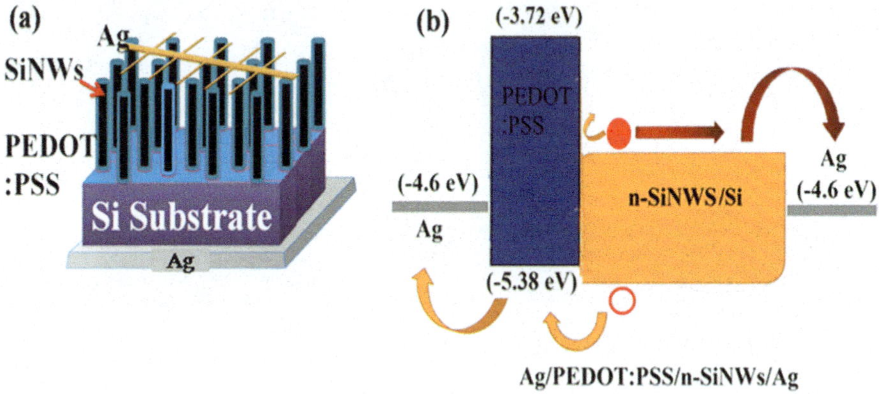

Fig. 6 **a** Schematic of SiNWs hybrid solar cell. **b** Bandgap diagram of SiNWs hybrid solar cell

1.6 SiNWs Hybrid Solar Cells

SiNWs hybrid solar cells are mainly composed of SiNWs as the inorganic material and the organic part constitutes organic semiconducting materials such as PEDOT: PSS, P_3HT:PCBM, MEH-PPV, spiro-(OMeTED), etc. [9]. The idea of the hybrid solar cell is very new and only in the past two decades, this area has become an essential subject of research. Scientists from all over the world are working extremely hard to obtain efficient hybrid solar cells because of their other properties such as lightweight, cost-effectiveness, ability to overcome Schokely-Queisser limit, flexibility as well as capability to produce the photovoltaic device on a large scale [10]. The reported work shows that the maximum efficiency that could be achieved by this hybrid solar cell lies in the vicinity of 15% and the theoretical efficiency that could be attained is more than 20% as reported in many simulation results.

The energy band diagram of SiNWs hybrid solar cell with the carrier transport mechanism is shown in Fig. 6b. In hybrid solar cells at first, the incoming photons generate a pair of charge carriers in the SiNWs, which are then separated by the SiNWs/PEDOT:PSS heterojunction. The separated electrons were then transported through Si to Ag rare contact and the holes get transported to the PEDOT:PSS/Ag terminal.

1.7 Required Materials for SiNWs Hybrid Solar Cell

1.7.1 Silicon and Silicon Nanowires

Silicon is a non-metallic semiconducting element that belongs to the carbon family in the periodic table. The atomic number of Si is 14 and is therefore tetravalent with the electronic configuration of [Ne] $3s^2 3p^2$. Si covers approximately 27.7% of the

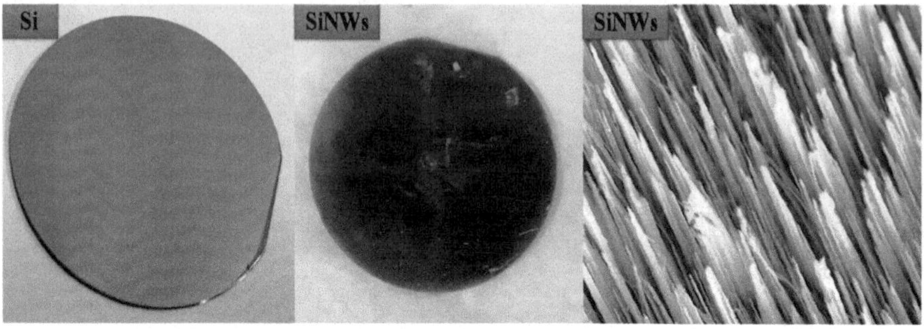

Fig. 7 Optical and SEM images of Si and SiNWs

earth's crust. Therefore, it is the second most abundant element on the earth. The bandgap of Si is 1.12 eV and is indirect (Fig. 7).

SiNWs is a one-dimensional structure of Si generally formed by etching of Si precursor or vapour-liquid phase growth on a substrate using different catalysts like gold, silver, etc. The properties of SiNWs are quite different as compared to their bulk counterpart due to their unusual quasi-one-dimensional electronic structure. Because of its one-dimensional nature, light trapping by SiNWs is many folds greater as compared to bulk Si. The reflectance of SiNWs is as low as 3% in the wavelength range of 400–1100 nm. The use of SiNWs over Si favours a reduction in material cost as well as increases the PCE of the solar cell, owing to its excellent electrical and optical properties.

1.7.2 Organic Conducting Polymers

Organic conducting polymers, also known as intrinsically conducting polymers, are polymers that are capable of conducting electricity. These polymers are different from extrinsically conducting polymers, which become conductive due to the presence of some external additive. These polymers are highly conjugated and therefore consist of a backbone of a long carbon chain bonded by σ bond with three of its kind and the remaining p_z orbital forms a delocalized π-bonded system. In other words, one can say that in such polymers, carbon atoms contain an alternating single and double bond between them. As the π-orbital of such a system is in the half-filled state, they must possess metallic behaviour; instead, they are slightly semiconducting, which is explained by Peierls theorem. In such polymers, the electronic states are rearranged, forming an alternating double bond between the carbon atoms of the system. This process results in a bandgap forming between a filled and unfilled π-band, making the system semiconducting instead of metallic. The initial conductivity of such a system is very low. Therefore, doping of ions on such conjugated polymers helps to increase the conductivity [11]. This process creates defects and deformation of the polymeric conjugated chain. The charge generated during this deformation process is the main reason for their excellent conductivity. As the doping concentration increases, more charges are formed in the polymer, resulting in higher conductivity.

1.7.3 Poly (3,4-Ethylene Dioxythiophene) Polystyrene Sulfonate (PEDOT:PSS)

PEDOT:PSS is a polyelectrolyte complex, which is a combination of two different intrinsically conducting polymers PEDOT and the polymer PSS which acts as a negative counterion. PEDOT:PSS is one of the most useful and commonly used intrinsically conducting polymers for photovoltaic devices because of its tunable electrical conductivity, high flexibility, and excellent optical transparency in the visible region. The structural model of PEDOT:PSS is shown in Fig. 8, where the oligomeric chain of PEDOT is tightly bonded with the much higher molecular weight chains of PSS. The high conductivity of PEDOT:PSS was achieved when the stacked PEDOT chain is encapsulated with the cross-linked and tangled highly swollen PSS gel particles [12, 13]. The conductivity of PEDOT:PSS can be increased from as small as 5–500 S cm^{-1}. This can be achieved by the incorporation of additives into the PEDOT:PSS solution. One of the crucial additives commonly used to increase conductivity is high boiling solvents such as ethylene glycol and dimethyl sulfoxide. The only problem with this conducting polymer is its fast degradation process due to the presence of oxygen and moisture in the atmosphere, which reduces the solar cell's lifetime. Therefore, recent research is being carried out to cease degradation to some extent.

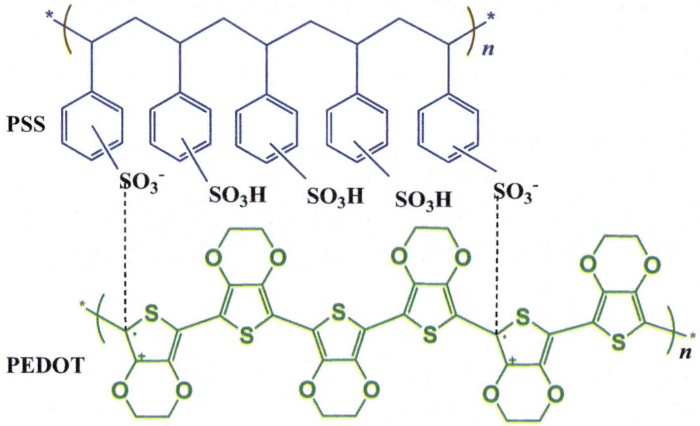

Fig. 8 Structure of PEDOT:PSS [14]

1.7.4 Graphene or Reduced Graphene Oxide

Graphene or reduced graphene oxide (rGO) is studied extensively by researchers all over the world due to its applications in various fields such as supercapacitors [15], sensors [16], photovoltaics [17], electrical devices [18], dye-degradation [18], catalysis [19], and many more. Graphene consists of a single sheet of hexagonally arranged sp^2 hybridized carbon atoms containing various oxygen functionalities like carbonyls, hydroxyls, etc. It is also called chemically obtained graphene as it is obtained by the chemical reduction of graphene oxide. Graphene's color is black and highly hydrophobic; therefore, it can be easily separated from an aqueous solution. The properties of graphene entirely depend upon the presence of different oxygen-containing functional groups. Therefore, it is possible to tune the properties of graphene by simply varying the oxygen functionalities.

2 Synthesis of SiNWs by Electroless Metal-Assisted Chemical Etching Method

EMACE is one of the most extensively used synthesis procedures for the easy growth of SiNWs. This method has gained much interest amongst researchers because of its facile synthesis procedure and cost-effectiveness. The EMACE technique also allows the tuning of various properties of SiNWs by varying the parameters like length and diameter of SiNWs for their application in different fields. The overall detailed explanation of the procedure is explained below (Fig. 9).

(a) **Cleaning of Si Wafer**

The purchased Si wafers are normally impure and therefore, before undergoing the etching process, their cleaning is necessary. The Si wafers were cleaned by following the Radio Corporation of America (RCA) cleaning procedure in the following sequence.

i. Cleaning of Si wafers in acetone for 15 min.
ii. Ethanol for 15 min.
iii. After that, clean the wafers in DI water 3–4 times.
iv. Cleaning in piranha solution (H_2SO_4:H_2O_2 = 3:1) for 15 min.
v. Cleaning in DI water for 15 min.
vi. Cleaning in HF (40%) for 1 min.

(b) **Electroplating**

After cleaning the Si wafers, the next step is to deposit silver nanoparticles (AgNPs) on the surface of the substrate as it acts as a catalyst for the etching of the Si substrate. The electroplating solution contains HF (4.8 M) and $AgNO_3$ (5–50 mM) in a 1:1 ratio. The cleaned Si wafers are dipped into the solution for 1 min at room temperature. The surface of the Si wafer appears to be shiny yellow due to the AgNP's electroplating.

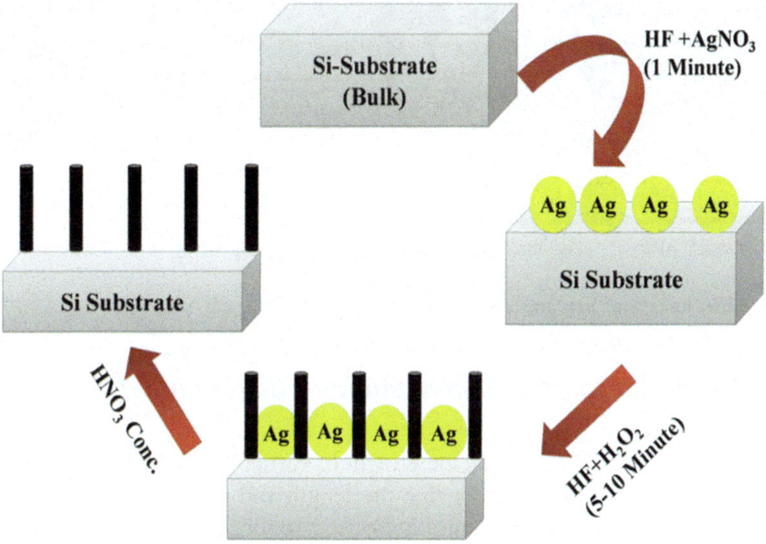

Fig. 9 Synthesis of SiNWs via EMACE method

The reaction is carried out in a dark room. AgNPs mainly help in the catalytic reduction of H_2O_2 at the AgNPs/Si interface resulting in the generation of holes (h+) in the Ag catalyst. These generated holes are then injected into the Si wafer lying beneath the AgNPs, consequently weakening the Si–Si bond. These bonds are further attacked by the F^- ion supplied by HF and form a water-soluble compound H_2SiF_6 and thus resulting in the etching of Si.

(c) **Etching**

After the deposition of AgNPS, the Si samples are then dipped into an etching solution containing HF (4.8 M) and H_2O_2 (0.808 M) in the dark at room temperature. The etching time is varied from 5–10 min to vary the length of SiNWs.

(d) **Removal of Silver**

The etched Si wafers are then put in concentrated nitric acid for 30 min to remove the AgNPs.

(e) **Tapering**

The SiNWs obtained after the etching is highly dense and get agglomerated on the tip. The agglomeration of SiNWs gives rise to bundle-like structures. The bunching of SiNWs deteriorates the various properties of SiNWs. Therefore, KOH tapering of SiNWs is done to reduce the density and make the SiNWs more conical such that they can absorb a wide range of solar radiations. The tapering solution contained IPA and KOH (20%) solution. Tapering time was varied from 10 to 100 s. The overall synthesis procedure is shown in Fig. 10.

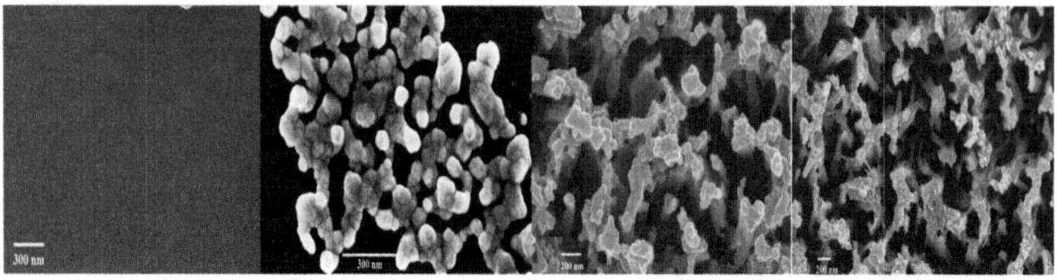

Fig. 10 SEM images of Si wafer undergoing etching to produce SiNWs

2.1 Morphological Characterization of SiNWs by SEM

The SEM image of SiNWs shows the growth of vertically aligned SiNWs with high density. It has been studied earlier that the length and diameter of SiNWs depend upon the etching time and concentration of the silver nitrate solution used for the electroplating of AgNPs on the surface of polished Si surface. With the increase in the size of AgNPs the diameter of SiNWs decreases. For the etching time, it has a direct relation with the length of SiNWs. As the etching time increases the length of SiNWs also increases. Figure 11 shows the SEM images of the microstructures of SiNWs synthesized by taking 20 and 25 mM of $AgNO_3$ concentration and performing etching for 5–10 min. The length of SiNWs obtained was ~ 2 μm when the etching time was 5 min and was increased to 5 μm when the wafers were etched for 10 min. Therefore, from the results obtained, we conclude that both $AgNO_3$ concentration and etching time has a profound effect on the diameter and length of SiNWs. It has also been observed that as the AgNP size increases the diameter of SiNWs decreases, which favours the obtained results as given above. The diameter of SiNWs synthesized by taking 20 mM $AgNO_3$ concentration lies in the range of 40–125 nm whilst in the case of 25 mM $AgNO_3$ concentration, the diameter is in the range of 30–70 nm.

Another important aspect we observed is the agglomeration in nanowires with the increasing etching time. As the length of the SiNWs increases, the nanowires tend to form bundle-like structures due to the capillary forces on SiNWs acting on each other. The surface of SiNWs contains many dangling bonds due to the unfulfilled valency of Si. As the length increases, these forces become more dominant, resulting in an agglomeration in SiNWs, giving rise to bundled structures. This can be seen from the SEM images of SiNWs, as shown in Fig. 12.

The agglomeration of SiNWs will harm its various properties such as optical, electrical, and photovoltaic properties. Moreover, the agglomeration of SiNWs and the low filling ratio will cause difficulty whilst preparing the core-shell heterojunction. Therefore, KOH tapering is chosen to tune the volume-filling ratio by increasing the space between SiNWs. The high filling ratio enables the penetration of foreign materials between the nanowires, thereby increasing the chance of getting the optimal quality of the heterojunction. Moreover, the KOH tapering on SiNWs also provides a chance to reduce the reflection phenomenon. It has been proven already that the

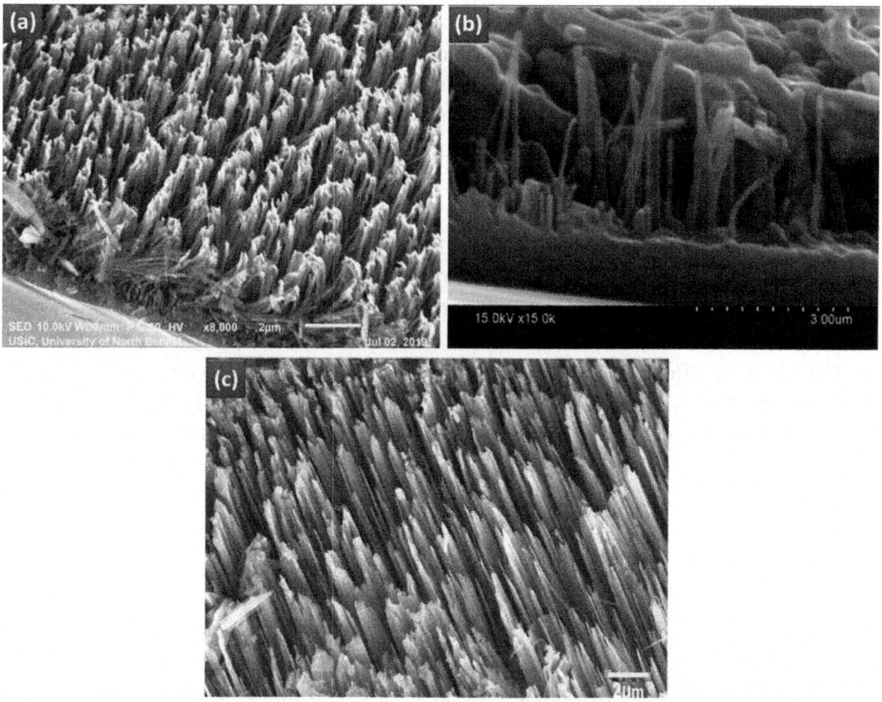

Fig. 11 SEM images of SiNWs synthesized by taking 20 mM of $AgNO_3$ concentration and etched for **a** and **b** for 5 min and **c** 10 min

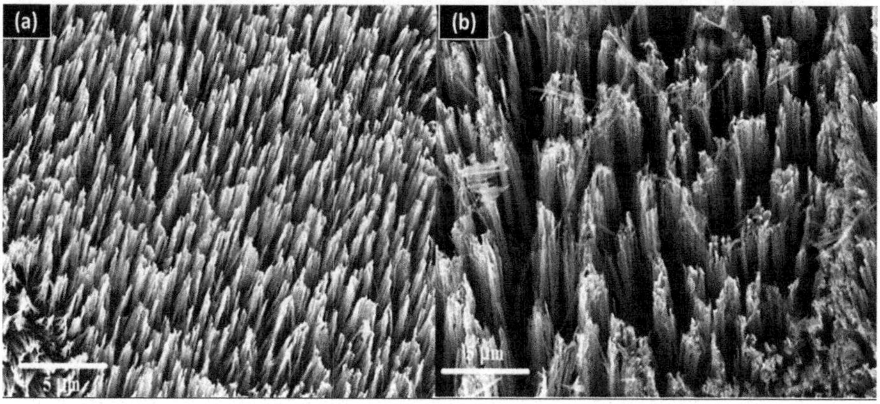

Fig. 12 SEM images of SiNWs with the $AgNO_3$ concentration of 25 mM and etched for **a** 5 and **b**10 min

KOH tapering on SiNWs makes the tip of nanowires sharper; consequently, the areal density of SiNWs gradually increases from top to bottom. This further leads to an increase in the refractive index from top to bottom. This is one of the crucial reasons for the reduction in reflectance of KOH tapered SiNWs. The next significant hurdle is choosing the tapering solution. If the tapering is performed only in KOH solution (e.g., 30% KOH solution), it has been observed that in 10 s the *n*-type SiNWs obtained by 2-min etching disappear completely. This might be due to the high defect sites present on the sidewalls of SiNWs. So it is hard to control the tapering results by precisely controlling the time. To overcome this problem, 31.2 g KOH is taken with 25 ml IPA in 100 ml of DI water to retard the etching process to some extent. The weight percentage of KOH is now 20%. The tapering experiments were performed on SiNWs only after completing the SiNWs synthesis and its cleaning in HNO_3. The tapering duration varied from 0–50 s. Figure 13 shows the SEM images of KOH tapered SiNWs. It is observed that by increasing the tapering time from 0–50 s, the density of SiNWs is reduced. From the SEM image of 30 s tapered SiNWs, it is clear that after tapering, some of the SiNWs appear to come out from the Si substrate due to an increase in the spacing between the nanowires. Moreover, the tip agglomeration is also reduced after performing the KOH tapering on SiNWs.

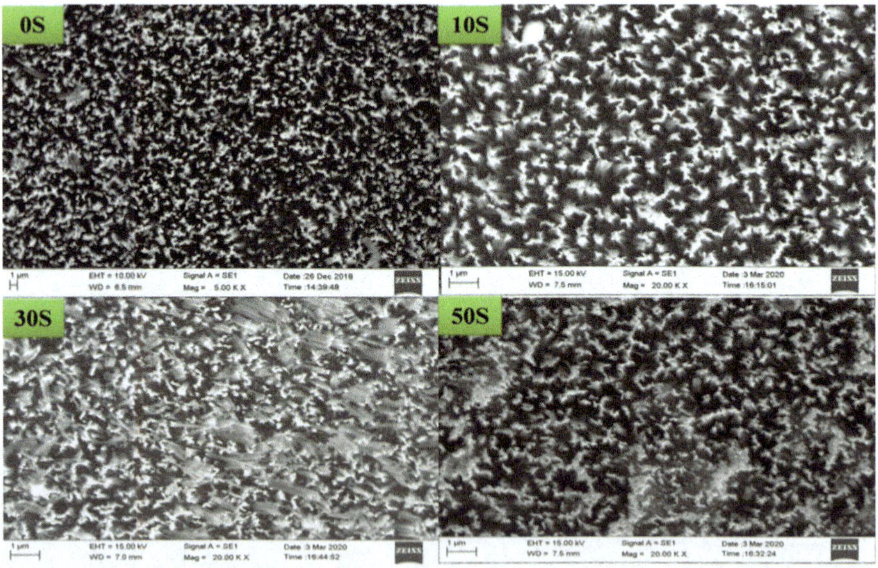

Fig. 13 SEM images of KOH tapered SiNWs

3 Chemical Synthesis of Reduced Graphene Oxide (rGO)

Synthesis of rGO via chemical route is one of the most popular and widely used methods due to its versatility, simplicity, and cost-effectiveness. This method allows the tuning of various graphene properties by changing the chemicals used and controlling the synthesis parameters. In this method, graphite is first oxidized to GO using oxidizing agents followed by its reduction to get the desired rGO sheets. The entire procedure is explained by the schematic as shown in Fig. 14.

(I) **Synthesis of GO**

GO was synthesized by following Hummer's method [20]. Graphite powder (1 g), potassium nitrate (0.5 g), and concentrated sulfuric acid (25 mL) were added to a 500 mL beaker. It was kept in an ice bath and continuously stirred for 5 min, which was followed by the slow addition of potassium permanganate (3 g). The reaction mixture was kept in an ice bath for 2 h. Then, the temperature was slowly raised to 35 °C, and vigorously stirred for 30 min. DI water (50 mL) was added dropwise to the mixture and stirred for 30 min. The reaction was completed by adding DI water (125 mL) and Hydrogen peroxide (10% v/v, 10 mL). The brown-colored product was separated by vacuum filtration. The GO thus formed was washed with dilute HCl (5%, 150 mL) solution and warm DI water.

(II) **Synthesis of rGO**

At first GO (50 g) was subjected to dispersion in 50 mL DI water in an ultrasonicator. When graphene oxide was fully dispersed, 10 ml of hydrazine hydrate was added to it. The reaction mixture was then stirred for an hour at a temperature of 90 °C. Finally, the black precipitate of rGO was collected using filter paper and washed several times with ethanol and DI water.

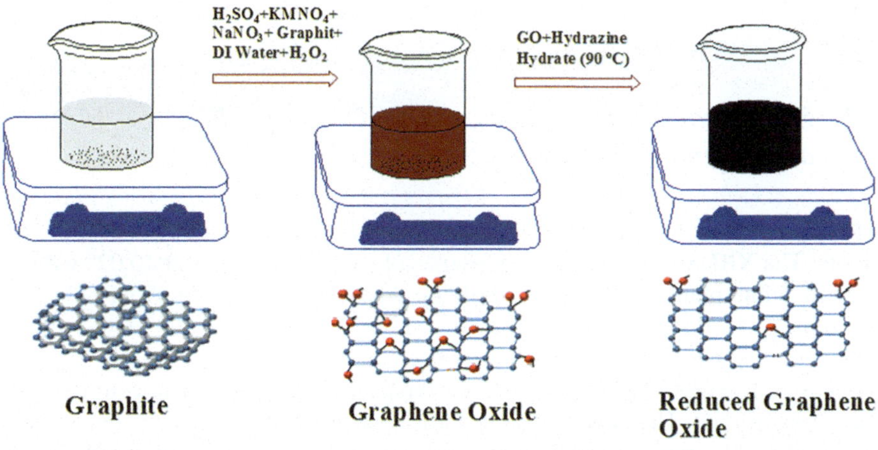

Fig. 14 Synthesis of rGO via chemical route [20]

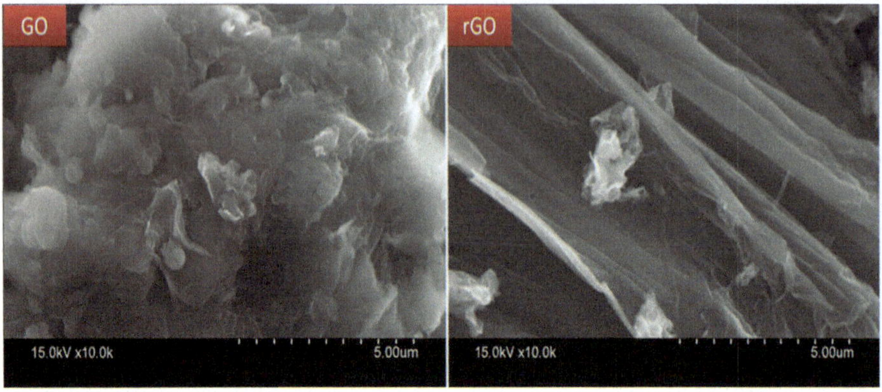

Fig. 15 SEM images of GO and rGO [21]

3.1 Morphological Characterization of rGO Sheet Using SEM

The SEM results of GO and rGO were synthesized by following Hummer's method as shown in Fig. 15. The SEM images illustrate the formation of a transparent sheet-like structure for both GO and rGO. The GO sheets have a wrinkled shape and are stacked one upon another as seen clearly in Fig. 15. Whilst in the case of rGO, the sheets are highly smooth, transparent, and wrinkle-free. The stacking and curling of the carbon sheets are caused due to the presence of an excessive number of oxygen-containing functional groups in GO. These functional groups can form bonds with the nearest functional groups present within the same or different GO sheets. This interaction between the functional groups present in different sheets causes stacking and interaction within the same sheet because of the twisting and curling of GO sheets.

3.2 Structural Characterization of rGO Using XRD and Raman Characterization

XRD and Raman are important tools to confirm the successful synthesis of rGO sheets. The XRD of GO shows the presence of a diffraction peak at 10.02° with the corresponding plane of (001) and a definite *d*-spacing of 0.85 nm. After reduction, the peak of GO disappears and a new band at ~ 24° appears for rGO with the corresponding plane of (002) and the reduced interlayer distance of ~ 0.36 nm. The reduction of the (001) plane and the appearance of the (002) plane confirmed the formation of rGO. The decrease in the lattice spacing of rGO after the reduction of GO is caused by removing the oxygen-containing functional groups bonded at the

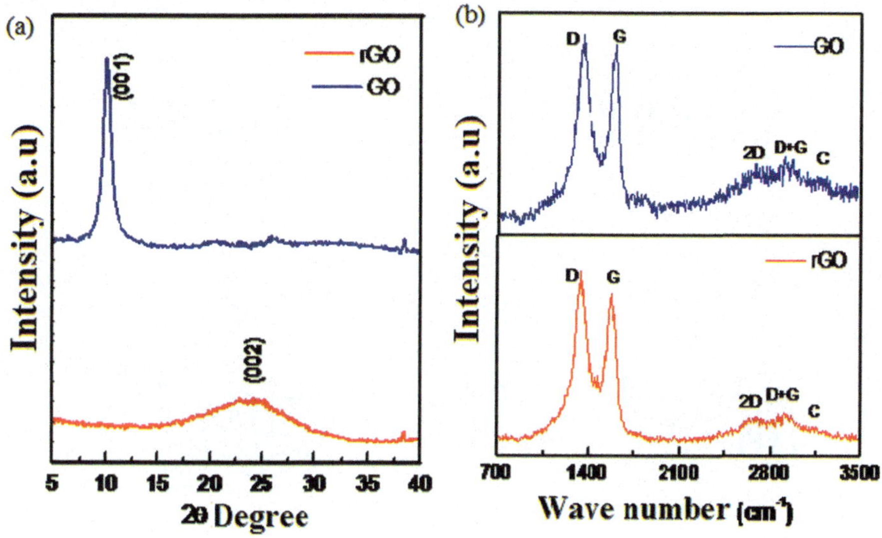

Fig. 16 **a** XRD pattern, and **b** Raman spectra of GO and rGO [23]

surface of the graphene layers. Similarly, the appearance of *D*, *G*, and 2D bands in the Raman spectra confirms the successful formation of rGO. The D band corresponds to the extent of disorder present in the graphene sheets and the in-plane vibration of sp^2 hybridized *C* atoms of the hexagonal *C* ring within the graphene sheets is shown by the G peak, which arises due to the vibration of LO phonon modes at the Γ point [22]. The second-order peaks are the result of combination mode frequencies of primary phonons. The double resonant process of *D* (LO) phonons gives rise to 2D bands (Fig. 16).

4 Fabrication of SiNWs Hybrid Solar Cell

Hybrid solar cells are a combination of organic and inorganic parts. In SiNWs hybrid solar cell, the inorganic part constitutes SiNWs and the organic part contains various organic conducting polymers such as PEDOT:PSS, P_3HT:PCMB, rGO- PEDOT:PSS composite, etc., in our study. We chose the drop-cast method for the deposition of an organic layer on the surface of SiNWs. The contacts of the solar cell were made of silver or aluminium (Fig. 17).

(I) Deposition of Organic Layer

The dimension of the substrate used was $1 \times 1\ cm^2$. The organic layer was deposited on the cleaned substrate by the drop-casting method. The thickness of the layers was varied by changing the number of drops on the substrate. The layer was then dried

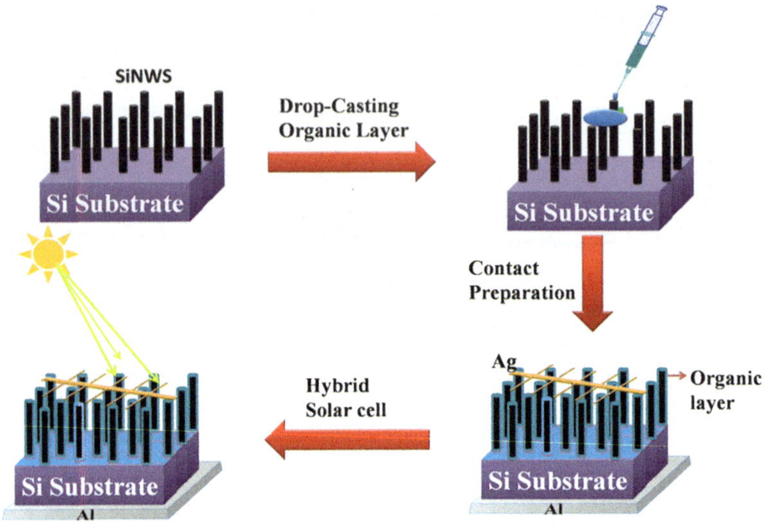

Fig. 17 Schematic representation for the fabrication of SiNWs hybrid solar cell

and annealed at a temperature of 110 °C for 30 min. The organic layer used was PEDOT:PSS and rGO incorporated PEDOT:PSS.

(II) **Contact Preparation**

The back contact of the heterojunction was made by depositing the Ag/Al layer by thermal evaporation method and the front contact was made using Ag paste. The overall schematic of the heterojunction preparation is shown in Fig. 3.5

4.1 J-V *Characterization of SiNWs Hybrid Solar Cells*

The electrical characterization of the solar cells is done using the Current Density–Voltage (*J*-*V*) curves. At first to study the effect of SiNWs in solar cells two heterojunctions Ag/PEDOT:PSS/*n*-Si/Al and Ag/PEDOT:PSS/SiNWs/*n*-Si/Al were prepared and their *J*-*V* characterization was done. The solar cells are represented in Fig. 18a and b. The solar cell dimension was $1 \times 1\ cm^2$.

The photo *J*-*V* characteristic curve of Ag/PEDOT:PSS/*n*-Si/Al and Ag/PEDOT:PSS/SiNWs/*n*-Si/Al heterojunctions is taken under the illumination of 1.5 AM and solar radiation of $100\ mW/cm^2$. Since the sample size was minimal ($1\ cm^2$), the obtained efficiencies of the heterojunctions are also very low. The obtained power conversion efficiency for Ag/PEDOT:PSS/*n*-Si/Al is found to be 0.0001%, whilst for Ag/PEDOT:PSS/SiNWs/*n*-Si/Al, the efficiency is found to be 0.01%. Therefore, from the above results, it is clear that the synthesis of SiNWs is essential to enhance solar cell efficiency. The enhancement is mainly caused due to the lower reflectivity

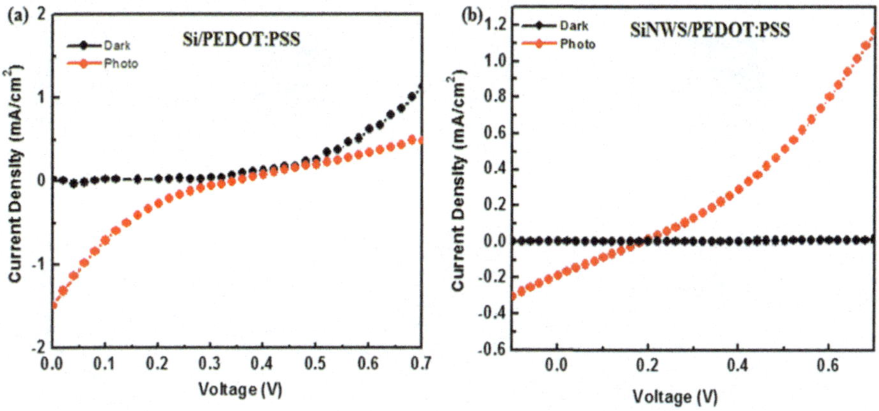

Fig. 18 Dark and Photo *J*-*V* characteristic curves for **a** Si/PEDOT:PSS **b** SiNWs/PEDOT:PSS hybrid solar cell

in SiNWs than that of bulk Si wafer. This is also due to the increased conductivity in SiNWs due to the conversion of the indirect bandgap of Si to direct bandgap in SiNWs.

4.2 *Effect of rGO Incorporation in PEDOT:PSS as the Hole Transport Layer in SiNWs/Organic Hybrid Solar Cell*

Figure 19c shows the *J*-*V* characteristic curve of Ag/rGO-PEDOT:PSS/SiNWs/*n*-Si/Ag and Ag/PEDOT:PSS/SiNWs/*n*-Si/Ag solar cells. In Table 6.2, the solar cell characteristics of device performance are given. The solar cell with only PEDOT:PSS as the conducting *p*-layer obtained V_{oc} of 0.22 V, J_{sc} of 2.46 mA/cm^2, FF of 0.19, and PCE of 0.1%. The solar cell fabricated with rGO-PEDOT:PSS composite as the *p*-type conducting layer led to a significant increase in the J_{sc} and the value of FF from 2.46 to 4.46 mA/cm^2, and 0.19–0.24, respectively. Hence, the PCE of Ag/rGO-PEDOT:PSS/SiNWs/*n*-Si/Ag solar cells increases from 0.1 to 0.22%. The incorporation of rGO in PEDOT:PSS efficiently enhances the efficiency of the solar cell due to the collection of more electrons from the junction; thereby reducing the recombination phenomenon. The incorporation of rGO provides an external electron transfer pathway by improving the carrier separation and transportation processes [24]. It is observed that the PCE of the device is low, which is due to the small sample size (1 cm^2) and also poor band edge matching of the different layers in the as-fabricated solar cells.

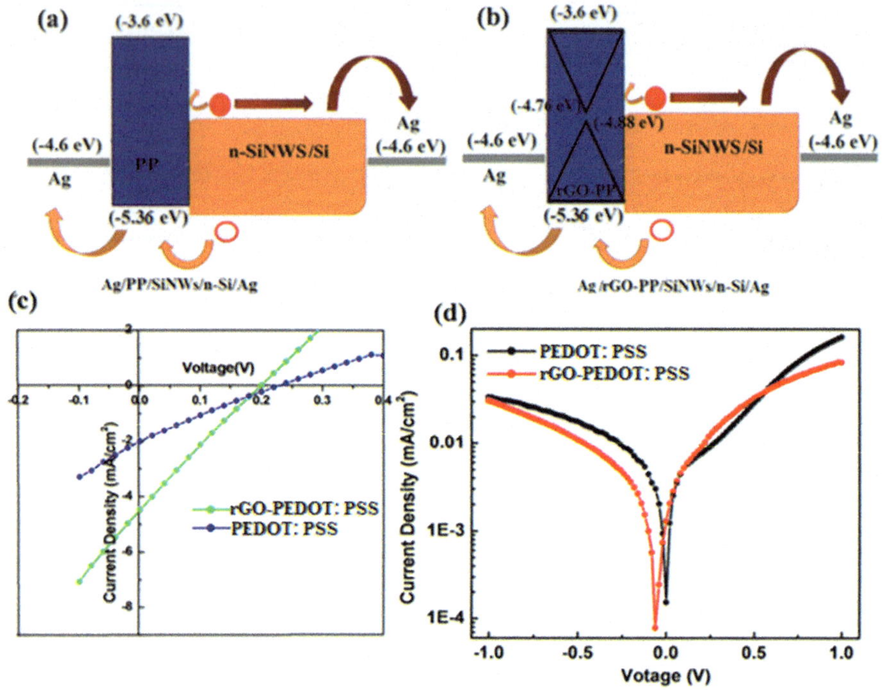

Fig. 19 Band structure diagram of **a** Ag/*n*-Si/SiNWs/ PEDOT:PSS (PP)/Ag, **b** Ag/*n*-Si/SiNWs/ rGO-PEDOT:PSS/Ag solar cells, **c** photo and **d** dark *J*-*V* characteristic curves of solar cells [25]

5 Conclusion

In this book chapter, we have studied the fabrication and characterization of different types of SiNWs hybrid solar cells. The conducting polymers used were PEDOT:PSS and rGO-PEDOT:PSS. The morphological characterizations of the samples were done by SEM characterization. The SEM images of SiNWs show growth of vertically aligned SiNWs with high density. KOH tapering on SiNWs is performed to reduce the density of SiNWs which further helps improve the solar cell efficiency. SEM images of rGO show the appearance of transparent sheets of GO and rGO. The XRD and Raman are important tools for the confirmation of rGO formation. *J*-*V* characterization is done to obtain solar cell efficiency. It was observed that an efficiency of 0.22% was obtained for the rGO-PEDOT:PSS composite whilst for PEDOT:PSS the efficiency was just 0.1%. Hence the incorporation of rGO in PEDOT:PSS can improve the photovoltaic properties by increasing the conductivity of the hole transport layer, making a good interface between organic–inorganic heterojunctions as well as by reducing the recombination of electron–hole pairs.

References

1. Wu HW, Emadi A, De Graaf G, Leijtens J, Wolffenbuttel RF (2011) Design and fabrication of an albedo insensitive analog sun sensor. Procedia Eng 25:527–530. https://doi.org/10.1016/j.proeng.2011.12.131
2. No Title (2012)
3. 08- 267 (2002) 53706
4. Zhu M, Zhu M (2014) Silicon nanowires for hybrid solar cells to cite this version : HAL Id : tel-00945787 Silicon nanowires for hybrid solar cells Nanofils de silicium pour les cellules solaires hybrides
5. Shockley W, Shockley W, Queisser HJ (1961) The Shockley-Queisser limit 1–5
6. de Santiago F, Trejo A, Miranda A, Carvajal E, Pérez LA, Cruz-Irisson M (2017) Band-gap engineering of halogenated silicon nanowires through molecular doping. J Mol Model 23(11). https://doi.org/10.1007/s00894-017-3484-8
7. Fonash SJ (2010) Solar cell device physics. Solar Cell Device Phys. https://doi.org/10.1016/C2009-0-19749-0
8. Dittrich T (2018) Basic characteristics and characterization of solar cells. Mater Concepts Solar Cells 3–43. https://doi.org/10.1142/9781786344496_0001
9. He LR, Jiang C, Wang H, Lai D (2011) Simple approach of fabricating high efficiency Si nanowire/conductive polymer hybrid solar cells. IEEE Electron Device Lett 32(10):1406–1408. https://doi.org/10.1109/LED.2011.2162222
10. Liu R, Wang J, Sun T, Wang M, Wu C, Zou H, Song T, Zhang X, Lee ST, Wang ZL, Sun B (2017) Silicon nanowire/polymer hybrid solar cell-supercapacitor: a self-charging power unit with a total efficiency of 10.5%. Nano Lett 17(7):4240–4247. https://doi.org/10.1021/acs.nanolett.7b01154
11. Liao SH, Jhuo HJ, Yeh PN, Cheng YS, Li YL, Lee YH, Sharma S, Chen SA (2014) Single junction inverted polymer solar cell reaching power conversion efficiency 10.31% by employing dual-doped zinc oxide nano-film as cathode interlayer. Sci Rep 4:4–10. https://doi.org/10.1038/srep06813
12. Eom SH, Senthilarasu S, Uthirakumar P, Yoon SC, Lim J, Lee C, Lim HS, Lee J, Lee SH (2009) Polymer solar cells based on inkjet-printed PEDOT:PSS layer. Org Electron 10(3):536–542. https://doi.org/10.1016/j.orgel.2009.01.015
13. Thomas JP, Leung KT (2014) Defect-minimized PEDOT:PSS/planar-si solar cell with very high efficiency. Adv Func Mater 24(31):4978–4985. https://doi.org/10.1002/adfm.201400380
14. Sun K, Zhang S, Li P, Xia Y, Zhang X, Du D, Isikgor FH, Ouyang J (2015) Review on application of PEDOTs and PEDOT:PSS in energy conversion and storage devices. J Mater Sci: Mater Electron 26(7):4438–4462. https://doi.org/10.1007/s10854-015-2895-5
15. Uma DGV, Gopalakrishnan SR (2019) Supercapacitor behavior and characterization of RGO anchored V 2 O 5 nanorods. J Mater Sci: Mater Electron 0123456789. https://doi.org/10.1007/s10854-019-01984-9
16. Manuscript A (2010) Nanoscale 207890
17. Sarkar A, Chakraborty AK, Bera S (2018) NiS/rGO nanohybrid: an excellent counter electrode for dye sensitized solar cell. Solar Energy Mater Solar Cells, 182(August 2017):314–320. https://doi.org/10.1016/j.solmat.2018.03.026
18. Ates M, Bayrak Y, Ozkan H, Yoruk O, Yildirim M, Kuzgun O (2019) Synthesis of rGO/TiO 2 / PEDOT nanocomposites, supercapacitor device performances and equivalent electrical circuit models. J Polym Res 26(2):1–16. https://doi.org/10.1007/s10965-018-1692-2
19. Krishna R, Fernandes DM, Ventura J, Freire C, Titus E (2016) Novel synthesis of highly catalytic active Cu@Ni/RGO nanocomposite for efficient hydrogenation of 4-nitrophenol organic pollutant. Int J Hydrogen Energy 41(27):11608–11615. https://doi.org/10.1016/j.ijhydene.2015.12.027
20. Yu H, Zhang B, Bulin C, Li R, Xing R (2016) High-efficient synthesis of graphene oxide based on improved hummers method. Sci Rep 6(November):1–7. https://doi.org/10.1038/srep36143

21. Das S, Ghadai R, Krishna A, Trivedi A, Bhujel R, Rai S, Shivakoti I, Kalita K (2020) A comparative analysis of structural and electrochemical properties of graphene oxide (Go) and reduced graphene oxide (rgo) synthesized by using hummer's and modified hummer's method. Iranian J Mater Sci Eng 17(4):25–32. https://doi.org/10.22068/ijmse.17.4.25
22. Ferrari AC (2007) Raman spectroscopy of graphene and graphite : disorder , electron – phonon coupling, doping and nonadiabatic effects 143:47–57. https://doi.org/10.1016/j.ssc.2007.03.052
23. Rai S, Bhujel R, Swain BP (2018) Electrochemical analysis of graphene oxide and reduced graphene oxide for super capacitor applications. In: Proceedings of international conference on 2018 IEEE electron device Kolkata conference, EDKCON 2018, pp 489–492. https://doi.org/10.1109/EDKCON.2018.8770433
24. Hilal M, Han JI (2018) Interface engineering of G-PEDOT: PSS hole transport layer via interlayer chemical functionalization for enhanced efficiency of large-area hybrid solar cells and their charge transport investigation. Sol Energy 174(September):743–756. https://doi.org/10.1016/j.solener.2018.09.031
25. . Bhujel R, Rai S, Deka U, Sarkar G, Biswas J (2021) Bandgap engineering of PEDOT : PSS/rGO a hole transport layer for SiNWs hybrid solar cells. Bull Mater Sci 0123456789. https://doi.org/10.1007/s12034-021-02376-8

Chapter 6
Proton Mobility in Solid Electrolyte: The Heart of Fuel Cell

Bibek Kumar Sonu, Gayatri Dash, and Ela Rout

1 Introduction

Most of the energy systems today are based on utilizing fossil fuels, which is a significant contributing factor to a wide variety of environmental issues, including air pollution, emissions of greenhouse gases, and acid rain. In addition, a significant quantity of energy is used up in the production of power (electricity) and the movement of people/goods, both of which are the foundations and necessities of contemporary existence. These two areas are inextricably linked to the massive use of energy. The rise in industrial production has resulted in a corresponding rise in the need for energy, the excessive use of energy resources, and emissions of carbon dioxide (CO_2) [1, 2]. Every area of the global economy is impacted when the production and delivery are inefficient of energy. The need for more energy is a direct consequence of significant technological advances. In the search for a solution to the growing energy dilemma, the issue of how energy might be utilized on earth more effectively while still maintaining a clean environment has become more important. When working to find solutions to these issues, every endeavor should prioritize the conservation of energy resources and net zero emissions of greenhouse gas. This issue may be expressed in terms of reduced energy usage in developed nations and increased usage of improved energy systems on a mass scale. These needs might be stated as "extremely efficient technologies in power generation and transportation."

Solid-state ionic devices, e.g. Solid Oxide Fuel Cells (SOFCs), are potential energy conversion and storage devices that have the potential to tackle certain environmental challenges while also reducing resource consumption and creating job opportunities. The SOFCs are anticipated to play a significant role both in stationary and portable energy production besides mobility in the future decades.

B. K. Sonu · G. Dash · E. Rout (✉)
Department of Physics, Birla Institute of Technology, Mesra, Ranchi 83521, India
e-mail: elasinha@bitmesra.ac.in

B. P. Swain (ed.), *Energy Materials*, Materials Horizons: From Nature to Nanomaterials,
https://doi.org/10.1007/978-981-99-3866-7_6

The solid oxide fuel cell is presently drawing a great deal of attention due to the enormous potential it has to improve the efficiency of energy conversion, as well as dependability and security, and to lessen the detrimental effects on the environment. In addition to this, one of the key benefits of SOFCs over other kinds of fuel cells is that they can use a wide variety of fuels. The usage of specialized high-temperature ceramic materials brings with it significant financial burdens, which is the primary barrier to the commercialization of SOFCs.

The high operating temperature that is necessary to achieve excellent ionic conductivity with standard electrolytes is a critical factor in the utilization of SOFCs. Additionally, working temperatures that are greater than 800 °C urge the utilization of costly ceramic interconnects that are made on lanthanum chromite, while also considering shortening the lifespan of these devices. By utilizing electrolytes with higher conductivity levels even at low temperatures, it is possible to obtain practical working temperatures ranging from 500 to 800 °C or even below. This study aims to provide an overview of the current research on proton diffusion that is currently being conducted in the field of electrolytes for Intermediate Temperature Solid Oxide Fuel Cells (IT-SOFCs) or Proton Conducting Solid Oxide Fuel Cells (PC-SOFCs).

2 Operating Principles of Fuel Cell

A Fuel Cell is a device for the conversion of energy that generates electricity by the electrochemical combination of a fuel (such as hydrogen, methanol, carbon monoxide, gasoline, etc.) and an oxidant in the presence of an ionic conducting electrolyte. A dense electrolyte is the fundamental component of a fuel cell. This electrolyte is located between two porous electrodes, which are known as the anode and the cathode, respectively is referred to as a single cell as shown in Fig. 1. It takes use of the fact that oxygen and hydrogen have a natural inclination to react to generate water. The fundamental operating principle of a fuel cell involves three-step process. The first stage is the oxidation of fuel at the anode, which then releases electrons into the external circuit. The movement of ions (H^+, OH^-, O^{2-}, and CO_3^{2-}) across the electrolyte is the second step to consider. And thirdly, the oxidant is sent to the cathode, where it participates in a reduction process by taking electrons from the external circuit and undergoing reduction. The movement of electrons from the anode to the cathode in the external circuit is what generates the electric current.

When it is put into action, the SOFC has the flexibility to make use of either an electrolyte that conducts oxygen ions or one that conducts protons. The majority of recent research efforts have been concentrating on the PC-SOFC that uses the proton (H^+) conducting electrolyte as opposed to the OC-SOFC that uses the oxygen ion (O^{2-}) conducting electrolytes. A driving force for the migration of the ions inside the electrolyte is provided by the differential in ion activity that exists between the two electrodes. In the case of OC-SOFC, oxide ions are produced as a result of the dissociation of oxygen at the cathode under the influence of electron consumption.

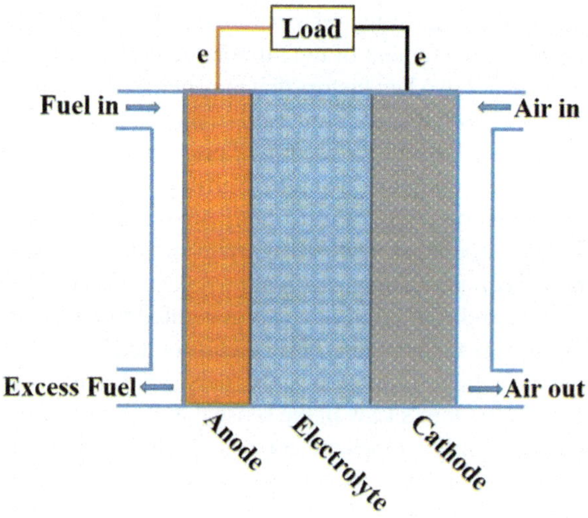

Fig. 1 A single-cell fuel cell

These oxide ions then migrate through the electrolyte to the anode, where they react with the oxidation products to produce water, heat, and carbon dioxide. In the case of PC-SOFC, protons (H^+) are produced at the anode, and then they migrate through the electrolyte to the cathode.

As long as the reactants are supplied to the electrodes, the cell will continue to generate current under the operating parameters that it was designed for. When the cell is unloaded and the ionic transport number is approximately 100%, the open-circuit voltage will be somewhat higher than one volt. This will be the case when the cell is not loaded. The ideal potential of an open-circuit cell is denoted by the symbol E, which stands for the Nernst potential. The upper limit or maximum voltage that can be achieved by a fuel cell is determined by this potential. When a fuel cell is coupled to an external load, the potential difference that exists between the fuel cell's electrodes is what propels the electrons through the external circuit. Nevertheless, the nature of the processes involved in the electrochemical system is such that they are highly complicated. Even though a great number of research have been conducted on the topic, the process of charge transfer across the interfaces created by solid electrolytes and electrodes continues to be one of the elements of electrochemistry that are least well understood.

The functioning of the cell is linked to several irreversibilities and results in several possible losses. When it comes to electrodes, the overall resistance is made up of the internal resistance, as well as the contact resistance, the activation resistance, and the mass transport resistance. The term "internal resistance" refers to the electron transport in the electrode as well as the ionic transport in the electrolyte, both of which are determined by the electronic conductivity in the electrode as well as the ionic conductivity in the electrolyte as well as the thickness of the electrode and electrolyte. The term "contact resistance" refers to the inadequate connection that exists between the electrode and the structure of the electrolyte. The local current density is a function that governs all of the resistive losses. A suitable selection of electrode material and careful management of the microstructural characteristics during the manufacturing process are two key factors that may help reduce the overpotential losses. The voltage output of a SOFC as a function of the electrical current density dissipated drawn in the external circuit is used as a performance index for SOFCs. This curve, which reflects a divergence from the Nernst potential, is known as the polarization curve.

As can be seen in Fig. 2, there are three distinct kinds of polarizations as well as zones of operation in which a certain loss mechanism is predominant in comparison to the other polarizations. Activation polarization is the primary loss term at open-circuit voltage and near it because it is generated by a slower response. Open-circuit voltage or OCV is associated with low current densities. At current moderate densities, the ohmic voltage reduction is the most important loss mechanism. Ohmic losses are subject to Ohm's Law in the sense that they proportionately increase with increasing current. Ohmic losses are correlated with the resistance, also known as area-specific resistance or ASR. This resistance is mostly brought on by the electrolyte and the interfacial surfaces. When there is a greater demand for current, concentration polarization becomes the primary source of overpotential. The measured voltage, E, may be expressed as

$$E = E_{\text{eq}} - E_L - \eta_{\text{act}} - \eta_{\text{ohm}} - \eta_{\text{diff}} \tag{1}$$

where E_{eq} represents the Nernstian voltage (at equilibrium), E_L represents the voltage loss attributable to leakage across the electrolyte, η_{act} represents the activation overpotential owing to a slower reaction rate at the electrode, η_{ohm} represents the overpotential as a result of ohmic resistances developed within the cell, and η_{diff} represents the overpotential caused by mass diffusion restrictions. Figure 2 illustrates how each of the elements in Eq. 1 contributes to the overall effect.

Fig. 2 Polarization in fuel cell (*V-I* curve) and power density

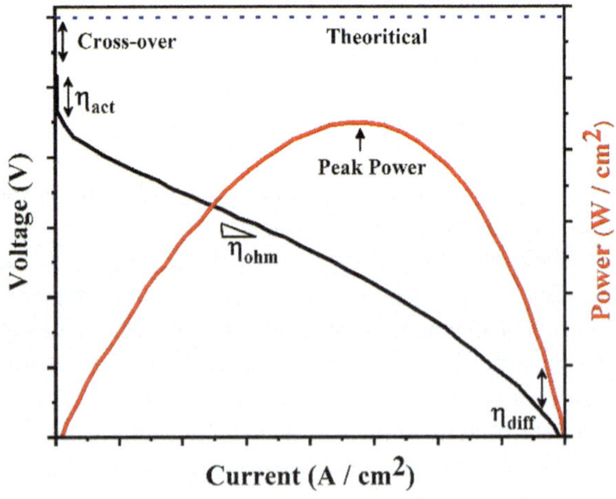

SOFCs provide several advantages over conventional methods of energy conversion, particularly high efficiencies (in the range of 50–60%), durability, flexibility, and exceptionally fewer emissions of localized vital air pollutants like CO, NO_x, and unburned hydrocarbons. These benefits may be attributed to the fact that in SOFCs, the chemical energy of the fuel is immediately converted into electrical energy. In this way, the typical losses that occur in traditional combustion systems throughout the transformation of fuel into heat, then into mechanical energy, and finally into electrical energy are avoided. SOFCs can be run on a variety of fuels, such as biomass and pipeline natural gas, without suffering a substantial decrease in efficiency or experiencing an increase in the complexity or expense of the system.

3 Fuel Oxidation

The oxidation of fuel (i.e. electrochemical reactions) occurs at the electrode–electrolyte interface, typically within the electrode at a distance less than 10–20 μm away from the surface of the electrolyte [3]. This section of the electrode is known as the functional/active layer. The portion of the electrode that is thicker than this threshold is essentially a current-collector structure that must be porous to ensure that gas can reach the active layer. Both the electrodes and the electrolyte are made of ceramic materials. To prevent direct gas mixing and their combustion, the electrolyte must be dense and gas-impermeable.

In the case of OC-SOFC, at the cathode-functional layer, oxygen (O_2) from the air is reduced to O^{2-}-ions. After that, these O^{2-}-ions migrate toward the anode passing through the electrolyte (O^{2-}-ion conductor). At the anode-functional layer, these ions combine with the hydrogen (H^+-ions) to form water via a chemical reaction, as can be seen in Fig. 3. As a consequence of this reaction, electrons are liberated and flow to the cathode via the external circuit.

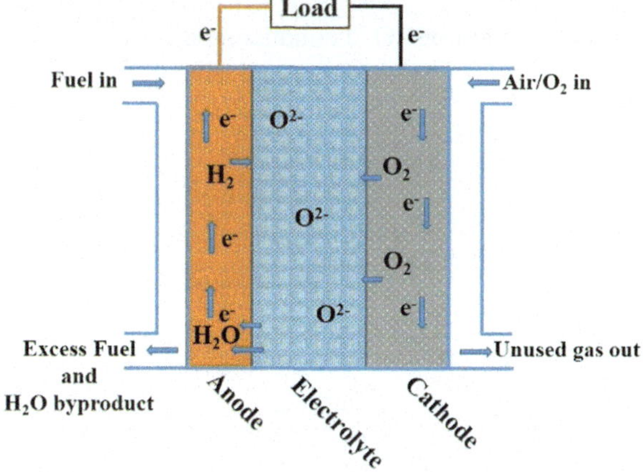

Fig. 3 Oxygen ion conducting SOFC

Fig. 4 Proton conducting SOFC

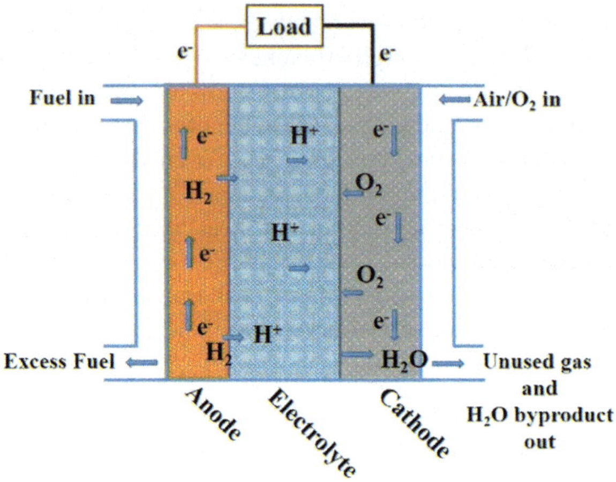

Similarly, in the case of PC-SOFC, at the anode-active layer, hydrogen (H_2) is oxidized to H^+-ions. Then, these H^+-ions move through the electrolyte (H^+-ions conductors) toward the cathode. At the cathode-active layer, these ions react with the oxygen (O^{2-}-ion) to form water as can be seen in Fig. 4. Consequently, electrons are liberated and flow to the cathode through the external circuit.

3.1 Ions Formation

The precise reactions that take place at the electrodes are determined by the types of species that are transferred/carried across the electrolyte (as discussed in the previous section).

The reaction mechanism is as follows:

When hydrogen (H_2) is used as fuel, the overall reactions are summarized below:

Case-I: Oxide-ions (O^-) conducting electrolyte

$$\text{At Anode} : 2H_2 + 2O^{2-} \rightarrow 2H_2O + 4e^- \tag{2}$$

$$\text{At Cathode} : O_2 + 4e^- \rightarrow 2O^{2-} \tag{3}$$

$$\text{Overall cell reaction} : 2H_2 + O_2 \rightarrow 2H_2O \tag{4*}$$

Case-II: Proton (H^+) conducting electrolyte

$$\text{At Anode}: 2H_2 \rightarrow 4H^+ + 4e^- \tag{5}$$

$$\text{At Cathode}: O_2 + 4e^- + 4H^+ \rightarrow 2H_2O \tag{6}$$

$$\text{Overall cell reaction: } 2H_2 + O_2 \rightarrow 2H_2O \tag{4*}$$

Case-III: Hydroxide OH^--ion conducting electrolyte

$$\text{At Anode: } 2H_2 + 4OH^- \rightarrow 4H_2O + 4e^- \tag{7}$$

$$\text{At Cathode: } O_2 + 4e^- + 2H_2O \rightarrow 4OH^- \tag{8}$$

$$\text{Overall cell reaction: } 2H_2 + O_2 \rightarrow 2H_2O \tag{4*}$$

Similarly, the total reactions when CO and hydrocarbons are utilized as fuels may be depicted as follows:

$$\text{At Anode: (CO as fuel)} 2CO + 2CO_3^{2-} \rightarrow 4CO_2 + 4e^- \tag{9a}$$

$$(H_2 \text{ as fuel}) 2H_2 + 2CO_3^{2-} \rightarrow 2H_2O + 2CO_2 + 4e^- \tag{9b}$$

$$\text{At Cathode: } 2CO_2 + O_2 + 4e^- \rightarrow 2CO_3^{2-} \tag{10}$$

$$\text{Overall cell reaction: } 2H_2 + O_2 \rightarrow 2H_2O \tag{4*}$$

$$\text{At Anode: } CH_4 + 4O^{2-} \rightarrow 2H_2O + 2CO_2 + 8e^- \tag{11}$$

$$\text{At Cathode: } 2O_2 + 8e^- \rightarrow 4O^{2-} \tag{12}$$

$$\text{Overall cell reaction: } 2H_2 + O_2 \rightarrow 2H_2O \tag{4*}$$

$$\text{At Anode}: 2CH_3OH + 2H_2O \rightarrow 2CO_2 + 12H^+ + 12e^- \tag{13}$$

$$\text{At Cathode: } 3O_2 + 12H^+ + 12e^- \rightarrow 6H_2O \tag{14}$$

$$\text{Overall cell reaction: } 2CH_3OH + 3O_2 \rightarrow 4H_2O + 2CO_2 \tag{15}$$

*(4) Since the overall cell reaction is the same, same number is used for all equations.

3.2 *The Fuel Cell Potential*

The OCV of a fuel cell is the highest working voltage (when no current is drawn) and is governed by the thermochemistry of the complete cell reaction. The Nernst equation establishes a relation between the OCV and the standard potential (E_T^o), for the chemical reaction occurring within the cell. In other words, if the standard potential at the desired temperature (marked by the subscript T) is known, the open-circuit voltage for various partial pressures of reactants and products at that temperature may be calculated.

The generalized Nernst equation used to calculate the OCV for the overall cell reaction of any fuel cell may be given as

$$E_T = E_T^o + \left(\frac{\mathrm{RT}}{\mathrm{nF}}\right)\ln\frac{\prod\left[(P_{\mathrm{reactants}})^x\right]}{\prod\left[(P_{\mathrm{products}})^y\right]} \tag{16}$$

where,

E_T = open-circuit voltage (OCV).
E_T^o = standard potential of the reaction.
R = universal gas constant, 8.314 J K^{-1} mol^{-1}
T = absolute temperature of the cell, K
F = Faraday's number, 94,487 C/equiv.
n = no. of electrons involved (equiv/mol).
$P_{\mathrm{reactants}}, P_{\mathrm{products}}$ = partial pressure or activity of the species involved.
x, y = stoichiometric coefficients.

Conventionally, the OCV (E_T) of the cell is calculated as the potential difference between the cathode and anode potential drops at a certain temperature T.

$$E = E_C - E_A \tag{17}$$

Now, using Eq. (16) we can estimate the OCV for PC-SOFC (Case II) i.e. for hydrogen and oxygen reaction that takes place at the anode and cathode respectively, as.

Using equilibrium reaction Eq. 5, we get

$$E_A = E_A^o + \left(\frac{\mathrm{RT}}{4F}\right)\ln\frac{(P_{H^+})^4}{(P_{H_2})^2} \tag{18}$$

To simplify, we have

$$E_A = E_A^o + \left(\frac{\mathrm{RT}}{4F}\right)\left[4\ln(P_{H^+}) - 2\ln(P_{H_2})\right] \tag{19}$$

Similarly, using equilibrium reaction Eq. 6, we get

$$E_C = E_C^o + \left(\frac{\mathrm{RT}}{4F}\right)\ln\frac{(P_{O_2})^1(P_{H^+})^4}{(P_{H_2O})^2} \tag{20}$$

Simplifying, we have

$$E_C = E_C^o + \left(\frac{\mathrm{RT}}{4F}\right)\left[4\ln(P_{H^+}) + \ln(P_{O_2}) - 2\ln(P_{H_2O})\right] \tag{21}$$

Now, from Eq. (17) E is the difference between Eqs. (21) and (19). Thus, we have

$$\begin{aligned} E = &\left[E_C^o + \left(\frac{\mathrm{RT}}{4F}\right)\left[4\ln(P_{H^+}) + \ln(P_{O_2}) - 2\ln(P_{H_2O})\right]\right] \\ &- \left[E_A^o + \left(\frac{\mathrm{RT}}{4F}\right)\left[4\ln(P_{H^+}) - 2\ln(P_{H_2})\right]\right] \end{aligned} \tag{22}$$

Simplifying, we get

$$E = E^o + \left(\frac{\mathrm{RT}}{4F}\right)\left[2\ln(P_{H_2}) + \ln(P_{O_2}) - 2\ln(P_{H_2O})\right] \tag{23}$$

This is equivalent to

$$E = E^o + \left(\frac{\mathrm{RT}}{2F}\right)\ln\frac{(P_{H_2})(P_{O_2})^{1/2}}{(P_{H_2O})} \tag{24}$$

Therefore, the Nernst equation may also be used to calculate the open-circuit voltage E between the electrodes of an ideal fuel cell from the cell reaction if the number of electrons is known.

3.3 Standard Potential

The basic process in which we may determine a cell's standard potential at any temperature is described as follows.

(i) When the activity ratio of all species is equal to unity $\left(\frac{P_{\text{reactants}}}{P_{\text{products}}} = 1\right)$, then the OCV is equal to the cell's standard potential.

$$\text{i.e. } E_T = E_T^o \tag{25}$$

(ii) When the species involved are not at unit activity, the standard potential at the required temperature may be utilized in the Nernst equation to calculate the OCV at that temperature.

In the overall cell reaction, the ion absorbs "n" moles of an electron against standard potential E_T^o. Then, the work done in electron movement is given by

$$W_T^o = \text{nFE}_T^o \tag{26}$$

The Gibbs free energy (ΔG) change indicates spontaneity and equals the maximum useful work (except volume expansion) in a process.

$$\text{i.e. } W_T^o = -\Delta G_T^o \tag{27}$$

$$\text{Or, } \Delta G_T^o = -\text{nFE}_T^o \tag{28}$$

$$\text{Therefore, } E_T^o = -\frac{\Delta G_T^o}{\text{nF}} \tag{29}$$

The free energy change can be calculated from the Gibbs–Helmholtz relation (beyond the scope of this chapter).

4 Electrolyte: The Core of Fuel Cells

The primary function of an electrolyte is to act as a medium that enables a particular ion to move freely between two electrodes in order to complete the electrochemical reaction of the cell. No significant current could pass through the fuel cell without that particular ion conducting, leaving just potential. The total performance of SOFCs is eventually constrained by the conductivity of the materials used for the electrolyte as fabrication technology advances. An electrolyte should ideally be both an electrical insulator and an ionic conductor. The most challenging conditions for SOFC electrolytes to function in are high temperatures, oxygen on the cathode side, hydrocarbons or hydrogen on the anode side, and the formation/presence of water at elevated temperatures.

Since electrolytes are at the heart of SOFCs, there are certain requirements that the materials must fulfill to function effectively as components in SOFCs: It must be an electrical insulator but a good ionic conductor at operating temperatures, conducting ions but not electrons (which pass through an external circuit), being extremely dense or be free of porosity to prevent gases from passing through it from one side to the other (i.e. gas impermeable), being stable in reducing and oxidizing environments, having both long-term chemical and thermal stability, mechanical adaptability with the electrode materials, being resistant to thermal shocks, and having an extensive surface area for maximum current capacity [4]. Additionally, to work with little ohmic loss, the electrolyte with these required qualities must ultimately be placed as thinly as feasible.

The following are some basic requirements of a solid electrolyte material that is intended for use in a SOFCs.

- Simple and easy manufacturing of dense, mechanically strong, and uniformly thin with large surface area to decrease bulk resistance.
- Ionic conductivity (σ) greater than 10^{-3} Scm^{-1} at the working temperature of the cell.
- Excellent compatibility with electrode materials both chemically and mechanically to reduce the interfacial resistances and prevent the development of blocking interface phases.
- The compatibility of the Thermal Expansion Coefficients (TEC) between the working temperature of the cell and the ambient temperature of the electrodes, electrolytes, interconnects, and seals.
- Reasonably cost-effective including the raw materials, fabrication process, and the finished product.

4.1 Oxide Ions Conductors

Intensive research has been done during the last two decades to explore the ionic conduction properties of oxides with different crystal structures for SOFC electrolyte applications. In oxide ion conductors, the flow of current is caused by the oxide ions' motion across the crystal lattice. Nernst demonstrated over a century ago that oxygen ion vacancies account for the majority of this movement [5]. Which is caused by the superimposed effect of thermally-activated oxygen ion hopping within the crystalline lattice, with the drift superimposed in the direction of the electric field. Due to this strong temperature dependence, the ionic conductivity can reach values near to 1 S cm^{-1} at high temperatures, which is equivalent to the same order of ionic conductivity observed in liquid electrolytes. This is indeed a significant characteristic possessed by these solids, and in order to comprehend its causes, we must make a few crucial observations. The first thing to notice is that there must be vacant sites inside the crystal that are equal to those that are filled by lattice oxygen ions. Second, the energy required to migrate from one site to its corresponding empty site must be negligible, and it must be less than around 1 eV.

Since the size of oxygen ions is big, there are only a few specific structures that make it feasible for oxygen ions migration. The fluorite-structured Zirconia (ZrO_2), Ceria (CeO_2), and Thoria (ThO_2) doped with trivalent rare earth are the most traditional high-temperature oxide ion conductors [6–8]. The fluorite crystal structure is characterized by the presence of a cubic oxygen lattice, in which alternating body centers are occupied by eight coordinated cations. There exists a face-centered cubic close packing of cations while anions occupy all the tetrahedral sites. These materials possess a large concentration of octahedral interstitial free-volume sites which helps in achieving high oxide ion diffusivity. Yttria-Stabilized Zirconia (YSZ) is commercially used as an oxide ion conducting electrolyte in high-temperature SOFC, which along with high oxide ion conductivity (10^{-3}–10^{-2} S cm^{-1}) displays excellent mechanical and chemical stability. However, the material has several issues with electrode–electrolyte incompatibility and high-temperature operating conditions (800–1000 °C) [9–11]. On the other hand, doped ceria displays better oxide ion conductivity than its zirconia counterpart, but an unavoidable electronic conductivity is observed in certain conditions which is one of its major drawbacks for practical applications.

4.2 Proton Conductors

In recent decades, protonic conducting oxide electrolytes have received a lot of attention as a result of the extensive research that has been done on them due to their adequate protonic conductivity at intermediate temperatures. Numerous substances transport protons, however, the majority either disintegrate at temperatures higher than 300 °C or only do so above 1000 °C [12]. According to the investigation by Demin and Tsiakaras [13], PC-SOFC has superior thermodynamic efficiency than OC-SOFC for converting chemical energy into electrical power for systems supplied by hydrogen in the temperature range of 727 and 1027 °C, exceeding 80%. Another advantage of the proton conductors is producing water at the cathode when hydrogen is utilized as fuel, preventing the fuel from becoming diluted while the cell is operating [14]. This increases cell efficiency and practicality [15]. Ceramic proton conductors are more chemically compatible with standard SOFC materials than Lanthanum Strontium Gallates (LSGM) and additionally have a higher ionic transport number than Ceria-Doped Gadolinia (GDC). The fact that these oxides can function between 400 and 750 °C due to their lower activation enthalpies and conductivity makes them a potential replacement for the traditional SOFC electrolytes based on YSZ, which only works at higher temperatures.

Thus, ceramic electrolytes that transport protons compete with anionic electrolytes like CeO_2-Gd_2O_3 (CGO) [16] and LSGM [17] for use in solid oxide fuel cells that operate at intermediate temperatures. Although, these materials do not have any structural protons. The formation of protonic defects that have a suitably high mobility

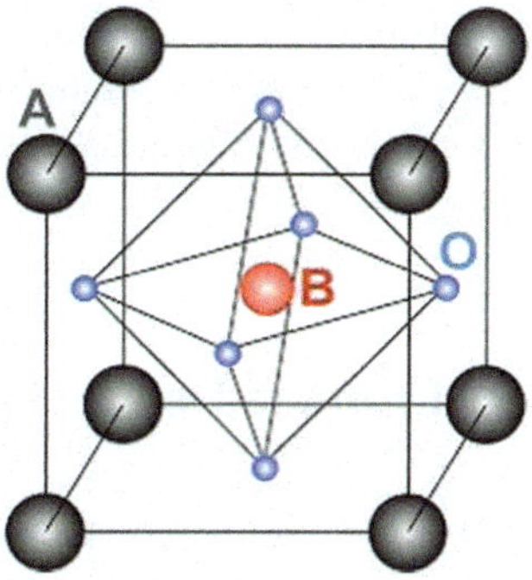

Fig. 5 Ideal representation of cubic perovskite structure, ABO_3

gives these materials their ionic conductivity. It was first reported by Thomas and Lander [18], that the creation of proton defects in ZnO at intermediate temperatures is due to the dissociative absorption of water during oxide preparation, characterization, or usage that necessarily requires the presence of oxide ion vacancies.

Therefore, developing highly efficient proton conducting SOFC electrolyte and simultaneously lowering its functioning temperature (below 600 °C) has been a major challenge. The compounds that have been discovered are acceptor-doped perovskite-type oxides with the formula $AB_{1-x}M_xO_3$, where A and B are the primary elements, and dopant M is a trivalent rare earth element, with a dopant level x lower than 1. The undoped perovskite-type structure has an orthorhombic crystal structure and the generic formula ABO_3. Since the first perovskite $CaTiO_3$ discovered by Lev Perovski [19] has an orthorhombic crystal structure. However, the structure of an ideal perovskite is cubic as shown in Fig. 5. The perovskite lattice has an atomic arrangement having larger A-cations at the corner of the cubic unit cell, with the smaller B-cations at the body center and the oxygen at the face center.

Doping this fundamental structure with a rare earth element (M^{3+}), causes a few small B-cations to be displaced. Because of this replacement, there are now vacant sites in the structure that were previously occupied by oxygen ions. At high temperatures, these types of materials frequently exhibit significant proton conductivity in the presence of a water and/or hydrogen environment. Conductivity may be associated with protonic defects, and their mobility can be attributed to the concentration of those defects, which can only fall within a specific range. However, whenever the symmetry is reduced, it results in a decrease in the concentration as well as the mobility of the protonic defects. Among different known solid-state proton conductors, perovskite materials based on doped alkaline earth zirconate and cerate have shown the most potential for protonic conducting electrolytes [20, 21], owing to their low activation energy and high protonic conductivity at moderate temperatures (400–700 °C) [22]. Where $SrZr_{1-x}M_xO_3$, $BaZr_{1-x}M_xO_3$, and other rare-earth-doped zirconate proved to be stable electrolytes with lower proton conductivity. On the other hand, rare-earth-doped cerate-based electrolyte such as $SrCe_{1-x}M_xO_3$, $BaCe_{1-x}M_xO_3$ showed high protonic conductivity with lower chemical/phase stability [23, 24]. Some latest class of perovskite oxides based on niobates [25, 26], tantalates [27, 28], titanates [29, 30], stannates [31, 32] also prove to have low activation energy and high protonic conductivity at intermediate temperatures. Different experiential methods were adopted to chemically modify the cerate to improve their stability [33–36], it appeared to exhibit that stability and high proton conductivity are contrasting characteristics [23].

5 Protonic Defects in Perovskite

The perovskite-based materials are crystalline because these have long-range atomic order. However, it is impossible to find crystalline materials that are completely free of defects, and there will always be some deviations in the ideal arrangement regardless of the temperature or other environmental factors. The defects are created due to atomic deviation from the mean lattice position, different size and/or charge ratios, vacancy at one particular lattice site etc. In defect chemistry, the Kröger-Vink notation [37] is widely used to describe the conventions used for electric charge and lattice position of point defects in crystals.

$$H_2O\ (g) + V_O^{\cdot\cdot} + O_O^X \leftrightarrow 2OH_O^{\cdot} \tag{30}$$

where, H_2O (g) is water in the gaseous phase, $V_O^{\cdot\cdot}$ is an oxygen vacancy site, O_O^X is neutral oxygen at its respective lattice site, and $OH_O^{\cdot}$ is a protonic defect.

As discussed above, the creation of proton defects is due to the dissociative absorption of water during oxide formation. The water in the gas phase decomposes into a proton (H^+) and a hydroxide ion (OH^-). The OH^--ion fills an oxygen ion vacancy, while the H^+-ion forms a covalent bond with one lattice of oxygen that has a unit positive effective charge, thus contributing to the conductivity of the material [38]. This moisture-induced conductivity needs to be significantly higher than that of electronic defects (electrons or holes) or other ionic species like oxide ion vacancies for the material to be a good protonic conductor. In dry conditions, these proton conducting oxides have a large concentration of oxygen vacancies, which get partially filled under hydrating atmosphere. Due to their extremely short ionic radius, protons are incapable of occupying a lattice or a regular interstitial site. Instead, they are permanently trapped within the electron cloud of an oxygen ion, which results in the formation of a hydroxide defect. The movement of ions as light as these requires a technique known as hopping. The transport features of such a model might be facilitated by lattice vibration and interaction, and the result would be that the protons exhibit small activation energy for conduction [39, 40].

5.1 Diffusion Mechanism in Solid Electrolyte

The diffusion process is referred to as the movement of matter across space as a result of the thermal motion of atoms or molecules in the solid, liquid, or gaseous phase [41]. Of course, the nature of the diffusion medium and diffusing species determines the transport mechanism. Theoretically, the most natural method to express matter displacement is to define a flux (**J**) with physical dimensions of matter per unit of time and surface expressed by

$$\mathbf{J} = c.v \tag{31}$$

where c and v denote the concentration and average displacement velocity of the diffusing species respectively. Fick initially formulated diffusion law, assuming a concentration gradient-driven effect between two adjacent sample volumes

$$\mathbf{J} = -D\nabla c, \quad D = D(c) \tag{32}$$

where ∇c is the concentration gradient and D is a proportionality factor.

Now, assume any arbitrary function $f = f(x, t)$ is the conserved quantity; for any instance, diffusion coordinate and time might be represented as x and t respectively. In some specific case, let f denotes the concentration of a given element, i.e. $f \equiv c$. Then, Fick's equation can be written for the present case as [42]

$$\nabla.\mathbf{J} = -\frac{\mathrm{d}c}{\mathrm{d}t} \quad \mathbf{J} = \mathrm{vc} \tag{33}$$

The previously stated equations seem to remain true independently of the specific sort of material or phase constituting the diffusion system, according to the thermodynamic base of the Fick equations and the generic nature of the underlying notions. The diffusion coefficient, a macroscopic number that, like pressure and temperature, has a statistical physical meaning and summarizes tiny details of transport processes, is the primary factor that regulates the movement of matter. The Fick equations allow one to bypass a full understanding of local interactions between diffusion species and diffusing medium, while the value of D that fits actual diffusion data reveals the transport processes. Solving Eq. 33 is problematic since D as a function of temperature, concentration, and microstructure is typically unknown. Solids have several diffusion processes. In crystal lattices, direct and indirect interstitial processes are important for small atoms and ions (e.g. H or B) that fit interstitial sites in metals, especially near phase transitions [43]. The most plausible diffusion mechanisms for thermally stimulated atom and ion jumps are vacancy and divacancy processes, which involve numerous jumps on neighboring unoccupied sites in the lattice. Proton conduction determines fuel cell performance.

There are several mechanisms of point defect migration through the crystalline sample depending upon the differences in effective mass, charge, and the electronic arrangement of the solid and defect species. The diffusion of a defect is also influenced

$$H_3O^+ + H_2O \Leftrightarrow H_2O + H_3O^+$$

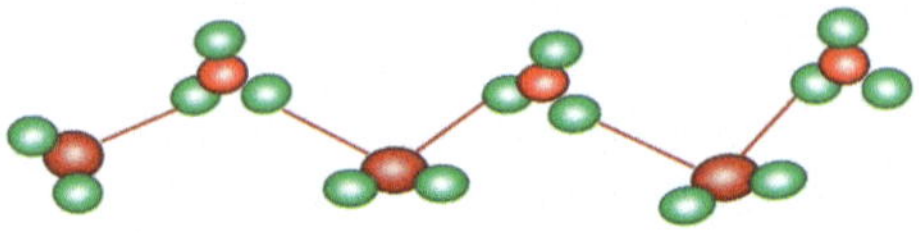

Fig. 6 Vehicle mechanism in the aqueous medium

by temperature, pressure, crystal structure, and morphology of the composition. Proton (H^+) is unique due to its extremely small size (8.768 fm) and having bare nucleus. The proton resides within the electron shell of some anion with which it is attached. In case of oxides, the ionic radius of oxygen is ~ 1.4 Åand the O–H distance is ~ 0.9 Å . Hence it is most unlikely that a proton can jump from one crystallographic site to another all by itself. Hence it strongly associates with the electron cloud of oxygen (in oxides), when integrated into the lattice in the form of protonic defects. Two possible mechanisms are considered for the diffusion of protons in oxides namely, the "vehicle mechanism" [44] and "Grotthus type mechanism" [45].

5.2 The Vehicle Mechanism

A protonated solvent molecule functions as the "vehicle" in the movement known as "vehicular" motion. This mechanism is distinguished by its normally greater activation energy as well as its typically lower proton mobility. In the vehicle mechanism, the proton moves as a passenger on a larger ion like O^{2-} in the form of OH^- or H_3O^+ as can be seen in Fig. 6. This type of mechanism is mostly observed in aqueous solutions and some polymer membranes.

5.3 The Grotthuss Mechanism

However, the most accepted form of the proton transport mechanism in solid oxide materials is the "Grotthuss mechanism" or "free migration mechanism," where the proton migrates through a combination of molecular re-orientation around the oxygen and proton hopping between adjacent stationary host oxygen ions [30]. The proton transport in perovskites is believed to be coupled with the local oxygen dynamics due to large O–O nearest neighbor distance and strong local interaction of proton within the electron cloud of the Oxygen.

The proton needs the dynamics of the host oxygen ion sublattice to jump to the nearest neighbor oxygen when the OH–O bond is shortened. Figure 7 shows the schematic diagram of proton conduction in ABO_3 perovskites via Grotthuss type mechanism, where only BO_6 octahedra have been shown for clarity. As illustrated

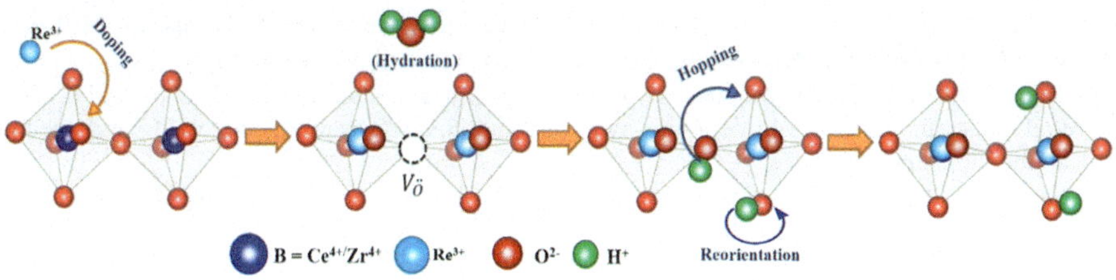

Fig. 7 Grotthuss mechanism in the solid electrolyte

in the figure, the principal features of the proton transport mechanism involve two steps, i.e.

- (the protonic defect first reorients about the host oxygen atom which directs the H^+ close to adjacent oxygen ion) followed by the second step, and
- (proton hops to the next nearest neighbor oxygen atom breaking its original bond and simultaneously forming a new bond).

The same process continues further throughout the lattice as a long-range conduction mechanism. Some of the atomistic simulation studies have rated the rotational period of a proton on a time scale of 10^{-12} s for a complete turn, while the proton hop/jump time of 10^{-9} s which is the rate-limiting step of this conduction mechanism.

6 Concluding Remarks

Understanding the mechanisms involved is essential for the development of novel or improved electrolytes for solid oxide fuel cells. The current constraints might be circumvented by operating at lower temperatures. In this article, we have discussed the chemistry, crystal structure, and electrical characteristics of numerous different electrolytes. These electrolytes all have their own unique set of advantages and disadvantages. When considering the mechanics of ion migration, the electrolytes were separated into three distinct classes: anionic, protonic, and mixed ions. In order to determine whether or not electrolyte materials are suitable for use in fuel cells and certain kinds of batteries, it is essential to have information on the proton conductivity as well as the proton diffusion rates.

References

1. Faro ML, La Rosa D, Antonucci V, Arico AS (2009) Intermediate temperature solid oxide fuel cell electrolytes. J Indian Inst Sci 89:363–380
2. Karuppiah K, Ashok AM (2019) Review of proton-and oxide-ion-conducting perovskite materials for SOFC applications, in, Thomas Telford Ltd

3. Brown M, Primdahl S, Mogensen M (2000) Structure/performance relations for Ni/Yttria-stabilized zirconia anodes for solid oxide fuel cells. J Electrochem Soc 147:475
4. Taroco H, Santos J, Domingues R, Matencio T (2011) Ceramic materials for solid oxide fuel cells. In: Advances in ceramics-synthesis and characterization, processing and specific applications, pp 423–446
5. Sunarso J, Baumann S, Serra JM, Meulenberg WA, Liu S, Lin YS, Diniz da Costa JC (2008) Mixed ionic–electronic conducting (MIEC) ceramic-based membranes for oxygen separation. J Membr Sci 320:13–41
6. Badwal S, Ciacchi F, Milosevic D (2000) Scandia–zirconia electrolytes for intermediate temperature solid oxide fuel cell operation. Solid State Ionics 136:91–99
7. Kharton VV, Naumovich EN, Yaremchenko AA, Marques FM (2001) Research on the electrochemistry of oxygen ion conductors in the former Soviet Union: IV. Bismuth oxide-based ceramics. J Solid State Electrochem 5:160–187
8. Lakshmi VV, Bauri R (2011) Phase formation and ionic conductivity studies on ytterbia co-doped scandia stabilized zirconia (0.9 ZrO2–0.09 Sc2O3–0.01 Yb2O3) electrolyte for SOFCs. Solid State Sci 13:1520–1525
9. Badwal S, Ciacchi F, Drennan J (1999) Investigation of the stability of ceria-gadolinia electrolytes in solid oxide fuel cell environments. Solid State Ionics 121:253–262
10. Seo DJ, Ryu KO, Park SB, Kim KY, Song R-H (2006) Synthesis and properties of Ce1−xGdxO2− x/2 solid solution prepared by flame spray pyrolysis. Mater Res Bull 41:359–366
11. Babu AS, Bauri R (2013) Rare earth co-doped nanocrystalline ceria electrolytes for intermediate temperature solid oxide fuel cells (IT-SOFC). ECS Trans 57:1115
12. Iwahara H (1995) Technological challenges in the application of proton conducting ceramics. Solid State Ionics 77:289–298
13. Demin A, Tsiakaras P (2001) Thermodynamic analysis of a hydrogen fed solid oxide fuel cell based on a proton conductor. Int J Hydrogen Energy 26:1103–1108
14. Iwahara H, Asakura Y, Katahira K, Tanaka M (2004) Prospect of hydrogen technology using proton-conducting ceramics. Solid State Ionics 168:299–310
15. Coors WG (2003) Protonic ceramic fuel cells for high-efficiency operation with methane. J Power Sources 118:150–156
16. Tianshu Z, Hing P, Huang H, Kilner J (2002) Ionic conductivity in the CeO2–Gd2O3 system ($0.05 \leq$ Gd/Ce$\leq$ 0.4) prepared by oxalate coprecipitation. Solid State Ionics 148:567–573
17. Yan J, Lu Z, Jiang Y, Dong Y, Yu C, Li W (2002) Fabrication and testing of a doped lanthanum gallate electrolyte thin-film solid oxide fuel cell. J Electrochem Soc 149:A1132
18. Thomas D, Lander J (1956) Hydrogen as a donor in zinc oxide. J Chem Phys 25:1136–1142
19. Rabe KM, Dawber M, Lichtensteiger C, Ahn CH, Triscone J-M (2007) Modern physics of ferroelectrics: essential background. In: Physics of ferroelectrics. Springer, pp 1–30
20. Fabbri E, Pergolesi D, Traversa E (2010) Materials challenges toward proton-conducting oxide fuel cells: a critical review. Chem Soc Rev 39:4355–4369
21. Kreuer KD (2003) Proton-conducting oxides. Annu Rev Mater Res 33:333–359
22. Fabbri E, D'Epifanio A, Di Bartolomeo E, Licoccia S, Traversa E (2008) Tailoring the chemical stability of Ba (Ce0. 8− xZrx) Y0. 2O3− δ protonic conductors for intermediate temperature solid oxide fuel cells (IT-SOFCs). Solid State Ionics 179:558–564
23. Kreuer K (1997) On the development of proton conducting materials for technological applications. Solid State Ionics 97:1–15
24. Iwahara H, Esaka T, Uchida H, Maeda N (1981) Proton conduction in sintered oxides and its application to steam electrolysis for hydrogen production. J Solid State Ionics 3:359–363
25. Animitsa I (2009) High-temperature proton conductors with structure-disordered oxygen sublattice. Russ J Electrochem 45:668–676
26. Liang K, Du Y, Nowick AS (1994) Fast high-temperature proton transport in nonstoichiometric mixed perovskites. Solid State Ionics 69:117–120
27. Irvine JT, Corcoran DJ, Lashtabeg A, Walton JC (2002) Incorporation of molecular species into the vacancies of perovskite oxides. Solid State Ionics 154:447–453

28. Corcoran DJ, Irvine JT (2001) Investigations into Sr3CaZr0. 5Ta1. 5O8. 75, a novel proton conducting perovskite oxide. Solid State Ionics 145:307–313
29. Shah MY, Lu Y, Mushtaq N, Rauf S, Yousaf M, Asghar MI, Lund PD, Zhu B (2022) Demonstrating the potential of iron-doped strontium titanate electrolyte with high-performance for low temperature ceramic fuel cells. Renew Energy 196:901–911
30. Malavasi L, Fisher CA, Islam MS (2010) Oxide-ion and proton conducting electrolyte materials for clean energy applications: structural and mechanistic features. Chem Soc Rev 39:4370–4387
31. Zvonareva IA, Medvedev DA (2022) Proton-conducting barium stannate for high-temperature purposes: a brief review. J Euro Ceram Soc
32. Exner J, Nazarenus T, Kita J, Moos R (2020) Dense Y-doped ion conducting perovskite films of BaZrO3, BaSnO3, and BaCeO3 for SOFC applications produced by powder aerosol deposition at room temperature. Int J Hydrogen Energy 45:10000–10016
33. Katahira K, Kohchi Y, Shimura T, Iwahara H (2000) Protonic conduction in Zr-substituted BaCeO3. Solid State Ionics 138:91–98
34. Ryu KH, Haile SM (1999) Chemical stability and proton conductivity of doped BaCeO3–BaZrO3 solid solutions. Solid State Ionics 125:355–367
35. Guo Y, Lin Y, Ran R, Shao Z (2009) Zirconium doping effect on the performance of proton-conducting BaZryCe0. 8− yY0. 2O3− δ (0.0≤ y≤ 0.8) for fuel cell applications. J Power Sour 193:400–407
36. Nasani N, Dias PA, Saraiva JA, Fagg DP (2013) Synthesis and conductivity of Ba (Ce, Zr, Y) O3− δ electrolytes for PCFCs by new nitrate-free combustion method. Int J Hydrogen Energy 38:8461–8470
37. Kröger F, Vink H (1956) Relations between the concentrations of imperfections in crystalline solids. In: Solid state physics. Elsevier, pp 307–435
38. Stotz S, Wagner C (1966) The solubility of water vapor and hydrogen in solid oxides. Ber Bunsenges Phys Chem 70:781788
39. Sonu BK, Sinha E (2021) Structural, thermal stability and electrical conductivity of zirconium substituted barium cerate ceramics. J Alloys Comp 860
40. Samgin A (2000) Lattice-assisted proton motion in perovskite oxides. Solid State Ionics 136:291–295
41. Kärger J, Heitjans P (2005) Diffusion in condensed matter: methods, materials, models. Springer
42. Di ML, Vona ES, Tosto S (2012) Diffusion in solid proton conductors: theoretical aspects and nuclear magnetic resonance analysis. In: Solid state proton conductors: properties and applications in fuel cells, vol 25
43. Ittermann B, Ackermann H, Stöckmann H-J, Ergezinger K-H, Heemeier M, Kroll F, Mai F, Marbach K, Peters D, Sulzer G (1996) Diffusion of boron in copper by direct-exchange mechanism. Phys Rev Lett 77:4784
44. Kreuer KD, Rabenau A, Weppner W (1982) Vehicle mechanism, a new model for the interpretation of the conductivity of fast proton conductors. Angew Chem, Int Ed Engl 21:208–209
45. Hempelmann R (2012) Structure and diffusivity in proton-conducting membranes studied by quasielastic neutron scattering. In: Solid state proton conductors: properties and applications in fuel cells 71–108

Chapter 7
Perovskite Manganite Materials: Recent Advancements and Challenges as Cathode for Solid Oxide Fuel Cell Applications

Paramananda Jena and Pankaj Kumar Patro

1 Introduction

Nowadays, researchers, scientists and entrepreneurs have great concern and attention towards the development of low-cost, eco-friendly, highly energy-efficient, renewable energy sources and technologies due to the uncertainty in the hydrocarbon reserve in the earth's crust and its depleting trend. Thus, the renewable energy technologies such as batteries, super-capacitors and fuel cells are the most suitable electrochemical energy resources [1]. Among these renewable electrochemical energy technologies fuel cells, solid oxide fuel cells (SOFCs) are the most promising future markets due to their non-polluting nature (zero emissions), modular construction, high energy efficiency and fuel flexibility. A solid oxide fuel cell consists of a solid electrolyte sandwich between cathode and anode electrodes. It converts the chemical energy of a fuel into electrical energy through redox reactions [2–6]. In this context, the novel materials play a crucial role for the enhancement of the electrochemical properties [7–9]. Hence, the engineering of novel materials has been more complicated than simply optimizing the known one for electrochemical properties. In particular, the materials used for cathode in SOFCs are most important owing to the catalytic role in oxygen reduction reaction of the oxidant. Concurrently, the

P. Jena
Centre of Excellence for Bioresource Management and Energy Conservation Material Development, Fakir Mohan University, Balasore, Odisha 756019, India

Department of Physics, S.K.C.G. (Autonomous) College, Paralakhemundi, Odisha 761200, India

P. K. Patro (✉)
Powder Metallurgy Division, Materials Group, Bhabha Atomic Research Centre, Vashi Complex, Navi Mumbai 400703, India
e-mail: pkpatro@barc.gov.in

Homi Bhabha National Institute (HBNI), Anushaktinagar, Mumbai 400094, India

B. P. Swain (ed.), *Energy Materials*, Materials Horizons: From Nature to Nanomaterials,
https://doi.org/10.1007/978-981-99-3866-7_7

cathode materials should be chemically stable in the oxidizing environment and also exhibit sufficient electronic conductivity during the operation. Since, the solid oxide fuel cell generally operates at high temperature ~ 1000 °C. So the cathode materials should be chemically as well as thermally compatible with other cell components starting from room temperature to the high operating temperature. Some of these key features of the cathode materials for SOFC applications are given below [10–12].

- Conductivity: It should have high electronic conductivity; i.e. the maximum number of electrons minimizes the ohmic loses as well as supports the electron flow in oxidizing atm. during the high-temperature operation.
- Catalytic activity: It should have more surface area for the ample catalytic activity, thereby minimizing the polarization for the electrochemical reduction of the oxidant.
- Stability: The cathode should be morphologically, chemically and dimensionally stable in the oxidizing environment. It should not have any phase transition during fabrication and operation temperatures. Also it should preserve its desired microstructure during the long-term operation.
- Chemical compatibility: The material should be chemically non-reactive against other cell components. In other words, the chemical reactions between the cathode and adjoining interface (electrolyte and interconnector) should be limited in order to minimize creation of secondary phase, alteration of thermal expansion, creation of the electronic conductivity in the electrolyte and stability reduction, etc.
- Coefficient of thermal expansion: The coefficient of thermal expansion should be well matched with the other cell components during cell fabrication as well as operation at high temperature in order to avoid cracking, delamination and thermal cycling.
- Porosity: It should possess adequate porosity to facilitate the transport of gases to the reaction sites in the cathode materials. The lesser porosity leads to the mass transport consideration.

In addition to these requirements, high strength and toughness ease of fabrication and low cost are the other desirable properties for applications in system level of SOFCs.

The various cathode materials were investigated such as lanthanum manganite, cobaltite, ferrite-based systems and also Ruddlesden-Popper series as well as double perovskites for high-temperature SOFCs [13–21]. Among these cathode materials, $LaMnO_3$ (lanthanum manganite, LM) belongs to perovskite family having ABO_3 formula that has been extensively studied due to its good chemical stability, possesses high electronic conductivity and thermally compatible to the other established solid electrolytes [22–24]. The normal cubic perovskite structure consists of three interpenetrating 3D networks. Out of these two are the different A and B cations, and the remaining third one is the corner-sharing octahedral oxygen (O_6). Thus, the ideal structure of lanthanum manganite ($LaMnO_3$) can be described as a $MnO_{6/2}$ scaffold of corner-shared octahedra that contains lanthanum cations within 12 coordinate sites. The presence of two types of oxidation sates, i.e. Mn^{3+} and Mn^{4+} in $LaMnO_3$, is responsible for the mixed valence semi-conductivity in it. In the pure

$LaMnO_3$ with increase Mn^{4+} stoichiometric content, composition leads to a structural phase transformation in it. The Mn^{4+} stoichiometric content of LM materials has been achieved by doping or substitution of the divalent cations on La site or by firing with different temperatures with different surrounding atmosphere [25–28]. In fact, $LaMnO_3$ perovskite in the A-site, i.e. La site, can tolerate considerable amount of vacancies, whereas in B-site, i.e. Mn site is for charge compensation by Mn^{3+} to Mn^{4+} formation. Conventionally, lanthanum manganite is considered as a anion excess compound ($LaMnO_{3+\delta}$), further the in-depth investigations were done by employing a high-resolution electron microscopy and other techniques show the existence of metal ion vacancies. A polaron hopping mechanism, i.e. charge carrier transport by the Mn^{3+}–O–Mn^{4+} network, has been suggested for alkaline earth substituted $LaMnO_3$ and $LaCrO_3$ [29–32]. Moreover proposed polaron hopping mechanism suggested that the maintenance of the homogeneity by various synthesis techniques inside the perovskite structure has been an essential criterion for thermal and chemical stability as well as high electrical conductivity at operating temperature. Further the maintenance of homogeneity leads to a better compatibility with other components like anodes, electrolytes and interconnectors for the solid oxide fuel cells.

2 Synthesis and Structural Phase Purity of LSM Materials

The synthesis techniques such as solid state, sol–gel, co-precipitation, sonication assists co-precipitation, sonochemical, citrate route, combustion, and pyrolysis were used to prepare $La_{1-x}Sr_xMnO_3$ (LSM) materials [33–42]. According to Conceicao et al., the citrate route showed major weight loss as compared to other synthesis techniques such as solid state combustion due to the decomposition of the polymeric network formed from citric acid and ethylene glycol. The sample synthesis from solid-state techniques showed large particle size and low surface area [42]. Bidrawn et al. have studied the influence of infiltrated dopants on the performance of SOFCs. According to them, the conductivity increases due to addition of dopants, which enhances the performance of SOFCs. However, the improved performance is related to the structural changes in the electrodes [43]. The synthesis routes have greatly influenced on the catalytic properties of the $La_{1-x}Sr_xMnO_3$ [41]. Mechanochemical synthesis is also one of the most important synthesis techniques as compared to the above-mentioned other preparative techniques. Zhang et al. synthesized the perovskite-type strontium manganite ($SrMnO_3$) and $La_{0.7}Sr_{0.3}MnO_3$ (LSM) powders with the grain size nano range by using planetary ball-mill in air [44]. In an attempt, LSM also synthesized by self-propagating high-temperature synthesis (SHS), relatively a new technique, which has been normally used to produce complex metal oxides. This SHS synthesis method is more economical for production of $La_{1-x}Mn_xO_3$ as compared to the existing commercial synthesis method [40].

$La_{1-x}Sr_xMnO_3$ ($x = 0$, 0.1, 0.16, 0.2 and 0.3) samples were synthesized by combustion techniques and observed phase changes from 36 mol% Mn^{4+} cubic to

28 mol% Mn^{4+} rhombohedral phase of $LaMnO_3$ on calcinations at 1000 °C [45]. The lifetime of the Sr-doped $LaMnO_3$ materials synthesized by wet chemical techniques of different stoichiometric compositions was investigated by applying various storage conditions such as low and high humidity in ambient temperature up to more than 350 days. It is observed that the electrode performance, chemical stability and adhesion of LSM materials highly depend upon the doping level of A-site rather than storage conditions. In this context, the LSM materials with A-site stoichiometric content, where the molar ratio, i.e. (La + Sr)/Mn ≤ 0.9, is found to be effectively suitable for achieving consistent electrode performance, high chemical stability and good adhesion as compared to the molar ratio (La + Sr)/Mn = 1 [46]. The complete phase formation conditions of $La_{1-x}Sr_xMnO_3$, i.e. utmost insertion of Sr into the perovskite structure, were investigated and reported in several literatures. However, the reported literature indicates that the substitutions of strontium only can be up to $x \leq 0.5$, and for $x \geq 0.6$ the presence of La_2O_3 has been detected. Further the maximum incorporation is influenced by the ionic radius of the substituted cations [34]. The co-ordination number in $LaMnO_3$ is 12 for La^{3+} with an ionic radius 1.46 Å, whereas the substituted ionic radius for Sr^{2+} is 1.54 Å, which is higher than Ca^{2+} (1.49 Å). Thus, the limit of substitution of Sr^{2+} is less in $LaMnO_3$ as compared to Ca^{2+} ($x = 0.8$) [47, 48]. The partial substitution of a Sr^{2+} which is lower valence as compared to the La^{3+} sites in $LaMnO_3$ leads to formation of Mn^{4+} species in Mn^{3+} matrix as per the ideal formula ${La_{1-x}}^{3+}{Sr_x}^{2+}{Mn_{1-x}}^{3+}{Mn^{4+}}_xO_3$. The formation of Mn^{4+} species varies linearly as per the Sr concentration [34]. The Mn^{4+} species concentration in $LaMnO_3$ oxides is fixed by substitution of Sr^{2+} in La^{3+} lattice sites and also by creation of cation vacancies [34, 48]. The over stoichiometric of $La_{1-x}Sr_xMnO_3$ compounds in air predominately disappears at $x \geq 0.3$ strontium content. It has been also reported that rhombohedral structure exists within the composition $0 \leq x \leq 0.3$, while it becomes cubic for the composition $0.3 \leq x \leq 0.5$. As reported by several literatures, the manganite structure symmetry has been largely dependent upon the level of substitution and therefore reasonably independent upon the preparation conditions. The stabilization of the cubic phase is due to the partial substitution of Sr^{2+} for La^{3+} sites leading to high Mn^{4+}content, which is accounted for Jahn–Teller distortion associated with decrease in Mn^{3+} concentration [49]. Aruna et al. also observed the phase transition from rhombohedral to cubic for 44 mol% of Mn^{4+} [50]. The formation of Mn^{4+} for the few stoichiometric amount of Mn^{3+} in the ideal formula $LaMnO_3$ leads to the distortion of the perovskite structure in terms of lattice shrinkage due to the smaller size of Mn^{4+}. Therefore, the structural symmetry of $LaMnO_3$ is very sensitive to the synthesis techniques and also the stoichiometric composition. However, the content of Mn^{4+} in $La_{1-x}Sr_xMnO_3$ oxides not only depends upon substitution of Sr^{2+} into the lattice but also by the non-stoichiometric composition ($La_{1-x}Sr_xMnO_3$), which depends upon the temperature, firing atmosphere and dwell time [30, 51]. Further the difference in Mn^{4+} content can be observed by the differences in cell parameters. The high-temperature studies are very much important for LM to use as cathode in SOFC applications. The influence of temperature on structural phase transition of LM has been reported by various authors. The two-phase structural evaluation of LM was observed using high-temperature XRD by van Roosmalen et al.

[26]. Also, Shu et al. found the two-phase transitions in the temperature ranges 373–473 K and 973–1273 K, respectively [52]. The first transition observed at 373–473 K from orthorhombic–rhombohedral is due to depression of Jahn–Teller distortion. The orthorhombic structure LM at room temperature, i.e. 373 K, is widely accepted. The second transition observed at 973–1273 K from rhombohedral to cubic is caused by the ionic oscillation. The second transition was not completely observed due to the slow phase transition rate. On the other hand, there is a cubic phase that exists in between orthorhombic and rhombohedral phase observed by Norby et al. [25]. They observed orthorhombic at room temperature and rhombohedral by heating at 1273 K under N_2 atmosphere.

3 Electrical, Thermal and Compatibility Properties of LSM as Cathode Materials

The intrinsic p-type conductivity in A- and B-site doped $LaMnO_3$ (LM) has been due to the creation of cation vacancies. Jonker et al. have prepared LM for the first time in 1950. They measured the electrical conductivity from room temperature to 1000 °C and observed that there is an appreciable electrical conductivity [53–55]. For solid oxide fuel cell applications, the electrical conductivity of LM materials can be enhanced by doping or substitution of lower valence cations either on A or B sites. With these encouraging findings, LM has been doped or substituted with various cations such as barium, strontium, calcium, cobalt, copper, lead, magnesium, potassium, rubidium, sodium titanium and yttrium by several researchers [56–60]. Apart from electrical conductivity, many of the dopants or substituents have been used to modify the thermal expansion properties of the materials [61]. Currently, strontium and calcium dopants or substituents have been used most frequently for solid oxide fuel cell applications, because $LaMnO_3$ substituted or doped with these two cations exhibits high electrical conductivity in oxidizing atm. and comparable thermal expansion to the other cell components such as anode and electrolyte [62–65].

The high-temperature thermodynamic stability in oxidizing atmosphere and compatibility with anode, electrolyte and interconnector in terms of activity of strontium is one of the major criteria for cathode and that has led to the preference of $La_{1-x}Sr_xMnO_3$ (LSM), as one of the trusted cathode materials [66]. First time Nakamura et al. observed that the decomposition pressure of LM material into La_2O_3 and MnO at 1000 °C and 1200 °C respectively [67]. According to this work, Shridharan et al. studied the decomposition pressure of LM material at temperature 1000 °C by using solid electrolyte galvanic cell [68]. Hildrum et al. also studied decomposition pressure of pure $LaMnO_3$ and $La_{0.8}Sr_{0.2}MnO_3$ at temperatures 1000 °C and 1100 °C respectively by using solid electrolyte galvanic cell [69]. Kuo et al. measured the decomposition pressure of pure $LaMnO_3$ at 1000 °C by employing an electrical conductivity measurement with the variation of oxygen partial pressure of the sample atmosphere [70]. The oxygen stoichiometry of pure $LaMnO_3$ and Sr-doped $LaMnO_3$

in oxidizing atmosphere has been widely changed from those under a reducing or inert environment [71–74]. The activity of SrO in $LaMnO_3$ under oxidizing atmosphere would facilitate a practical evolution of the compatibility of strontium-doped lanthanum manganite with the solid electrolyte and interconnector w.r.t transfer of SrO from region of higher to lower potentials [66].

In $LaMnO_3$, at the oxygen-rich end of the system for the symmetry has been rhombohedral while at the other end cubic [75]. Materials with perovskite structure have attracted the attention since they exhibited mixed ionic and electronic conductivities. The LSM being most common in this type has added advantage. The strontium has been preferred as a dopant due to its appropriate size match with the lanthanum. Although different studies have been dedicated towards the LSM materials, and their properties have been well studied, there seems to be persistence with a disagreement on the position of the electrochemical reaction site. This has especially been important for the intermediate temperature operation because the polarization of the cathode increases significantly with the decrease in temperature of SOFC. Operating of SOFC at about 650 °C using mixed conducting oxides has many advantages [76]. In other words, LSM shows high oxygen reduction reaction activity at the cathode and a best suitable candidate for a high-temperature solid oxide fuel cell [77]. Also, it fulfils the basic requirements of cathode such as high electronic conductivity and mechanical as well as chemical compatibility with interconnector and electrolyte materials [63, 64]. The comparison of ionic and electronic properties of $La_{0.8}Sr_{0.2}MnO_3$ at 900 °C is given in Table 1.

The $La_{1-x}Sr_xO_3$ ($0 \le x \le 0.5$, M = Mn, Co) cathode materials have been prepared by solid-state sintering (SSS) and deposited on yttria-stabilized zirconia (YSZ) electrolyte. The study on electrode layers using DC polarization measurements has revealed that the screen-printed electrodes performance was mainly affected by the development of a resistive reaction products with poor microstructure [81]. This limitation also severely found in $LaCoO_3$-based electrodes having strong reactivity towards conventional YSZ electrolyte, which requires a strong bonding in order to prevent failure because of the extreme coefficient of thermal expansion mismatch. The electrical conductivity of $La_{1-x}Sr_xMnO_{3+\delta}$ has been found very much dependent on temperature, density, microstructure and chemical composition. The oxygen

Table 1 Ionic and electronic properties of $La_{0.8}Sr_{0.2}MnO_3$ at 900 °C

Properties	Value	References
Electronic conductivity, σ_e S cm^{-1}	300	[78]
Ionic conductivity, σ_i S cm^{-1}	5.93×10^{-7}	[79]
Oxygen tracer diffusion, D_0^* cm^2 S^{-1}	1.27×10^{-12}	[79]
Oxygen self-diffusion, D^* cm^2 S^{-1}	3×10^{-12}	[80]
Oxygen surface exchange, K cm S^{-1}	9×10^{-8}	[80]

reduction reaction on LSM cathode has been represented by Eq. (1)

$$\frac{1}{2}O_2(g) + V_O^{\cdot\cdot}(\text{solid}_i) \rightarrow O_O^x(\text{solid}_i) + 2h^*_{\text{LSM}}, \quad (1)$$

where (solid$_i$) stands for ionically conducting material. There are two feasible schemes that have been proposed by Vanherle et al. [82]. The details of the reaction areas are as follows:

- Firstly where the cathode, solid electrolyte and oxygen-rich gas phase meet is called triple phase boundary; in this case, solid would be represented by YSZ.
- Secondly at the surface of the perovskite cathode making the reaction zone only two phase; for this case, solid represents the mixed conducting of Sr-doped LM (LSM).

Accordingly, in the former case, very large TPB has been achieved during the fabrication of finely porous cathode, whereas, in latter case the cathode layer is found to be dense.

The oxygen reduction reaction on the cathode site for SOFCs can be represented according to Kroger–Vink notation as follows:

$$O_2(g) + 2V^{\cdot\cdot}_{O,\text{electrolyte}} \rightarrow 2O^x_{O,\text{electrolyte}} + 4h^*_{\text{electrode}}, \quad (2)$$

where O_O^x and $V_O^{\cdot\cdot}$ represent an oxide ion and vacancy, respectively, in the crystal lattice of the electrolyte. The h^* represents the electron hole in the crystal lattice of the electrode. The h^* is an electron hole in the electrode crystal lattice. The transportation of the oxide ion away from the reaction sites towards the bulk of the electrolyte and excellent electronic conductivity as well as the occurrence of gaseous oxygen in the electrode materials have been found the most important criteria for the reaction. These criteria can be fulfilled in the locality of the triple phase boundary between the electrolyte, electrode and gases [83]. A triple phase boundary is found to be electrochemically active only when oxide ions, electrons and gases can percolate through the electrolyte, electrode and pores, respectively, to and fro from it. The extent of activeness of triple phase boundaries is determined by the microstructure which leads to the overall reaction rates in it [84–88].

In recent years, the compositions $La_{1-x}Sr_xMnO_3$ with $x \leq 0.2$ generally were used as cathode materials for SOFCs due to their high mixed electronic ionic conductivity, chemical as well as thermally stability during the operation temperature, and also it is the compatibility with conventional YSZ electrolyte [89]. Interestingly, maintaining compositional homogeneity has been the vital criterion for obtaining many of these properties in a reproducible manner. According to the conditions of oxygen partial pressure and temperature, the ratio of Mn^{4+}/Mn^{3+} is strongly modified in LSM material. It is essential to know the conditions used precisely to change in cationic composition that can occur as a function of composition. In other words, it has been always necessary to know the exact Mn^{4+}/Mn^{3+} ratio in the LSM system at 1000 °C in the air to select the best material for SOFC cathode application. The properties of the perovskite-type $La_{1-x}Sr_xMnO_3$ oxides ($0 \leq x \leq 1$) have been related to the reversible redox reaction of Mn^{3+} and Mn^{4+}. It has been shown that this reaction depends upon the experimental conditions as well as relative cationic concentrations and plays a key role in the interfacial transfer and transport properties. For these oxides by substitution of strontium on the lanthanum site (A-site), subsequently, there is a change in the valence state of the manganese site (B-site) which leads to enhancement of the electrical conductivity. Concurrently, finding the proper chemical composition of the compound for the excellent transport, chemical and mechanical properties, which is suitable for SOFC applications for the long life of the electrolyte–electrode interface, has been a challenging task. The complexities of these problems are due to the basic properties of the material which relates to the reversibility of the oxidizing–reducing reaction in the Mn^{3+}–Mn^{4+} [90]. It has been shown that for any given experimental conditions, the maximum equilibrium can exist between Mn^{3+} and Mn^{4+} ions, which can be controlled by adopting different synthesis techniques which have already been discussed in the synthesis part. The $La_{1-x}Sr_xMnO_{3+\delta}$ within the composition $0.3 \leq x \leq 0.5$ satisfies most of the criteria required for a good cathode for solid oxide fuel cell application based on YSZ electrolyte. The ratio of Mn^{4+}/Mn^{3+} is close to 1 and has a cubic symmetry of LSM which possesses good thermal stability [91].

The study of temperature-dependent resistivity (ρ) of $LaMnO_3$ and La (Sr) MnO_3 by Aruna et al. revealed the low concentration of Sr^{2+} at lanthanum sites ($x \leq 0.1$) and behaves like a semiconductor [92]. In contrast, at higher Sr^{2+} ($x \geq 0.16$) metallic behaviour has been observed at low temperatures. The metal-to-semiconductor phase transition that occurs at a particular temperature (T_t) was found to match well with ferromagnetic transition temperature [93]. The increase in resistivity below a transition temperature (T_c) manifested semiconductor behaviour [94]. The resistivity behaviour below T_c has been attributed to holes in an extended band and above T_c to the hopping motion of localized holes [95]. In general, log (σT) versus $1/T$, i.e. Arrhenius plot of $La_{1-x}Sr_xMnO_3$ for $x \geq 0.16$, exhibits linear behaviour [92]. This linear dependence has been the characteristic of the polaron hopping transport mechanism [96] represented by

$$\sigma = \left(\frac{A}{T}\right)\exp\left(\frac{-E_a}{kT}\right). \tag{3}$$

According to the literature, the pure LM materials sintered at 1200 °C show a transition in conductivity due to the structural phase transition from orthorhombic to rhombohedral of it, and with an increase in Mn^{4+} content, the transition temperature decreases [35, 97].

The electrical conductivity of $La_{1-x}Sr_xMnO_3$ has been measured using DC four-terminal methods at different temperatures by Lee [98]. According to him, $x = 0.5$ substitutions of Sr on the La site give the maximum conductivity. Shu et al. have reported that $La_{0.84}Sr_{0.16}MnO_3$ shows the highest conductivity in air, which has been higher than the value reported by Kertesz et al. [99] for the same composition at 900 °C. The activation energy decreases with increasing strontium content in $La_{1-x}Sr_xMnO_3$ composition. The activation energy decreases which is ascribed to the increase in Mn^{4+} content, leads to the reduction of Mn–Mn distance and brings the Mn–O–Mn angle closer to 180° [100]. The conductivity of LSM material with different synthesis methods at different temperatures is compared in Table 2.

The thermal expansion behaviour of $La_{1-x}Sr_xMnO_3$ ($x = 0, 0.16, 0.3$) has been studied by several researchers and found that the coefficient, α, increases with increasing substituted strontium content. Interestingly, the coefficient of thermal expansion (TEC) of $La_{1-x}Sr_xMnO_3$ has been reported closely matching with that of YSZ, a solid electrolyte with light doping [35]. A linear thermal expansion coefficient for various doping concentrations of LSM is depicted in Table 3.

One of the most striking requisite characteristics of cathode material has been the non-formation of an insulating secondary phase when in contact with the electrolyte. The reactivity with electrolyte has been investigated as a function of x in $La_{1-x}Sr_xMnO_3$ ($x = 0, 0.1, 0.3$ and 0.5) [98]. The study on electrolyte and $La_{1-x}Sr_xBO_3$ (B = Cr, Mn and Co) has revealed non-formation of the secondary

Table 2 Conductivity of $La_{1-x}Sr_xMnO_3$ at various temperatures with different synthesis methods

Composition	Method	T (°C)	σ (S cm^{-1})	References
$La_{0.8}Sr_{0.2}MnO_3$	Combustion	800	180	[96]
$La_{0.8}Sr_{0.2}MnO_3$	Solid state	900	0.146	[98]
$La_{1-x}Sr_xMnO_3$ $x = 0, 0.15, 0.30, 0.50$	Co-precipitation	1000	115, 175, 265, 300	[101]
$La_{0.8}Sr_{0.2}MnO_3$	Solid state	900	31.60	[98]
$La_{0.8}Sr_{0.2}MnO_3$	Solid state	900	96.98	[102]
$La_{0.8}Sr_{0.2}MnO_3$	Dip coating	800	07.00	[103]
$La_{0.8}Sr_{0.2}MnO_3$	Mechanochemical	600	0.139	[104]
$La_{0.84}Sr_{0.16}MnO_3$	–	1000	190	[105]
$La_{0.84}Sr_{0.16}MnO_3$	Solid state	1000	200	[106]
$La_{0.6}Sr_{04}MnO_3$	Sol gel	800	16	[107]
0.9 ($La_{0.8}Sr_{0.2}MnO_3$) + 0.1YSZ	Solid state	800	16	[108]
0.5 ($La_{0.8}Sr_{0.2}MnO_3$) + 0.5YSZ	Solid state	800	27	[108]
$La_{0.8}Sr_{0.2}MnO_3$	Solid state	800	200	[108]

Table 3 Linear thermal expansion coefficients of $La_{1-x}Sr_xMnO_3$

No.	Compound	$\times 10^{-6}$	References
1	$La_{0.8}Sr_{0.2}MnO_{3-\delta}$	13.1 K^{-1}	[109]
2	$La_{0.79}Sr_{0.2}MnO_{3-\delta}$	10.8 K^{-1}	[109]
3	$La_{0.69}Sr_{0.3}MnO_{3-\delta}$	11.7 K^{-1}	[109]
4	$La_{0.65}Sr_{0.3}MnO_{3-\delta}$	12.0 K^{-1}	[109]
5	$La_{0.8}Sr_{0.2}MnO_3$	11.7 °C^{-1}	[60]
6	$La_{0.6}Sr_{0.4}MnO_3$	10.8 °C^{-1}	[110]

phase, $La_2Zr_2O_7$, for the substitution of Sr up to 50 mol%. In recent years, in perovskites structure (ABO_3) on A-site with lanthanum, strontium and B-site with various combinations such as Fe/Mn, [110, 111] Fe/Co, [112, 113] Mn/Co [114, 115] has been widely investigated. The result of these investigations has been found that the electronic and ionic conductivity is the highest for the composition containing 50% Sr on the A-site and 50–100% Co on the B-site. Although these compositions have been advantageous for electrochemical performance but disadvantageous concerning the increase in the coefficient of thermal expansions and interactions with YSZ solid electrolytes, it has been leading to the formation of $SrZrO_3$ or $La_2Zr_2O_7$. A solution for the reduction of the coefficient of thermal expansion maintaining excellent electrical conductivities has been achieved by the substitution of other rare earth ions instead of lanthanum on the A-site. The coefficient of thermal expansion decreases with a decrease in the ionic radius of the lanthanide ions on the A-site of the perovskite structure. Introducing the proper lanthanide ion can compensate for slight changes in Co or Sr content. However, large amounts of Co or Sr always lead to high coefficients of thermal expansion which is not suitable for SOFC applications. A decrease in the coefficient of thermal expansion has been observed only in the case of substitution of Pr or Nd with La, and for other lanthanide elements, the coefficient of thermal expansion did not change considerably [109].

SOFCs with LSM as cathode and YSZ as electrolytes have been studied by several researchers at the operating ~ 1000 °C. At such high temperatures, degradation has been observed for a longer duration of operation due to the interaction of cell components chemically. Reducing the operating temperature of SOFC in the range of 700–850 °C has been suggested to overcome this problem [116]. But, at such intermediate temperature also LSM suffers from poor kinetics of electrode reaction [33]. In the process of developing suitable cathode materials for intermediate temperature operation, an attempt has been made to substitute either Mn by other transition elements or Sr by other alkaline earth metals [117]. Consequently, Cr, Co, Fe and Ni have been found to be suitable candidates for substitution on the B-site of LSM [118, 119]. Kuharuang has investigated the effect of Ni substitution up to $x = 0.3$ on electrical conductivity, microstructure and TEC [119]. LSM exists as single-phase compounds only up to the 40 mol% Ni [33]. Furthermore, symmetry of such sample was found to depend on composition as well as annealing temperature. Additionally, reduction in cell parameters has been found on annealing in air at 900 °C. Such reduction in cell parameters has been attributed to oxygen vacancies created by Ni^{2+} substitution for Mn^{3+}/Mn^{4+}. Thermal expansion coefficient was reported to change drastically on substitution of small amount of nickel for manganese.

The dense $La_{0.81}Sr_{0.09}MnO_3$ and $La_{0.6}Sr_{0.4}CoO_3$ (LSC) have been synthesized by laser ablation techniques on YSZ and samarium-doped ceria (SDC) substrates, respectively [120]. It is found that the interfacial electrical conductivity of the $La_{0.81}Sr_{0.09}MnO_3$ electrodes is inversely proportional to its thickness and independent on the oxygen partial pressure. However, for the $La_{0.6}Sr_{0.4}CoO_3$ electrodes the interfacial electrical conductivity found to be dependent upon the square root of oxygen partial pressure which indicates the surface reaction to the over-potential [41]. The effect of different synthesis techniques for LSM such as sol–gel, pyrolysis, solid state, and oxalate co-precipitation has been studied by Bell et al. [41]. According to them although the synthesized samples have alike compositions, they perform quite differently during catalytic characterization. This observation has been correlated more closely to the different synthesis techniques rather than impurity levels or physical properties of the powers. Finally, they concluded that the surface structures formed by different synthesis techniques execute very differently under catalysis conditions. However, the obtained structures from pyrolysis and co-precipitation synthesis techniques were found to be most efficient that facilitates oxidation-type reactions [41].

The transition metals such as Cr, Mn and Co have been partially substituted on B-site for $La_{0.5}Sr_{0.5}BO_3$, and the electrochemical characterizations have been studied by AC impedance spectroscopy [98]. The interfacial resistance and electrical conductivity have been found to be decreased with the increase in strontium substitution, and the value has been achieved for 50 mol%. According to a study, the reactivity of $La_{1-x}Sr_xBO_3$ (B = Cr, Mn, Co) cathode and electrolyte gives a secondary phase $La_2Zr_2O_7$ for 50 mol% Sr addition [121]. In accordance with many researchers, the $La_{0.9}Sr_{0.1}Ga_{0.8}Mg_{0.2}O_3$ (LSGM) and $La_{0.9}Sr_{0.1}MnO_3$ have been the suitable choices of the electrolyte and cathode materials, respectively. Since both the materials belong to perovskite structure, there is no possibility of reaction to occur in between them. Even different compositions of $LaSr(GaMg)_{1-x}Mn_xO_3$ or $La_{0.9}Sr_{0.1}MnO_3$ and LSGM ($x = 0$–1) have been identified without any formation of intermediate phase in spite of sintering at high temperatures [122]. The oxygen reduction reactions have been studied on the compositions LSM and (La, Sr)(Co, Fe)O_3 (LSCF) cathodes within the temperature range 700–900 °C in air atmosphere and also the absence of chromium species. Three reaction steps are identified for the oxygen reduction reaction such as (i) surface dissociative adsorption and diffusion, (ii) transfer of charge and (iii) migration of oxygen ion inside the zirconium electrolyte. The reaction is influenced by the dissociative adsorption and diffusion in LSM cathode surface at low temperature. Further the oxygen diffusion/migration inside the zirconium electrolyte occurs at high temperatures. It has been reported that bulk as well as surface diffusion plays a key role of the reaction kinetics for the oxygen reduction reaction on LSCF cathodes as compared to the LSM [63].

4 Other Cathode Materials for SOFC Applications

Apart from the LSM materials, some other cathode materials have also been studied such as $La_{1-x}Ca_xMnO_3$ and $Gd_{1-x}A_xCo_{1-y}MnYO_3$. The local structure of former has been determined from pair-distribution function (PDF) investigation from pulsed neutron diffraction. From the comparison of the effect on the local structure during the metal–insulator transition as a function of composition and temperature, it has been shown that crossing the metal–insulator boundary at low temperature by varying the composition yields similar type distortions as like LSM electrode materials with increase in temperature. Such behaviour indicates the occurrence of polarons in the insulating phase, which vanish in the metallic phase [123]. In an attempt, the $Gd_{1-x}A_xCo_{1-y}MnYO_3$ (A = Sr, Ca) system has been studied as a cathode for SOFC applications. The thermal expansion compatibility has been shown by Mn-rich compositions with the YSZ electrolyte. It has also been shown that this composition exhibits less reactivity with YSZ and the formation of very small amounts of $La_2Zr_2O_7$ on annealing at 1000 °C [124]. On the other hand, sintering, electrical conductivity and coefficient of thermal expansion for the composition $Ln_{1-x}Sr_xMnO_3$ (Ln = Pr, Nd and Sm; $x = 0, 0.16$ and 0.25) synthesized by combustion method have been studied as a cathode by Aruna et al. [50]. According to them, the ionic conductivity decreases with the decrease in ionic radii of rare earth ions. The highest electrical conductivity has been achieved in $(Pr/Nd)_{0.75}Sr_{0.25}MnO_3$ material. The coefficients of thermal expansion for $Pr_{0.75}Sr_{0.25}MnO_3$ and $Nd_{0.75}Sr_{0.25}MnO_3$ have been found to be 10.2×10^{-6} and 10.7×10^{-6} K^{-1}, respectively, which is closer to the conventional YSZ electrolyte [50]. Furthermore, Sakaki et al. have studied the perovskite-type manganite system such as $Ln_{1-x}Sr_xMnO_3$ (Ln = Sm, Nd, Pr, La and Gd; $0 \geq x \geq 0.5$) [86]. From this study, it is found that the compositions with smaller ionic radii lanthanide systems such as $Pr_{1-x}Sr_xMnO_3$ and $Nd_{1-x}Sr_xMnO_3$ succeeded in suppressing the development pyrochlore $Ln_2Zr_2O_7$ phase. Kenjo et al. prepared a series samples with different concentrations, i.e. $Pr_{0.6-x}Sr_{0.4}MnO_3$ ($x = 0, 0.05$ 0.1, 0.15, 0.2) by solid-state reaction method [85]. According to them, Pr scarcity at the A-site has an immense effect on the properties of $Pr_{0.6}Sr_{0.4}MnO_3$ from cathode material viewpoint. Also it has been compared with conventional used $La_{0.6}Sr_{0.4}MnO_3$ and $La_{0.55}Sr_{0.4}MnO_3$. It has been found that the $Pr_{0.6-x}Sr_{0.4}MnO_3$ cathode is better in terms of conductivity, over-potential and impedance. It has also been reported that the $Pr_{0.55}Sr_{0.4}MnO_3$ exhibited the best performance as compared to LSM [50]. But the oxygen ion conductivity of LSM materials was found very low (10^{-7} s cm^{-1}) at 900 °C, which has limited its use in the development of intermediate temperature SOFC (IT-SOFC), which operates at 600–800 °C [125]. Hence, mixed-ion–electron conductors (MIEC) (La, Sr) (Co, Fe)$O_{3-\delta}$ (LSCF) with electronic conductivity about 230 S cm^{-1} at 900 °C and oxygen ion conductivity about 0.25 S cm^{-1} have been preferred. However, LSCF reacts with YSZ to form $SrZrO_3$. Hence, instead of YSZ, gadolinia-doped ceria (GDC) has been used [63]. In an attempt to enhance the cathode performance in turn the power density, perovskites based on LSCF have been proposed [126]. $La_{0.58}Sr_{0.4}Fe_{0.8}Co_{0.2}O_3$ has emerged as a potential cathode in the light of reproducible power output of 1.2 W cm^{-2} at 800 °C and 0.7 V with hydrogen as fuel gas. Long-term cell tests, however, lasted up to 3000 h that suggested degradation of the cells between 0.5 and 1.5 for 1000 h of operation. The loss in performance has been found more than for conventional cathodes based on LSM [126].

Maguire et al. investigated the electrical conductivity and cathode performance of the ceramic materials such as $La_{0.84}Sr_{0.1}6Fe_{0.7}O_3$ ($LSCFe_7$) and $La_{0.84}Sr_{0.16}Co_{0.7}Fe_{0.3}O_3$ ($LSCFe_3$) synthesized by solid-state reaction method from oxides and carbonate [127]. They found that a large decrease in cell ohmic resistance with increase in cathodic over-potentials for both the cathode materials. According to them, the electrochemical reaction spreads from the TPB line to the electrode surface as per a characteristic feature of mixed ionic-electronic conducting electrode materials. Generally, the steady-state cathodic polarization measurements performed at 800 °C revealed that the LSCFe3 can be the best cathode material because of its high conductivity than LSCFe7 (in the range 200–800 °C) [127]. Though the LSM has been the best candidate material as compared to other one, in order to enhance the efficiency of the cell at intermediate temperatures, the other materials with mixed ionic-electronic conductivity have been also used. Mixed conductivity facilitates the exchange of oxygen not only at the TPB area, but can also be extended over a larger surface of the cathode. As per the literature, the $La_{1-x}Sr_xCoO_3$ (LSC) exhibits the high oxygen ion conductivity and, therefore, the high rate of oxygen surface exchange. Concurrently, it maintains higher current densities and lower cathode over-potentials [77]. However, its TEC that mismatches with the electrolyte restricted the utility as a cathode material for SOFC application [128]. As a matter of fact, the performance of conventional cathodes based on $La_{1-x}Sr_xMnO_{3-\delta}$ has not been satisfactory. Thus, electrode-solid electrolytes, e.g. LSM-$ZrO_2(Y_2O_3)$ composite perovskite electrodes, have been emerged as an promising alternative [129]. The electrical properties of manganite-based perovskite materials are not suitable for SOFC applications below 800 °C. The absence of oxygen vacancies at such lower temperature restricts the oxygen reduction at TPB. To avoid such situation, two approaches have been taken by researchers [10]. First approach has been to add ionically conducting phase to LSM. This approach extends the surface area where oxygen reduction can occur. Generally, GDC has been used as an electrolyte to avoid any intermediate phase formation at operating temperature. In the second approach, La has been replaced with other rare earth elements or LSM doped with a cation (Co, Fe or Ni) that promotes the oxygen vacancies while Sr at A-site. They behave effectively as the mixed conductors on the macroscopic scale similar to $La_{0.8}Sr_{0.2}FeO_{3-\delta}$ or $La_{1-x-y}Sr_xCo_{0.2}Fe_{0.8}O_{3-\delta}$ ($x = 0.2$ and 0.4; $y = 0$–0.05). In other words, it can be said that apart from high electronic conductivity, composite electrodes exhibit high oxygen ion conductivity, thereby enlarging the TPB. Most of the later studies have been focused on the electrochemical performance of LSM-$ZrO_2(Y_2O_3)$ composite cathodes, while a few researchers concentrated on the electrochemical performance of iron and cobalt-containing perovskite cathodes, e.g. $La_{1-x}Sr_xCo_{1-y}FeyO_{3-\delta}$ cathodes [130]. Furthermore, to enhance the cathodic performance at lower temperature, several perovskites-based mixed ionic-electronic systems like $La_{0.6}Sr_{0.4}Co_{0.2}Fe_{0.8}O_{3-\delta}$ [130], $Sm_{0.5}Sr_{0.5}CoO_{3-\delta}$ (SSC) [131] and $Ba_{0.5}Sr_{0.5}Co_{0.8}Fe_{0.2}O_{3-\delta}$ (BSCF) [132], has been studied in order to replace the LSM system. These MIECs have much higher oxygen ion conductivity, electronic conductivity and catalytic activity for oxygen reduction than LSM. Furthermore, the interfacial polarization resistance has been found to be lower as compared to LSM. TEC for $La_{0.6}Sr_{0.4}Co_{0.2}Fe_{0.8}O_{3-\delta}$ has been reported to 20.5×10^{-6} K^{-1}, which has been much higher than that for SDC 12×10^{-6} K^{-1}. Replacing Co with Fe has an effect on reducing TEC; 13.8×10^{-6} K^{-1} for $La_{0.6}Sr_{0.4}Co_{0.2}Fe_{0.8}O_{3-\delta}$, thus, minimizes the mismatch. Nevertheless, the conductivity of LSCF decreases with increasing Fe content [133]. The sharp discontinuity in coefficient of thermal expansion leads to delamination during thermal cycling; in order to avoid it, the composite functional-graded cathodes have been designed and fabricated for SOFCs based on

YSZ electrolytes [134–136]. The electronically conducting phase in these functional-graded cathode materials changed easily from LSM to LSC. Further enhanced electrochemical performances have been reported for the functional-graded cathodes deposited on YSZ electrolytes using screen-printing [134], slurry spray [135] and slurry coating [136]. It has been understood that, instead of a sudden modification microstructure and composition between the electrochemically active layer and the current collecting layer, functional-graded cathodes have a graded interface at which the composition slowly changes from one layer to another and, thus, minimizes the coefficient of thermal expansion mismatch leading to better cathodic performance [137–142].

Further, rare earth-doped cobaltites have attracted much attention because of their mixed conduction as well as relatively high ionic conductivity. According to Ishihara et al., the strontium-doped samarium cobaltite, $Sm_{1-x}Sr_xCoO_3$ (SSC), exhibits not only high conductivity, up to 10 S/cm when $x = 0.5$ but also good compatibility with ceria electrolyte as compared to that of YSZ [141]. However, the cathodic reaction mechanism of SSC has not been clearly understood in this case [143]. Interestingly, the effects of dopants on the electrical conductivity of the perovskite-type oxide $LaInO_3$ have been investigated. Accordingly, the replacement of La by Sr has been the most effective way to enhance the conductivity of $LaInO_3$; whereas, Ca substitution for In has been rather difficult due to the large difference in the ionic radii. The maximum conductivity has been reported for $La_{0.9}Sr_{0.1}InO_{3-\delta}$ composition [144].

Also Yasumoto et al. suggested (La, Sr)(Co, Cu)$O_{3-\delta}$ (LSCC) as the electrode material for SOFC [144]. According to them, LSCC has no oxygen excess region. Its behaviour has been found similar to that of (La, Sr)$FeO_{3-\delta}$ (LSF) but not with (La, Sr)$MnO_{3-\delta}$. However, the electrical conductivity was found lower as compared to other cathode materials. The electrode activity of LSCC has been better than that of LSC, in spite of lower electrical conductivity of former than latter [145]. The electrochemical properties at low temperatures (400–600 °C) in between the interfaces of the $Sm_{0.2}Ce_{0.8}O_{1.9}$ (SDC) solid electrolyte and composite porous cathodes consisting of $Sm_{0.5}Sr_{0.5}CoO_3$ (SSC)-SDC have been studied by forming anode-supported single cells [146]. This study has revealed that the firing temperature at 950 °C and the maximum concentration about 30 wt% of the SDC solid electrolyte in the composite cathode are required. It has also been found that adding up the adequate quantity of SDC to SSC noticeably improves the catalytic properties at the interface which leads to the reduction of interfacial resistance more than one order magnitude as compared with an SSC cathode without SDC [146].

5 Effect of Porosity on LSM and Other Cathode Materials

Porous microstructure (70% theoretical density and 30% open porosity) of any cathode material apart from electrical conductivity, compatible thermal expansion and compatibility with the electrolyte is one of the most important features for SOFC applications [147]. Hence, a porous composite cathode for SOFCs has been made from a mixture of LSM and YSZ. A simplified model of porous electrode response has been understood on optimization of fuel cell performance. The AC impedance technique has been first applied to porous electrodes by Levie et al. [148, 149]. Ostergard et al. also investigated the kinetics O_2 reduction at the $La_{0.85}Sr_{0.15}MnO_3$/ ZrO_2–Y_2O_3 electrode–electrolyte interface by AC impedance spectroscopy [150]. According to them, in a system there is three time constants present, every one corresponds to a rate-limiting step towards the whole reaction mechanism. At high frequencies, the time constant corresponds to the charge transfer of oxide ions from the electrode–electrolyte to the oxide ion vacancies in the electrolyte [150, 151].

The electrochemical performance of O_2 reduction of porous composite electrodes consisting of LSM-YSZ has been studied for better understanding of the O_2 reduction mechanism and enhancement of the cathode's performance. Half cell SOFCs consist of YSZ, and slurry-coated cathodes have been tested with the three electrodes configuration [152]. Since electrode resistance has been found to be dependent on the size, the process takes place in the bulk of the cathode [152]. The complex impedance spectroscopy (CIS) study has been carried on the electrochemical cell obtained by forming interface between dense YSZ and porous LSM followed by annealing at temperature in the range 1100–1400 °C for 200 h [153]. In the complex impedance spectra, the two semicircles have been represented by an equivalent circuit which consists of a series combination of the electrolyte resistance and two parallel combinations of a resistance and a constant phase element (CPE). The semicircle at high frequencies corresponds to the diffusion of oxide ions from the electrode–electrolyte interface LSM-YSZ to the oxide ion vacancies situated at the electrolyte surface, whereas at lower frequencies, the semicircle corresponds to the O_2 species adsorption and diffusion in the LSM cathode.

A composite cathode consists of a porous mixture of electronic conductor such as LSM and a solid electrolyte typically YSZ that have been used to enhance the cathodic performance of SOFCs. Several studies have been confirmed that the composite cathodes revealed very much less over-potentials than that of the single-phase electrodes by virtue of the parallel paths of ionic and electronic charge carriers [154]. By using a cathode, it has been possible for the reaction zone to extend from the electrode–electrolyte interface into the electrode and effectively results in a functionally diffuse interface where the charge transfer reaction takes place. As a matter of fact, the performance of composite cathodes in SOFCs is mainly governed by mass transport, interfacial kinetics and ohmic drop. A theoretical model based on activation and mass transport steps has been proposed for a porous composite cathode [155]. Pure and strontium-doped LM have not been used in the fuel cell as they lack in requisite porosity and, also, due to poor oxygen ion conductivity. Since the formation of the intermediate phase with the YSZ electrolyte, various scientists have employed a new electrolyte GDC. Reaction products of LSM and YSZ at various temperatures for are depicted in Table 4.

Practically, composite cathode has been obtained by mixing LSM with GDC. Also, it has been investigated that the LSM-GDC composite cathode although comparable to the usually used LSM-YSZ cathodes has great advantages since LSM has no reactivity with GDC and the higher ionic conductivity [158]. Ralphz et al. have

Table 4 Reaction products of $Ln_{1-x}Sr_xMnO_3$ and YSZ at various temperatures

x in $La_{1-x}Sr_xMnO_3$	1000 °C	1100 °C	1200 °C	1300 °C	1400 °C	References
0.0	$La_2Zr_2O_7$	$La_2Zr_2O_7$	$La_2Zr_2O_7$	$La_2Zr_2O_7$	$La_2Zr_2O_7$	[156]
0.1	$La_2Zr_2O_7$	$La_2Zr_2O_7$	$La_2Zr_2O_7$	$La_2Zr_2O_7$	$La_2Zr_2O_7$	[156]
0.2	$La_2Zr_2O_7$	$La_2Zr_2O_7$	$La_2Zr_2O_7$	$La_2Zr_2O_7$	$La_2Zr_2O_7$	[156]
0.3	–	–	–	–	$La_2Zr_2O_7$	[156]
0.4	–	$SrZrO_3$	$SrZrO_3$	$SrZrO_3$	$SrZrO_3$	[156]
0.5	$SrZrO_3$	$SrZrO_3$	$SrZrO_3$	$SrZrO_3$	$SrZrO_3$	[156]
0.0	–	–	–	–	$La_2Zr_2O_7$	[98]
0.1	–	–	–	–	$La_2Zr_2O_7$	[98]
0.3	–	–	–	–	$La_2Zr_2O_7$	[98]
0.5	–	–	–	–	–	[98]
0.2	–	–	–	–	$La_2Zr_2O_7$	[157]
0.35	–	–	–	–	$SrZrO_3$, $La_2Zr_2O_7$	[154]

suggested that the composite cathode provides a high density of TPB while having a continuous ionically conducting phase. Both these allow electrochemical reactions to occur within the electrode at a distance up to tens of micrometres from the electrode–electrolyte interface, thereby reducing overall resistance [155]. The absence of oxygen vacancies in LSM results in a conductivity that has been almost exclusively electronic. This restricts the reduction of oxygen to the TPB regions, whereas the cathode, air and electrolyte make an intimate contact. This constraint has been reported the primary reason for unacceptable performance of LSM at lower temperatures, where the entire cathode surface participates in the reduction and transport of O_2 to the electrolyte. While some uncertainty exists over the actual oxygen vacancy concentration in LSM and the mechanisms that introduce it, this concentration has been unquestionably low.

Also, research has been performed where both gallium and aluminium doping onto the B-site appear to enhance the performance of the cathodes [157]. Most of the candidate materials contain rare earth elements due to the geometric constraints placed on the A-site. Most of the published work on alternative A-site cations has been focused on YSZ electrolytes, but the theories can generally be adapted for GDC as well. The first approach has suggested improved interfacial resistance provided selected cation does not form a stable insulating phase with the YSZ electrolyte. Lanthanum-based cathodes have been commonly forming $La_2Zr_2O_7$ at the interface, which has a high resistance and rapidly degrades the performance of the cathode. Although making A-site deficient cathodes introduced some improvement in this situation, lanthanum zirconate has been still detected, especially in the non-manganite systems. Alternative cations that include Gd, Nd and Y have been studied. Among all these, Nd shows limited improvement and forms a more stable pyrochlore phase [159]. Gadolinium prevents the formation of a stable pyrochlore phase with zirconium, while yttrium being a constituent of the YSZ electrolyte does not cause any interfacial problems. The selection of a cation with multiple valences to improve the electronic conductivity of the cathode material has been one of the approaches. Cations like Sm, Pr, Ce and Tb have been studied extensively [143, 160]. The problem with a pyrochlore phase formation at the interface has not been seen in the case of the ceria-based electrolyte, and thus, one would expect an improvement from changing the A-site to be small [159]. Nowadays, researchers and scientists, the majority of their time, focused on the development of these perovskite materials, identifying novel mixed ionic-electronic conductors and developing improved cell designs for the improvement of operational performance by using the conventional existing materials.

6 Conclusions

Lanthanum strontium manganite (LSM)-based perovskite is the most significant cathode material for SOFC applications. The persistent enthusiasm and interest in this material arise because of its typical advantages of cathode for high-temperature SOFCs: its high mixed electronic and ionic conductivity, better thermal stability as well as compatibility with conventional electrolytes such as YSZ, GDC and LSGM and high electrochemical activity for O_2 redox reaction at operation conditions. The LSM cathode material is most important among other cathode materials because it has an excellent track record on long-term performance along with microstructural stability. The conventional approach to synthesizing the cathode materials desires high-temperature sintering to form an intimate electrode–electrolyte interface as well as a higher processing temperature, which leads to the limitation of the optimization for the microstructure of the cathode materials. Therefore, adapting different synthesis techniques for LSM-based cathode materials plays a crucial role in creating macroscopically homogeneous composite materials as well as nanostructure materials. Nano-structure materials would serve to substantially increase the thermal stability and electrical conductivity and also to increase the TPB. Hence, the nano-structured LSM-based composite cathode materials combined with higher surface area and adequate porosity possess superior electrochemical properties and have immense potential as cathodes for SOFC applications.

References

1. Winter M, Brodd RJ (2004) Chem Rev 104:4245–4270
2. Minh NQ (2004) Solid State Ionics 174:271–277
3. Chelmehsara ME, Mahmoudimehr J (2018) Int J Hydrogen Energy 43:15521–15530
4. Blum L, (Bert) de Haart LGJ, Malzbender J, Menzler NH, Remmel J, Steinberger-Wilckens R (2013) J Power Sour 241:477–485
5. Singhal SC (2000) Solid State Ionics 135:305–313
6. Yamamoto O (2000) Electrochim Acta 45:2423–2435
7. Mahato N, Banerjee A, Gupta A, Omar S, Balani K (2015) Prog Mater Sci 72:141–337
8. Hussain S, Yangping L (2020) Energy Transitions 4:113–126
9. da Silva FS, de Souza TM (2017) Int J Hydrogen Energy 42:26020–26036
10. Sun C, Hui R, Roller J (2010) J Solid State Electrochem 14:1125–1144
11. Harrison CM, Slater PR, Steinberger-Wilckens R (2020) Solid State Ionics 354:115410–211538
12. Kan WH, Samson AJ, Thangadurai V (2016) J Mater Chem A 4:17913–17932
13. Afroze S, Karim A, Cheok Q, Eriksson S, Azad AK (2019) Front Energy 13:770–797
14. Chen K, Li N, Ai N, Li M, Cheng Y, Rickard WDA, Li J, Jiang SP (2016) J Mater Chem A 4:17678–17685
15. Nagasawa K, Daviero-Minaud S, Preux N, Rolle A, Roussel P, Nakatsugawa H, Mentr O (2009) 21:4738–4745
16. Zou J, Park J, Yoon H, Kim T, Chung J (2012) Int J Hydrogen Energy 37:8592–8602
17. Fan W, Sun Z, Bai Y, Wu K, Chen Y, Appl ACS (2019) Mater Interfaces 11(26):23168–23179
18. Ni C, Zhou J, Zhang Z, Li S, Ni J, Wu K, Irvine JTS (2021) Energy Environ Sci 14:6287–6319
19. Amow G, Davidson IJ, Skinner SJ (2006) Solid State Ionics 177:1205–1210
20. Ding P, Li W, Zhao H, Wu C, Zhao L, Dong B, Wang S (2021) J Phys Mater 4:022002
21. Pelosato R, Cordaro G, Stucchi D, Cristiani C, Dotelli G (2015) J Power Sour 298:46–67
22. Tsai T, Barnett SA (1997) Solid State Ionics 93:207–217
23. Jiang SP (2008) J Mater Sci 43:6799–6833
24. Carpanese MP, Clematis D, Bertei A, Giuliano A, Sanson A, Mercadelli E, Nicolella C, Barbucci A (2017) Solid State Ionics 301:106–115
25. Norby P, Andersen IGK, Andersen EK, Andersen NH (1995) J Solid State Chem 119:191–196
26. van Roosmalen JAM, van Vlaanderen P, Cordfunke EHP, Ijdo WL, Ijdo DJW (1995) J Solid State Chem 114:516–523
27. Bogush AK, Pavlov VI, Balyko LV (1983) Crystal Res Technol 18:589–598
28. Shekhtman VSh, Sedykh VD (2007) J Surf Invest X-Ray Synchrotron Neutron Tech 5:513–521
29. Xu W, Apodaca N, Wang H, Yan L, Chen G, Zhou M, Ding D, Choudhury P, Luo H (2019) ACS Catal 9:5074–5083
30. Kuo JH, Anderson HU, Sparlin DM (1990) J Solid Sate Chem 87:55–63
31. Carini GF, Anderson HU, Nasrallah MM, Sparlin DM (1991) J Solid State Chem 94:329–336
32. Kertesz M, Riess I, Tannhauser DS, Langpape R, Rohr FJ (1982) J Solid State Chem 42:125–129
33. Wandekar RV, Wani BN, Bharadwaj SR (2005) Mater Lett 59:2799–2803
34. Gaudon M, Laberty-Robert C, Ansart F, Stevens P, Rousset A (2004) Solid State Sci 4:125–133
35. van Roosmalen JAM, Cordfunke EHP, Huijsmans JPP (1993) Solid State Ionics 66:285–293
36. Ghosh A, Sahu AK, Gulnar AK, Suri AK (2005) Scripta Mater 52:1305–1309
37. Pang G, Xu X, Markovich V, Avivi S, Palchik O, Koltypin Y, Gorodetsky G, Yeshurun Y, Buchkremer HP, Gedanken A (2003) Mater Res Bull 38:11–16
38. Das N, Bhattacharya D, Sen A, Maiti HS (2009) Ceram Int 35:21–24
39. Belardi RM, Deseure J, Brant MC, Matencio T, Domingues RZ (2009) Ionics 15:227–232
40. Ming Q, Nersesyan MD, Richardson JT, Luss D, Shiryaev AA (2000) J Mater Sci 35:3599–3606

41. Bell RJ, Millar GJ, Drennan J (2000) Solid State Ionics 131:211–220
42. da Conceicao L, Silva CRB, Ribeiro NFP, Souza MMVM (2009) Mater Charact 60:1417–1423
43. Bidrawn F, Kim G, Aramrueang N, Vohs JM, Gorte RJ (2010) J Power Sour 195:720–728
44. Zhang Q, Nakagawa T, Saito F (2000) J Alloy Compd 308:121–125
45. Alonso JA, Martinez-Lope MJ, Casais MT, Mac Manus-Driscoll JL, de Silva PSIN, Cohen LF, Fernandez-Diaz MT (1997) J Mater Chem 7(10):2139–2144
46. Jiang SP, Zhang JP, Ramorakash Y, Milosevic D, Wilshier K (2000) J Mater Sci 35:2735–2741
47. Duprat AM, Alphonse P, Sarda C, Rousset A (1994) Mater Chem Phys 37:76–81
48. Van Roosmalen JAM, Cordfunke EHP (1994) J Solid State Chem 110:106–108
49. Wold A, Arnott RJ (1959) J Phys Chem Solids 9:176–180
50. Aruna ST, Muthuraman M, Patil KC (1999) Solid State Ionics 120:275–280
51. Taguchi H, Yoshioka H, Matsuda D, Nagao M (1993) 104:460–463
52. Shu Q, Zhang J, Liu J (2005) 390:240–244
53. Jonker GH, Van Santen JH (1950) Physica XVI 3:337–349
54. Van Santen JH, Jonker GH (1950) Phys XVI 7–9:599–600
55. Jonker JH (1954) Phys XX 11:1118–1122
56. Kuo JH, Anderson HU, Sparlin DM (1990) J Solid State Chem 87:55–63
57. Hammouche A, Schouler EJL, Henault M (1988) Solid State Ionics 28–30:1205–1207
58. Goodenough JB (1955) Phys Rev 100(2):564–573
59. Takeda Y, Sakaki Y, Ichikawa T, Imanishi N, Yamamoto O, Mori M, Mori N, Abe T (1994) Solid State Ionics 72:257–264
60. Mori M (2004) Solid State Ionics 174:1–8
61. Scotti C, Gharbage B, Lauret H, Levy M, Hammou A (1993) Mater Res Bull 28:1215–1220
62. Kertesz M, Riess I, Tannhauser DS, Langpape R, Rohr FJ (1982) J Solid Sate Chem 42:125–129
63. Jiang SP (2002) Solid State Ionics 146:1–22
64. Jiang SP (2003) J Power Sour 124:390–402
65. Hansen KK, Vels Hansen K (2007) Solid Sate Ionics 178:1379–1384
66. Arul Antiny S, Swaminathan A, Nagaraja KS, Sreedharan OM (2001) J Alloy Compd 322:113
67. Nakamura T, Petzow G, Gauckler LJ (1979) Mater Res Bull 14:649
68. Shridharan OM, Pankajavalli R, Gnanammoorthy JB (1983) High Temp Sci 16:251
69. Hildrum R, Brustad M, Changzhen W, Johannesn O (1994) Mater Res Bull 29:851
70. Kuo JH, Anderson HU, Spaelin DM (1989) J Solid State Chem 83:52
71. Van Roosmalen JAM, Cordfunke EHP, Helmholdt RB (1994) Solid State Chem 110:100
72. Norby P, Krogh IG, Andersen EK, Krogh E, Andersen J (1995) Solid State Chem 119:191
73. Hammouche A, Siebert E, Hammou A (1989) Mater Res Bull 24:367
74. Jonker GH, Vansanten JH (1991) Physica 3:337
75. Harwood G (1990) Proc Phys Soc 68:586
76. Larminie J. Fuell cell systems explained (2nd edn). Andrew Dick
77. Takeda Y, Kanno R, Noda M, Tomida Y, Yamamoto O (1987) J Electrochem Soc 134:2656
78. Li Z, Behruzi M, Fuerst L, Stover D (1993) In: Singhal SC, Iwahara H (eds) SOFC III, PV 93-4. Eectrochemical Society, p 171
79. Yasuda I, Ogasawara K, Hishinuma M, Kawada T, Dokiya M (1996) Solid State Ionics 1197:86
80. Carter S, Selcuk A, Chater RJ, Kadja J, Kilner JA, Steele BCH (1997) Ionics 597:53
81. Labrlncha JA, Jian Meng L, Santos MP, Marques FMB, Frade JR (1993) Mater Res Bull 28:101
82. Vanherle J, Mcevoy AJ, Thampi KR (1994) Electrochim Acta 39(11/12):1675
83. Jurgensen MJ, Primdahl S, Mogensen M (1999) Electrochim Acta 44:4195
84. Kroeger FA, Vink HJ (1956) Solid State Phys 3:307
85. Kenjo T, Nishiya M (1992) Solid State Ionics 57:295
86. Sasaki K, Gauckler LJ (1995) In: Proceedings of the international symposium on struct. func. grad. mater., vol 3, p 651
87. Juhl M, Primdahl S, Manon C, Mogensen M (1996) J Power Sour 61:173
88. Estergaerd MJL, Clausen C, Bagger C, Mogensen M (1995) Electrochim Acta 40:1971

89. Gaudon M (2002) Solid State Sci 4:125
90. Decorse P, Caboche G, Dufour L (1999) Solid State Ionics 117:161
91. Van Roomsmalen JAM, Van Vlaanderen P, Corefunke EHP, Ijdo WL (1995) J Solid State Chem 114:516
92. Aruna TS, Muthuraraman M, Patil KC (1997) Mater Chem 7(12):2499
93. Mahendiran R, Mahesh R, Raychaudri AK, Rao CNR (1995) J Phys 44:393
94. Tanaka J, Umehara M, Tamura S, Tsukioka M, Ehara S (1982) J Phys Soc 51:1236
95. Karim DP, Aldrad AT (1979) Phys Rev B 20:2255
96. Kharton VV, Kovalevsky AV, Tikhonivich N, Naumivuch EN, Viskup AP (1998) Solid State Ionics 110:53
97. Jonker GH, Phys J (1959) Chem Solids 9:165
98. Lee HK (2002) J Mater Chem Phys 77:639
99. Kertesz M, Riess I, Tannhauser GS, Langpage R, Rahr FJ (1982) J Solid State Chem 42:125
100. Van Roosmalen JAM, Cordfunke EHP, Huijamans JPP (1993) Solid State Ionics 66:279
101. Van Roosmalen JAM, Huijsmans JPP, Plomp L (1998) Solid State Ionics 66:279
102. Haung T, Hong YS (2003) Mater Sci Eng B 103:207
103. Dessemond L, Stevens P (2004) J Power Sour 133:214
104. Nagde KR, Bhoga SS (2010) Ionics 16:361
105. Tanhauser DS, Langpape R, Rohr FJ (1982) J Solid State Chem 42:125
106. Sakaki Y, Takeda Y, Kato A, Imanishi N, Yamamoto O, Hattori M, Esaki Y (1999) Soid State Ionics 118:187
107. Gaudon M, Robert CL, Ansart F, Stevens P, Rousset A (2004) J Power Sour 133:214
108. Kening S, Jinhuo P, Naiqing Z, Xinbing C, Shen X, Derui Z (2008) Rare Met 27(3):278
109. Tietz F (1999) Ionics 5:129–139
110. Inoue T, Eguchi K, Setoguchi T, Arai H (1990) Solid State Ionics 40(41):407
111. Kindermann L, Das D, Bahadur R, Weir R, Nickel H, Hilpert K (1997) J Am Ceram Soc 80:909
112. Sekido S, Tachibana H, Yamamura Y, Kambara T (1990) Solid State Ionics 37:253
113. Tai LW, Nasrallah MM, Anderson HU, Sparlin DM, Sehlin SR (1995) Solid State Ionics 76:259
114. Roosmalen JAM, Cordfunke EHP (1992) Solid State Ionics 52:303
115. Tiff EI, Wersing W, Schiefl M, Greiner H, Bunsenges B (1990) Phys Chem 94:978
116. Uhlenbruck S, Tietz F (2004) Mater Sci Eng B Solid State Mater Adv Technol 107:211
117. Caboche G, Dufour LC, Morin F (2001) Solid State Ionics 144:211
118. Kuharuangrong S, Dechakupt T, Aungkavattana P (2004) Mater Lett 58:1964
119. Kuharuangrong S (2004) Cerem Int 30:273
120. Endo A, Fukunaga H, Wen C, Yamad K (2000) Solid State Ionics 135:353
121. Nagabushan BM, Sreekant RP, Ramesh KP, Shivkumara C, Chandrapp GT (2006) Mater Res Bull 42:1735
122. Yi JY, Choi GM (2002) Solid State Ionics 148:557
123. Steele BCH (1997) Solid State Ionics 94:239
124. Phillipps MB, Sammes NM, Yamamoto O (1999) Solid State Ionics 123:131
125. Huang X, Liu J, Lu Z, Liu W, Pei L, He T, Liu Z, Wenhui S (2000) Solid State Ionics 130:195
126. Tietz F, Haanappel VAC, Mai A, Mertens J, Stover D (2006) J Power Sour 156:20
127. Maguire E, Gharbage B, Marques FMB, Labrincha JA (2000) Solid State Ionics 127:329
128. Kostogloudis GC, Ftikos C, Ahmad-Khanlou A, Naoumidis A, Stover D (2000) Solid State Ionics 134:127
129. Bebelis S, Kotsionopoulos N, Mai A, Tietz F (2007) J Appl Electrochem 37:15
130. Dusastre V, Kilner JA (1999) Solid State Ionics 126:163
131. Xia CR, Chen FL, Liu ML (2001) Electrochem Solid-State Lett 4(5):A52
132. Shao Z, Haile SM (2004) Nature 43:170
133. Ullmann H, Trofimenko N, Tietz F (2000) Solid State Ionics 138:79
134. Hart NT, Brandon NP, Day MJ, Shemilt JE (2001) J Mater Sci 36:1077
135. Holtappels P, Bagger C (2002) J Eur Ceram Soc 22:41

136. Xia CR, Rauch W, Wellborn W, Liu ML (2002) Electrochem Solid State Lett 5:A217
137. Yamada H, Nagamoto H (1993) In: Proceedings of the 3rd international symposium on solid oxide fuel cells, Honolulu, Hawaii (USA). Electrochemical Society, p 213
138. Yamamoto O, Takeda Y, Kanno R, Kojima T (1989) In: Yamamoto O, Dokiya M, Tagawa H (eds) Proceedings of the international symposium on solidoxide fuel cells, Nagoya. Science House, Tokyo, p 87
139. Teraoka Y, Nobunaga T, Okamoto K, Miura N, Yamazoe N (1991) Solid State Ionics 48:207
140. Xu X, Xia C, Xiao G, Peng D (2005) Solid State Ionics 176:1513
141. Ishihara T, Honda M, Shibayama T, Minami H, Nishiguchi H, Takita Y (1998) J Electrochem Soc 145:3177
142. Fukunaga H, Koyama M, Takahashi N, Wen C, Yamada K (2000) Solid State Ionics 132:279
143. He H, Huang X, Chen L (2000) Solid State Ionics 130:183
144. Yasumoto K, Inagaki Y, Shiono M, Dokiya M (2002) Solid State Ionics 148:545
145. Xia C, Rauch W, Chen F, Liu M (2002) Solid State Ionics 149:11
146. Minh NQ (1993) J Am Ceram Soc 76:563
147. Selman JR, Lin YP (1993) Electrochim Acta 38(14):2063
148. Levie RD (1967) Adv Electrochem and Electrochem Eng 6:329
149. Levie RD (1963) Electrochemica acta 8:751
150. Ostergard MJL, Mogensen M (1993) Electrochim Acta 38(14):2015
151. Barbucci A, Viviani M, Carpanese P, Vladikova D, Stoynov Z (2006) Electrochim Acta 51:1641
152. Brant MC, Matencio T, Dessemond L, Domingues RZ (2006) Solid State Ionics 177:915
153. Choi JH, Jang JH, Oh SM (2001) Electrochim Acta 46:867
154. Deseure J, Bultel Y, Dessemond L, Siebert E (2005) Electrochim Acta 50:2037
155. Ralphz JM, Schoeler AC, Krumpelt M (2001) J Mater Sci 36:1161
156. Hildrum R, Aasland S, Johannesen O (1993) Solid State Ionics 66:207
157. Kostogloudis GC, Fertis P, Ftikos C (1998) J Euro Ceram Soc 18:2209
158. Kamata H, Hosaka A, Mizusakia J, Tagawa H (1998) Solid State Ionics 106:237
159. Yoon HS, Choi SW, Lee D, Kim BH (2001) J Power Sour 93:1
160. Kim J, Kim GD, Moon J, Park Y, Lee W, Kobayashi K, Nagai M, Kim C (2001) Solid State Ionics 143:379

Chapter 8
Silicon Nanowires/Graphene Oxide Heterojunction for Photovoltaics Application

Sadhna Rai, Rabina Bhujel, Joydeep Biswas, and Bibhu P. Swain

1 Introduction

Over the years, nanomaterials (NMs) and nanotechnology have made tremendous development in science and technology. The advantage of tuning the properties of NMs in comparison with bulk counterpart makes them more desirable. Such an advantage opens up new possibilities in different applications. NMs are used in energy harvesting and storage devices [1, 2], sensors [3], detectors [4], medicine [5], photocatalysis [6], etc. In day-to-day life, the NMs have been put to applications in cosmetics (TiO_2, ZnO), paints (TiO_2, SiO_2), clothes (TiO_2, ZnO and carbon nanotubes) and sports (carbon nanotubes). Due to the extraordinary property shown by the NMs, the researchers are working hard to explore them to better the world and humankind. In the upcoming section, the categorization of the NMs in different dimensions and the formation of heterojunctions between them have been discussed. The use of the NMs in energy generation is discussed in the following section, emphasizing photovoltaics (PV) and its principle. Further, the uses of silicon nanowires (SiNWs) and graphene oxide (GO) have been discussed.

S. Rai (✉) · R. Bhujel
Department of Chemistry, Sikkim Alpine University, Kamrang, Namchi, Sikkim 737126, India
e-mail: sadhnarai15@gmail.com

J. Biswas
Department of Chemistry, Sikkim Manipal Institute of Technology, Sikkim Manipal University, Majitar, East Sikkim 737136, India

B. P. Swain
Department of Physics, National Institute of Technology, Manipur, Langol, Manipur 795004, India
e-mail: bibhuprasad.swain@gmail.com

B. P. Swain (ed.), *Energy Materials*, Materials Horizons: From Nature to Nanomaterials,
https://doi.org/10.1007/978-981-99-3866-7_8

1.1 Dimensions of Nanomaterials

NMs are of four types based on dimensions, namely 0D, 1D, 2D and 3D NMs. Examples of 0D NMs are graphene quantum dots, carbon quantum dots, inorganic quantum dots, polymer dots [7] and perovskites [8]. 1D NMs include nanowires, nanorods, nanotubes and nanofibres. The 2D NMs exist as nanosheets, nanoflakes and nanofilms, viz. graphene, metal dichalcogenides [9] and hexagonal boron nitride (hBN). Further, 3D NMs consist of multiple nanosheet layers, nanoparticle aggregates, nanorod bundles and nanowires with nanoscale features [10].

1.2 Classification of Heterojunctions Based on the Dimension

The heterojunction is formed by combining two semiconductor materials of different bandgaps. The properties of the heterojunction depend primarily on the arrangement of the bands along the junction. "Based on the alignment of bandgaps at the interface of the semiconductors, heterojunctions are of three types: Straddling gap (Type I), Staggered gap (Type II) and Broken gap (Type III)", as shown in Fig. 1.

However, since the discovery of 2D materials, a new type of heterojunction known as van der Waals (vdWs) heterojunction was introduced. The 2D NMs can be layered on top of one another like Lego blocks to create heterojunctions [11]. Examples of 2D materials which can be considered for designing vdWs heterojunction are hBN, MoS_2, WS_2, WSe_2, TiO_2, bismuth strontium calcium copper oxide (BSCCO), graphene, etc. [11]. The weak vdWs forces hold the layered structures together. Hence, vdWs heterostructure may be defined as a heterostructure held together by weak vdWs forces and not strong chemical bonds [12]. The vdWs heterostructure has advantages, viz. (1) no requirement of chemical reactions for the integration of

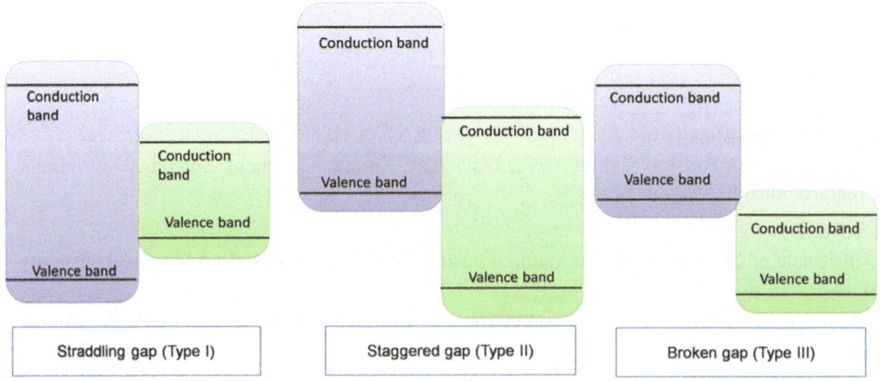

Fig. 1 Types of heterojunctions

heterostructure, (2) strict lattice matching is not required and (3) materials with different electronic or chemical properties can be integrated to create materials with new properties. The low-dimensional materials provide an extensive library for designing different heterostructures. The 2D materials are the most studied materials for forming vdWs heterojunctions. Taking 2D NMs as one of the component heterojunctions, viz. 2D-3D, 2D-2D and 2D-0D, can be integrated.

1.3 Density of States

"The density of states (DOS) function describes the number of states that are available in a system and is essential for determining the carrier concentrations and energy distributions of carriers within a semiconductor" [13]. Based on the confinement of the carriers, the materials can be 3D (no confinement), 2D (1D confinement), 1D (2D confinement) and 0D (3D confinement). The equation for the DOS for these materials is given below [14]

$$g(E)_{3\mathrm{D}} = \frac{1}{2\pi^2}\left(\frac{2m^*}{\hbar^2}\right)^{3/2} E^{1/2} \tag{1.1}$$

$$g(E)_{2\mathrm{D}} = \frac{m^*}{\pi\hbar^2} \tag{1.2}$$

$$g(E)_{1\mathrm{D}} = \frac{1}{\pi\hbar}\left(\frac{m^*}{2E}\right)^{1/2} \tag{1.3}$$

$$g(E)_{0\mathrm{D}} = 2\delta E \tag{1.4}$$

Here E, m^*, $\hbar$ and δ are the energy, mass of electron, reduced Planck's constant and delta function, respectively. A schematic for the DOS in 3D, 2D, 1D and 0D materials is given in Fig. 2.

1.4 Current Scenario of Energy in the World

Energy and power are significant to every nation in the world. They have an important role in the growth and development of a country. Energy is a key factor in achieving the global sustainable agenda. To date, the international energy demand is fulfilled mainly by fossil fuels. In 2018, fossil fuel, hydroelectric, nuclear and renewable energy accounted for 84.7, 6.8, 4.4 and 4% of global energy utilization. The major factors influencing energy demand are population growth, economic growth, technological advancement and urbanization [15].

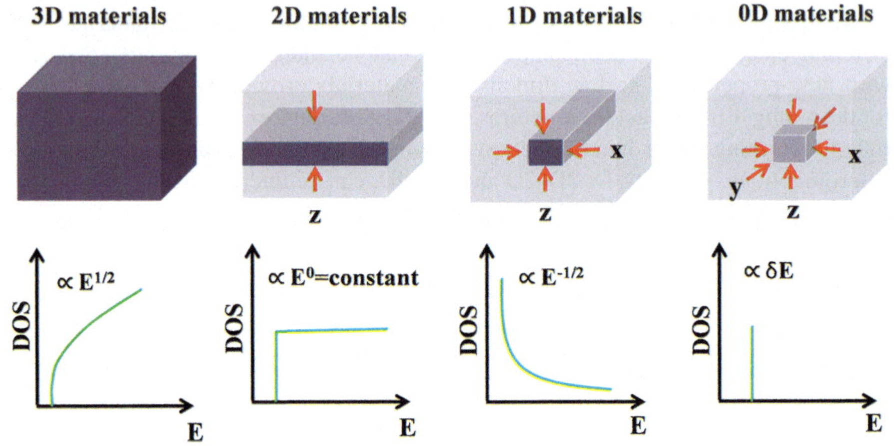

Fig. 2 Schematic showing DOS in 3D, 2D, 1D and 0D materials

There are four major sections for global energy demand: transportation, residential, industry and electricity. The energy requirement in the transportation division increases with the growth of a nation. The energy is required in aviation, railways, roads and maritime transport. Thus, the energy demand in transportation has a crucial part in the future energy demand. Industries also have a significant part in the development of a country. So, the demand for energy in this sector also increases as a country develops. Similarly, the requirement for energy in the residential area also increases. The electricity sector is highly significant because it causes less carbon contamination and is efficient. The electricity is mainly generated by fossil fuels (hydrocarbons), viz. natural gas, coal and oil. However, the supply of fossil fuels is limited because they are used more rapidly than they are being created. Also, fossil fuels are detrimental as they lead to significant concerns such as climate change, environmental pollution and health hazards. Alternative to handling these concerns are renewable sources such as solar, wind, geothermal, hydroelectricity, biofuels and hydrogen [16].

To control the threat of global warming, different countries in the world are shifting towards zero-emission. According to the Paris Agreement, the countries must reduce greenhouse gas emissions substantially by 2030 and reach net zero by 2050. The nations are shifting toward green energy sources to meet this goal. According to BP 2022, fossil fuel accounted for ~ 80%, while renewable energy accounted for ~ 10% of the global primary energy. However, the share of fossil fuels is gradually decreasing and may lie between 60 and 20% by 2050. Further, renewable energy is increasing, and its share may lie between 35 and 65% by 2050 [17].

1.5 Solar Energy

Solar energy is an essential source of sustainable energy. It is a profuse source of energy for the earth. Figure 3 presents the yearly average solar energy intensity over the earth's surface. "The worldwide variation of annual effective solar irradiance is ~ 60 to 250 W m^{-2}" [18]. Solar energy has a high capacity to solve the energy demand; it is environment-friendly, cost-free and an unlimited energy source [19]. These features make it a promising sustainable energy resource. The sunlight can be used mainly for the generation of heat and electricity.

1.5.1 Photovoltaics

The process of conversion of sunlight (photons) into electricity (voltage) is called photovoltaics (PVs). In PVs, the sunlight is concentrated on the PV cells and then converted into electrical energy. Since its discovery, the PV sector has started quick development. To reach the global energy requirement, the efficiency of PV cells plays a significant role. The researchers have dedicated their work to increasing solar cell efficiency, and some have obtained brilliant results. Based on the reports from the literature [20–23], Fig. 4 displays the historical development of PV cells.

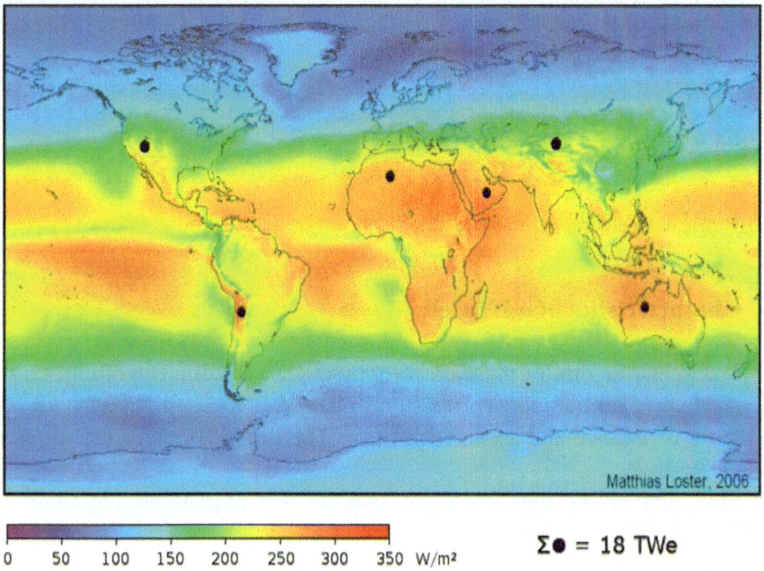

Fig. 3 Annual average intensity of solar energy over the surface of the earth [18]

Fig. 4 Historical development of photovoltaic cells

1.5.2 Characteristics of a PV Cell

Figure 5 presents the circuit diagram for a practical PV cell. The circuit consists of a power source, a diode and two resistors. Here, I_{ph}, R_{sh} and R_{s} refer to the input power, shunt resistance and series resistance. The diode is connected in parallel to R_{sh} and series to R_{s}.

The equivalent circuit diagram shows that the current generated by the PV cell is given by Eq. 1.5 [21].

$$J = J_{\text{ph}} - J_{\text{s}} - J_{\text{sh}} \tag{1.5}$$

Here J, J_{ph}, J_{s} and J_{sh} refer to the output current, photo-generated current, diode current and shunt current, respectively.

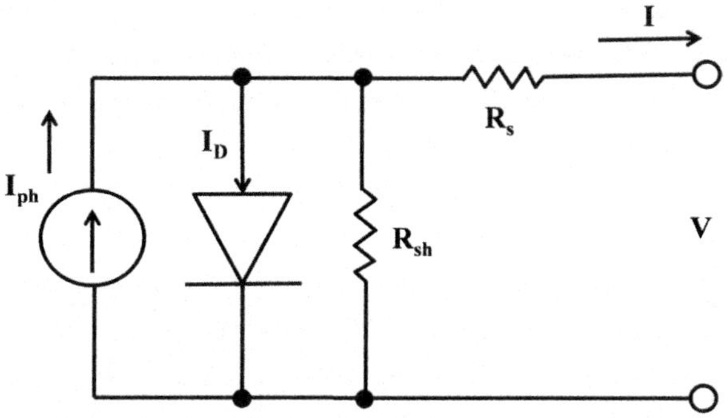

Fig. 5 Circuit diagram of a PV cell

The voltage across both the diodes (V_j) is given by

$$V_j = V + JR_s \tag{1.6}$$

The current that passes through the diode (J_s) is given by the Shockley diode equation

$$J_s = J_o\left[\exp\left\{\frac{q(V + JR_s)}{\eta k_B T}\right\} - 1\right] \tag{1.7}$$

Here J_o, q, V, η, k_B and T refer to the reverse saturation current of the diode, electronic charge, voltage, ideality factor, and Boltzmann's constant and absolute temperature, respectively.

Also, the shunt current J_{sh} is given by

$$J_{sh} = \left(\frac{V + JR_s}{R_{sh}}\right) \tag{1.8}$$

From Eqs. 1.7 and 1.8, Eq. 1.9 was obtained.

$$J = J_{ph} - J_o\left[\exp\left\{\frac{q(V + JR_s)}{\eta k_B T}\right\} - 1\right] - \left(\frac{V + JR_s}{R_{sh}}\right) \tag{1.9}$$

The important factors that control the power conversion efficiency (PCE) of a PV cell are maximum power point (P_m), open circuit voltage (V_{oc}), short circuit current density (J_{sc}) and fill factor (FF). The current density versus voltage (J–V) curve of a PV cell showing these parameters is presented in Fig. 6. The PCE of a PV cell is represented by Eq. 1.10.

$$\eta = \frac{P_{out}}{P_{in}} = \frac{V_{oc} J_{sc} \text{FF}}{P_{in}} \tag{1.10}$$

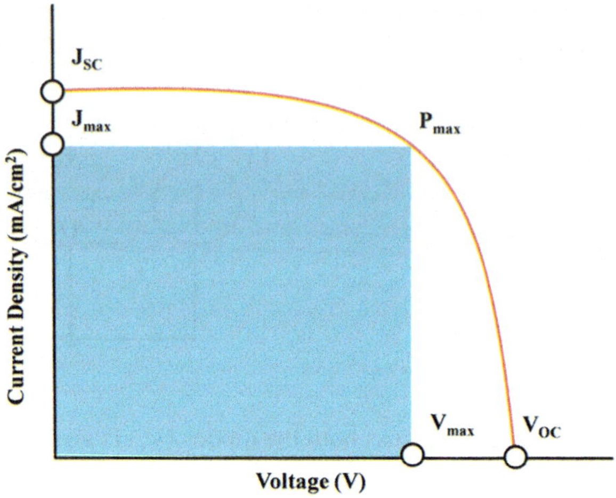

Fig. 6 *J–V* characteristics of a solar cell

1.6 Silicon Nanowires

Generally, silicon (Si) is the material used in PV cells. It is widely used because it is abundantly available and is safe for the environment. Greater than 90% of commercial PV cells comprise Si. However, Si suffers from the drawback of indirect bandgap and high reflectivity. This is tackled using thick Si (100–500 μm), which adds to the solar cell production cost. Also, the thick solar cells suffer loss in PCE due to non-radiative recombination losses [22]. However, an adequate solar cell with a higher PCE must be less expensive and smaller. According to the laws of thermodynamics, the maximum possible PCE of a mono junction Si PV cell at room temperature under one sun is 32.33% [24]. Nanomaterials are used in PV cells to address these issues. In this regard, silicon nanowires (SiNWS) have revealed great prospective for application in PVs since the nanowires (NWs) have a high aspect ratio with smaller dimensions. The optical image for Si and SiNWs is shown in Fig. 7. The researchers have much interest in studying SiNWs for PV application because the nature of the SiNWs can be tuned by varying the morphology, density and porosity of the NWs.

The size of the bulk Si is decreased when NWs are formed, which creates more surface atoms. The reflectance in the SiNWs is decreased to a great extent compared to that of the planar Si [25, 26]. In SiNWs, multiple reflections take place, as a result of which more light is trapped, as shown in Fig. 8. This increases the absorbance in SiNWs, thus enhancing the light-harvesting property. The length, diameter and roughness affect the antireflective properties. The size, growth direction and surface morphology affect the electronic property of the SiNWs. Owing to the quantum confinement effect (QCE), the motion of the particles is restricted, which in turn affects the energy bandgap. The bandgap depends on the size of the SiNWs. Bulk Si has an indirect bandgap of 1.12 eV, whereas SiNWs have a direct bandgap [27]. The bandgap can be regulated by controlling the diameter of the NWs. As the diameter

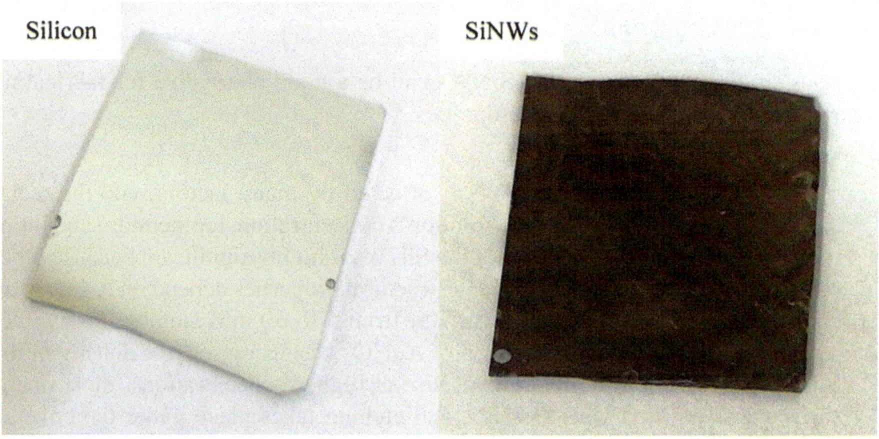

Fig. 7 Optical image of silicon and SiNWs

of the NWs decreases, the value of the bandgap increases [28]. Thus, the factors contributing to the change in the properties of the SiNWs must be considered.

Fig. 8 Schematic showing multiple reflections in SiNWs

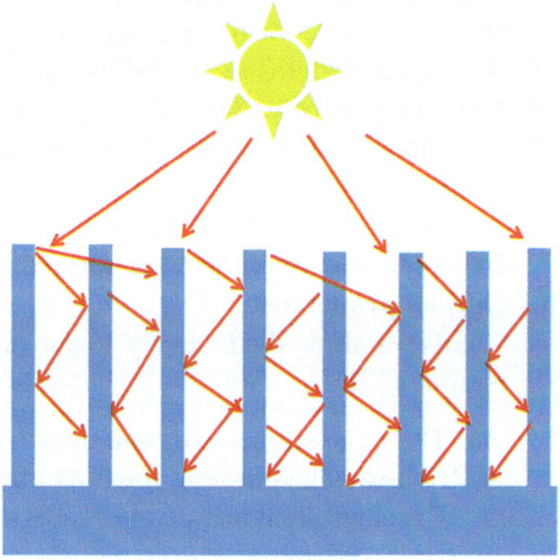

1.6.1 Properties of SiNWS

Previous studies have shown that SiNWs can be a good alternative for fabricating solar cells due to their distinctive properties.

a. **Morphological properties**

   The morphology of the SiNWs is affected by many factors, viz. the metal catalyst's concentration, etching solution's concentration, temperature and duration of etching. The morphology of the SiNWs is highly significant because other properties like optical, electronic and structural properties depend on it. Cao et al. considered the concentration of $AgNO_3$ from 5 to 30 mM suitable for synthesizing SiNWs. As the concentration of $AgNO_3$ was increased, the density of the SiNWs decreased. This may be because at higher concentrations, most of the surface was covered with Ag NPs, and etching takes place under the covered region [29]. Omar et al. demonstrated the effect of the concentration of H_2O_2 on the growth of SiNWs. The SiNWs were formed at concentration 0.3–0.6 M. However, at their highest concentration, i.e., 0.7 M, only rough surface and no SiNWs was obtained. Additionally, at low (0.3 M) and high (0.6 M) concentrations, the SiNWs were not vertically aligned. The vertically aligned SiNWs were obtained only at a concentration of 0.5 M [30]. Rahmani et al. varied the etching time from 30 to 300 s (Fig. 9). It was observed that at low etching duration, along with the NWs, pores were made on Si. At longer etching duration, the pores decreased, and more SiNWs were formed. Also, the bundling of the SiNWs increased with the etching duration, possibly due to the presence of vdWs forces. The lengths of the SiNWs were obtained as 1.3 and 1.7 μm for 30 and 300 s, respectively [31].

   Fellahi et al. deliberated on the effect of temperature on the MACE of Si. Their group observed that the well-aligned vertical SiNWs were formed in the temperature range of 30–50 °C. In the temperature range of 30–50 °C, porous and columnar structures were formed [32].

b. **Optical properties**

   SiNWs show distinctively lower reflectance than bulk Si for the following reasons [33]

   i. The large surface area and density of SiNWs
   ii. Sub-wavelength light trapping effects of SiNWs and
   iii. Excellent light harvesting properties due to the internal reflections.

   Gonchar et al. observed that SiNWs show reflectance values of 5–10% in the UV–Vis range which was much lower than bulk Si, i.e., 50% (Fig. 10). The reflection peaks observed at 280 and 370 nm were ascribed to the direct bandgap of the SiNWs. The intense light scattering and absorption that caused the light

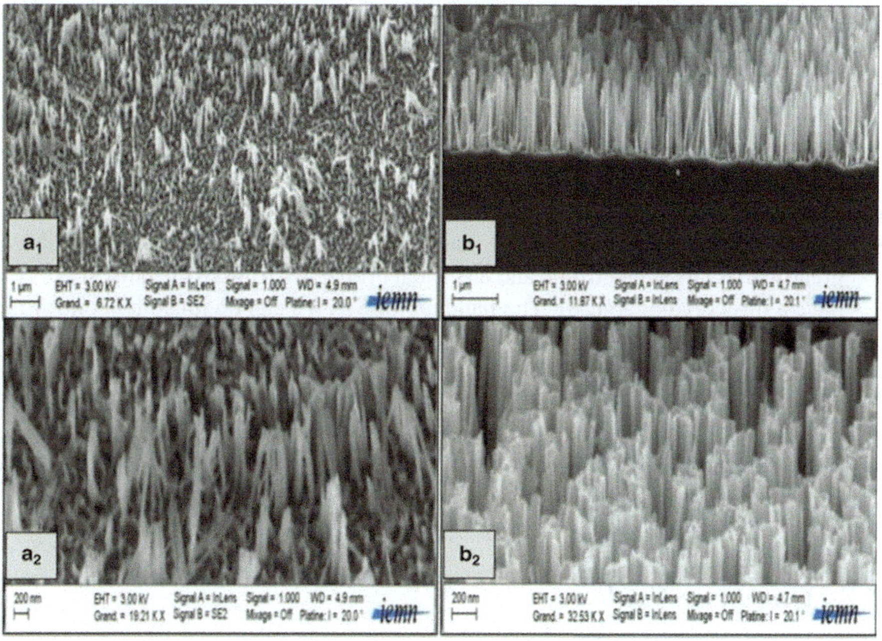

Fig. 9 SEM images of SiNWs synthesized at $\mathbf{a_1}$, $\mathbf{b_1}$ 30 s and $\mathbf{a_2}$, $\mathbf{b_2}$ 300 s [31]

localization were considered the factors that led to decreased reflectance values for SiNWs [34].

Ge et al. studied the variation in the length of the SiNWs for the manufacture of PEDOT:PSS/SiNWs solar cells. They observed that as the length of the SiNWs

Fig. 10 Total reflectance spectra of SiNWs synthesized by varying the pH of H_2O_2:NH_4F [35]

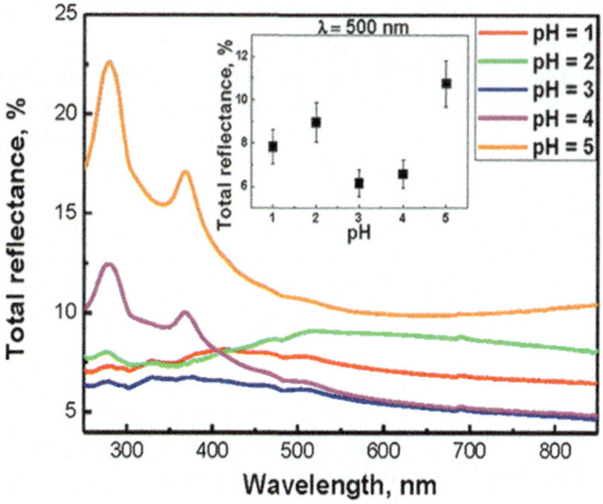

increased, the reflectance decreased. The longer SiNWs show increased light trapping compared to shorter ones resulting in lower reflectance [36].

Hutagalung et al. showed that SiNWs show different reflectance values in different wavelength ranges. The minimum reflectance of 3.5% (200–400 nm), 12.1% (400–750 nm) and 9.8% (750–1000 nm) was observed. The reflectance for SiNWs was much lower than the bulk Si [37].

c. **Electronic properties**

The morphology, size, surface roughness and direction of growth of SiNWs greatly affect their electronic and electrical properties. Unlike the bulk Si, SiNWs have a direct bandgap [28]. The bandgap of SiNWs can be tuned by controlling the diameter of the NWs. The bandgap increases with the decrease in the diameter of the NWs. Sacconi et al. demonstrated that the bandgap of the hydrogenated NWs increased from 1.56 to 2.66 eV when the diameter was decreased from 3.2 to 1.6 nm [38]. Leu et al. studied the surface passivation of SiNWs with halogens, viz. chlorine, bromine and iodine. This study observed that iodine passivation gave the smallest bandgap, followed by bromine and chlorine termination [39].

1.7 Graphene Oxide

The discovery of graphene in 2004 brought a revolution in the area of 2D materials. Due to the exceptional properties of graphene, viz. brilliant optical transparency, excellent electrical conductivity (1×10^8 S m^{-1}), flexibility and carrier mobility (2×10^5 cm^2 (V s)$^{-1}$), it may be used in PV applications [40]. However, due to the zero bandgap in graphene, its application in nano-electronic devices is restricted [41]. Additionally, graphene is hydrophobic, making it difficult to solution processability. The synthesis of pristine graphene involves complex procedures, and the yield is low, which makes the material expensive [42].

The drawbacks of graphene can be overcome by using other members of the graphene family, viz. GO and rGO. The structures of graphene, GO and rGO are displayed in Fig. 11. By extensive oxidation of graphite, GO can be synthesized. It contains various oxygen functional groups (OFGs) like –OH, –C=O and C–O–C along the base and periphery. GO is polar and hydrophilic. The process for the synthesis of GO is simpler, and the yield is high. GO consists of sp^3 carbons atoms, which disrupt the otherwise planar structure of graphene resulting in low electrical conductivity.

The problem can be addressed by controlling the presence of OFGs. The OFGs can be removed by chemical reduction of GO which gives rGO or chemically modified graphene. The properties of rGO are close to that of graphene and possess additional benefits of non-zero bandgap and solution processability [43]. The reduction of GO to rGO can be achieved by various methods, viz. chemical, thermal, hydrothermal and electrochemical, which are later discussed in the literature survey. Also, green-reducing agents are used for the partial removal of OFGs. The green-reducing agents provide an easy and eco-friendly route for the preparation of rGO

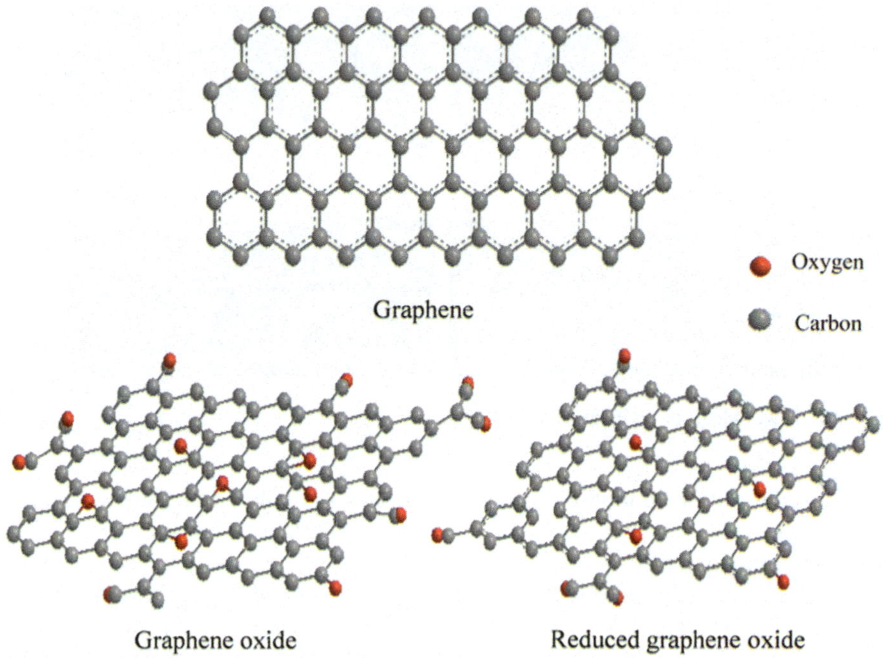

Fig. 11 Structure of graphene, GO and rGO

without the involvement of stabilizers [44]. The challenging factor for using GO in solar cells is obtaining thin and transparent conducting film.

1.7.1 Properties of GO and rGO

a. **Morphological properties**

The morphological properties of GO and rGO are studied using SEM and TEM characterization techniques. The GO and rGO mostly show wrinkled and crumpled structures. In the study by Sharma et al., SEM images (Fig. 12) revealed that the wrinkles in rGO were more than that of GO. The corrugation increased when GO was reduced to rGO because of the removal of OFs [45]. Lei et al. reported the preparation of rGO paper. The TEM images showed a few layers which were overlapped together. The FESEM images showed silk wave-like morphology. It was suggested that the aggregation might be because of the layers in rGO of solvent evaporation [46].

Aziz et al. studied the morphological properties of GO and rGO using TEM (Fig. 13). The thick flat layers of GO had irregular shapes with a rough surface and minimum creases. The presence of OFs and high exfoliation during the oxidation led to a disordered and smooth structure. On the other hand, thin flat

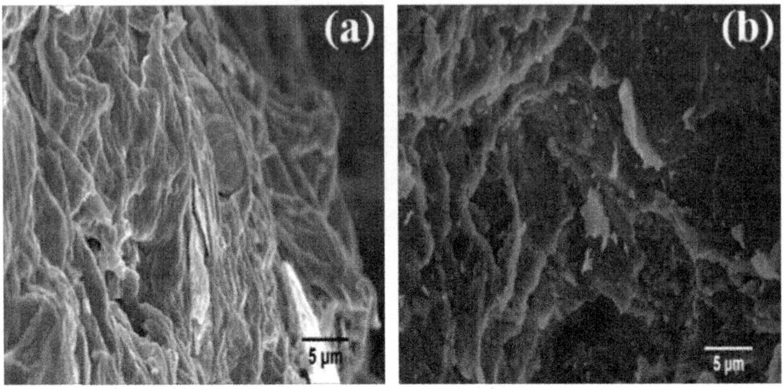

Fig. 12 SEM images of **a** GO and **b** rGO [45]

layers of rGO with multiple layers and creases were observed. This was caused by removing OFs between the layers during the reduction process [47].

b. **Optical properties**

The optical properties of GO and rGO are studied by employing UV–Vis spectroscopy. The UV–Vis spectra obtained by Johra et al. showed a peak at 235 nm corresponding to the $\pi \rightarrow \pi^*$ transition because of the presence of the C=C bonds in GO. After reduction, the peak shifted to 265 nm. The peak shift to a longer wavelength shows that the sp^2 carbons are restored when GO is reduced to rGO [48]. Similar results were obtained by several other groups [49–52].

c. **Electronic properties**

GO is a wide bandgap material. The bandgap is decreased when GO is reduced, i.e., removal of OFs. Velasco-Soto et al. used eco-friendly reagents (glucose,

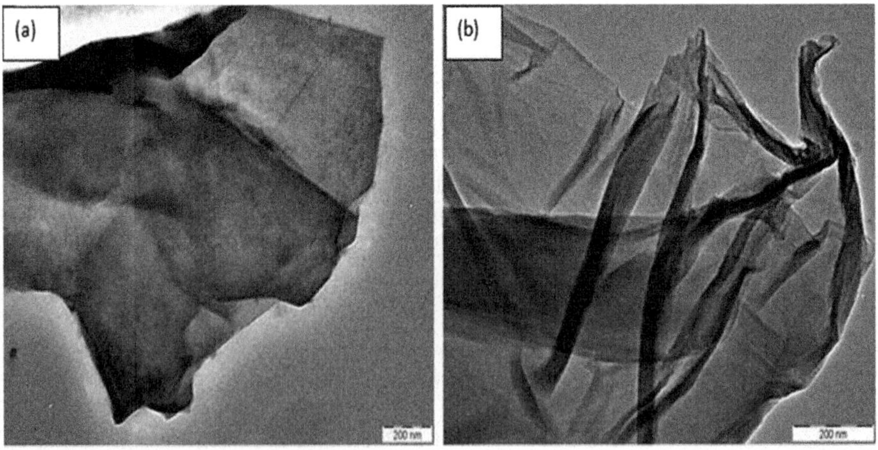

Fig. 13 TEM images of **a** GO and **b** rGO [47]

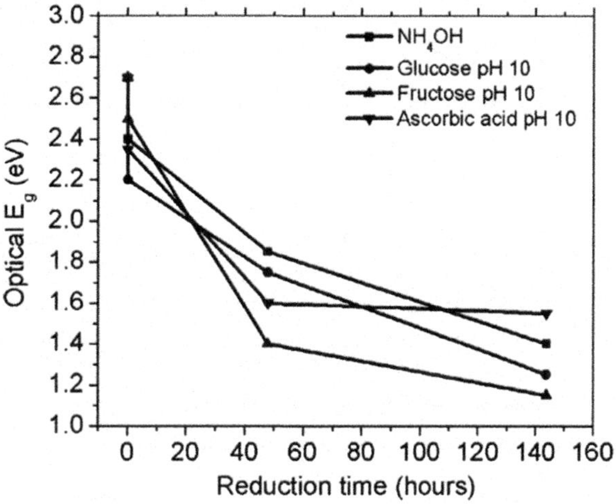

Fig. 14 Variations in the bandgap of GO obtained by varying the reducing agents and reduction time [53]

fructose and ascorbic acid) to reduce GO and tune the bandgap of GO. The bandgap was also tuned by controlling the reduction of GO over a varied time, as shown in Fig. 14. They successfully reduced the bandgap from 2.7 eV in GO to 1.15 in rGO [53].

Kimiagar et al. showed that the bandgap of GO varied from 3.3 to 4.2 eV. When MgO was incorporated in GO at temperatures 180 and 210 °C, the bandgap increased to 4.2–5.9 and 4.6–6 eV, respectively [54]. Jin et al. tuned the bandgap of GO by treating the thermally reduced GO with HNO_3. For the untreated rGO, the E_g value was 0.37 eV which increased gradually when the percentage of HNO_3 was increased. Finally, at 40% HNO_3, an E_g value of 1.49 eV was obtained [55].

d. **Electrical properties**

GO has low electrical conductivity because of the disruption of the planar structure due to the presence of OFs. The electrical conductivity can be improved by restoring sp^2 carbons on the removal of OFs. This can be achieved by the reduction of GO. However, even in rGO, the OFs are not completely removed, and some sp^3 carbons still exist. Thus, the movement of charge carriers is still disrupted in rGO. Several methods have been employed to increase the electrical conductivity of the GO. Sun et al. used HI and ascorbic acid combined to reduce GO. The electrical conductivity of GO was increased to 1115 S/m. The as-synthesized rGO films were uniform and highly transparent [56]. Sun et al. used hydrohalic acids for the reduction of GO. The electrical conductivity of rGO depended on the hydrohalic acid. The minimum resistivity was shown by hydroiodic acid-reduced GO. The resistivity lay in the range of 0.003–0.641 Ω cm [56]. Park et al. demonstrated that the increase in pressure leads to increased conductivity of rGO. Additionally, thermal reduction further increased the electrical conductivity. The rGO synthesized at 1000 °C showed the highest electrical conductivity [57].

1.8 Recent Status of SiNWs and Graphene Heterojunction

Li et al. first made the graphene/silicon solar cell in 2010, showing 1.5% PCE. Graphene sheets (GS) were deposited on *n*-type Si, and the active area of the device was 0.5 cm^2 [58]. The advancement in this field was extremely fast, and only after 5 years Song et al. successfully achieved a PCE value of 15.6% for a device with 0.11 cm^2 active area. They studied the impact of the silicon oxide layer between the graphene and Si interface on the PCE value. High efficiency was achieved after optimizing the oxide layer and doping graphene with $AuCl_3$ [59]. Rehman et al. used Al_2O_3 as the electron-blocking layer to decrease the recombination at the graphene/Si interface. The oxide layer's thickness was optimized to obtain an efficiency of 8.4%. Graphene was not doped, and the device was stable even after nine months [60]. Ma et al. introduced MoS_2 between graphene and silicon. It acted as a hole transport layer and a recombination suppressor at the interface. A high PCE of 15.8% was obtained for this graphene/MoS_2/Si solar cell [61].

Graphene/silicon heterojunction solar cells possess excellent properties and great potential for industrial application. However, they suffer from several drawbacks. Firstly, transferring CVD-grown graphene to silicon is difficult for practical applications. Secondly, the area of the device is relatively small, which hinders its commercial production. Thirdly, though the efficiency of the device is increased by doping, the stability of the device is low.

To address these problems, researchers have adopted alternate measures, viz. the use of GO and rGO as an alternative for graphene and texturization of silicon. GO and rGO are cheaper compared to graphene. GO is hydrophilic, and it can be dispersed effortlessly in any solvent. Thus, it can be simply applied to the substrate by simple procedures, viz. spin coating, drop casting, dip-coating and spray coating. As discussed previously, the OFs on GO can be removed by various measures to increase electrical conductivity. As mentioned earlier, the properties of rGO are similar to that of pristine graphene. Hence, using GO and rGO instead of graphene may help achieve higher PCE values with the benefits of easy device assembly and cost-effectiveness. Mahala et al. fabricated GO/Si heterojunction by coating GO on Si with the help of a spin coater. This device showed a PCE value of 0.07% with series and shunt resistance of 150 Ω and 450 MΩ, respectively [62]. Han et al. studied the thermal stability of nitrogen-doped rGO (NrGO)/Si solar cells. They found that the PCE of the device decreased when rGO was doped with nitrogen (4–2.5%). However, the thermal stability of the NrGO/Si was greater than rGO/Si [63]. Hu et al. drop-casted $FeCl_3$ functionalized GO over single-walled carbon nanotube/Si heterojunction to obtain a PCE of 17.5%. Here, GO acted as an antireflective coating, photocurrent generator and carrier transport bridge. The $FeCl_3$ increased the electrical conductivity, work function and carrier charge transport. This device showed 90% PCE retention after exposure to air for 15 days [64].

Usually, in silicon solar cells, an antireflective coating is applied. However, texturing Si helps decrease surface reflectance. Fan et al. compared the light-harvesting properties of graphene/Si and graphene/SiNWs for PV application.

Graphene was doped with $SOCl_2$ at varying time duration. They demonstrated that graphene/SiNWs showed superior properties with a PCE value of 2.86% [65]. Yu et al. created a pyramid structure on the surface of Si. GO/CNT/Si heterojunction was formed using a floating transfer method. The device showed a high-efficiency value of $13.01 \pm 0.32\%$, with a 99.9% Si surface covered by GO [66]. Fallahazad et al. deposited rGO on flat and pyramid Si using electrophoretic deposition. They demonstrated that the device with a pyramid structure (1.76%) showed better efficiency compared to flat Si (0.94%) [67]. Similar results were obtained by Abdullah et al. (PCE = $5.3 \times 10^{-3}\%$), Lee et al. (PCE = 5.20%) and Zhang et al. (PCE = 8.6%) [68–70].

GO is also used for interface engineering in graphene/Si heterojunction solar cell to reduce the recombination at the interface and increase the V_{oc}. These devices are found to be more stable than doped graphene/Si solar cells [71, 72].

From the literature, a summary of the important parameters for graphene/Si solar cell is presented in Table 1.

Table 1 Summary of important parameters for different types of graphene/Si solar cell

Device	Doping (G or rGO)	Area (cm^2)	V_{oc} (V)	J_{sc} (mA/ cm^2)	FF (%)	η (%)	Stability	References
GS/*n*Si	–	0.5	0.42–0.48	4–6.5	45–56	1.5	–	[58]
TiO_2/G/SiO_2/*n*Si	–	0.11	0.595	36.7	72	15.6	–	[59]
Cr/Au/G/Al_2O_3/*n*Si/Al	–	0.32	0.556	33.4	45.21	8.4	9 months	[60]
GOCNT/spiro-OMeTAD/Si	$AuCl_3$	0.087	0.575	28.1	75	12.83	15 days, ~ 90%	[73]
GO-SWCNT/Si	$FeCl_3$	0.09	0.637	36.7	74.6	17.5	15 days, 90%	[64]
G/SiNWs	$SOCl_2$	1.44	0.503	11.24	50.6	2.86	14 days, 41.95%	[65]
rGO/*n*Si pyramid	–	0.3	0.37	7.54	63.2	1.76	–	[67]
GO/CNT/*n*Si pyramid	$SOCl_2$	0.087	0.573	33.75	68	13.08	–	[66]
rGO/*n*Si pyramid	–	0.35	0.44	18.26	39.61	5.20	–	[68]
G/pSi	$POCl_3$	–	0.606	20.49	–	8.6	–	[69]
G/GO/*n*Si	–	0.81	0.48	26.4	41	5.2	–	[71]
G/GO/*n*Si	–	0.085	0.44–0.45	25.3–26.6	44–52	5.10–6.18	7 days, 95%	[72]

2 Conclusion

In the present world, the main energy source is a fossil fuel. For the development of any country, energy plays a crucial role. So, in the race to get the tag of a developed country, every nation leaps toward urbanization, economic growth and technological advancement. In doing so, the limited fossil fuel sources are depleting daily. In addition, they produce harmful gases like CH_4, CO, CO_2, SO_2, NO, etc., which lead to pollution and global warming. The countries have agreed to reduce greenhouse gas emissions by 2030 and reach net zero by 2050 through the Paris Agreement. Thus, renewable energy sources have emerged as a knight in shining armour to protect the world from the threat of energy crisis and global warming. Among them, solar energy is highly significant because sunlight is widely available. It is an environment-friendly, cost-free and unlimited source of energy. By using the principle of photovoltaics, light energy can be directly converted into electric energy. Currently, the commercial solar cell market is dominated by Si-based solar cells. However, SiNWs show better optical and electrical properties than bulk Si. Also, GO is solution processable, which helps to coat it on any substrate easily. The GO/SiNWs heterojunction can be fabricated by simple and easy methods because the two materials can hold each other due to the vdWs force of attraction without worrying about the lattice mismatch. The advancement in this field may contribute to producing cost-effective solar cells with high efficiency.

References

1. Santangelo S (2019) Electrospun nanomaterials for energy applications: recent advances. Appl Sci 9(6):1049. https://doi.org/10.3390/app9061049
2. Zhang Q, Uchaker E, Candelaria SL, Cao G (2013) Nanomaterials for energy conversion and storage. Chem Soc Rev 42(7):3127. https://doi.org/10.1039/c3cs00009e
3. Bochenkov VE, Sergeev GB (2007) Nanomaterials for sensors. Russ Chem Rev 76(11):1084–1093. https://doi.org/10.1070/RC2007v076n11ABEH003735
4. Chua CK, Ambrosi A, Pumera M (2011) Graphene based nanomaterials as electrochemical detectors in lab-on-a-chip devices. Electrochem Commun 13(5):517–519. https://doi.org/10.1016/j.elecom.2011.03.001
5. Lehner R, Wang X, Marsch S, Hunziker P (2013) Intelligent nanomaterials for medicine: carrier platforms and targeting strategies in the context of clinical application. Nanomed Nanotechnol Biol Med 9(6):742–757. https://doi.org/10.1016/j.nano.2013.01.012
6. Xiong J, Di J, Xia J, Zhu W, Li H (2018) Surface defect engineering in 2D nanomaterials for photocatalysis. Adv Func Mater 28(39):1801983. https://doi.org/10.1002/adfm.201801983
7. Wang Z, Hu T, Liang R, Wei M (2020) Application of zero-dimensional nanomaterials in biosensing. Front Chem 8(April):1–19. https://doi.org/10.3389/fchem.2020.00320
8. Mohammed OF (2019) Outstanding challenges of zero-dimensional perovskite materials. J Phys Chem Lett 10(19):5886–5888. https://doi.org/10.1021/acs.jpclett.9b00175
9. Cai S, Yang R (2020) Two-dimensional nanomaterials with enzyme-like properties for biomedical applications. Front Chem 8. https://doi.org/10.3389/fchem.2020.565940
10. Kebede MA, Imae T (2019) Low-dimensional nanomaterials. In: Advanced supramolecular nanoarchitectonics. Elsevier, pp 3–16. https://doi.org/10.1016/B978-0-12-813341-5.00001-2

11. Geim AK, Grigorieva IV (2013) Van der Waals heterostructures. Nature 499(7459):419–425. https://doi.org/10.1038/nature12385
12. Wang P, Jia C, Huang Y, Duan X (2021) Van der Waals heterostructures by design: from 1D and 2D to 3D. Matter 4(2):552–581. https://doi.org/10.1016/j.matt.2020.12.015
13. Kao KC (2004) 7 Electrical conduction and photoconduction. Part I: Electrical conduction. In: Dielectric phenomena in solids, pp 503–504
14. Edvinsson T (2018) Optical quantum confinement and photocatalytic properties in two-, one- and zero-dimensional nanostructures. R Soc Open Sci 5(9):180387. https://doi.org/10.1098/rsos.180387
15. Ghasemian S, Faridzad A, Abbaszadeh P, Taklif A, Ghasemi A, Hafezi R (2020) An overview of global energy scenarios by 2040: identifying the driving forces using cross-impact analysis method. Int J Environ Sci Technol0123456789. https://doi.org/10.1007/s13762-020-02738-5
16. Holt A, Pengelly IJ (2008) ITS and renewable energy. In: IET road transport information and control conference and the ITS United Kingdom members' conference (RTIC 2008), vol 6, pp 55–55. https://doi.org/10.1049/ic.2008.0789
17. bp Energy Outlook 2022 (2022) Edition 2022 explores the key uncertainties surrounding the energy transition
18. Kabir E, Kumar P, Kumar S, Adelodun AA, Kim K-H (2018) Solar energy: potential and future prospects. Renew Sustain Energy Rev 82(September 2016):894–900. https://doi.org/10.1016/j.rser.2017.09.094
19. Gong J, Li C, Wasielewski MR (2019) Advances in solar energy conversion. Chem Soc Rev 48(7):1862–1864. https://doi.org/10.1039/C9CS90020A
20. Green MA, Dunlop ED, Hohl-Ebinger J, Yoshita M, Kopidakis N, Hao X (2021) Solar cell efficiency tables (version 58). Prog Photovoltaics Res Appl 29(7):657–667. https://doi.org/10.1002/pip.3444
21. Zaidi B (2018) Introductory chapter: introduction to photovoltaic effect. In: Solar panels and photovoltaic materials. InTech, pp 1–8. https://doi.org/10.5772/intechopen.74389
22. Bhattacharya S, John S (2019) Beyond 30% conversion efficiency in silicon solar cells: a numerical demonstration. Sci Rep 9(1):12482. https://doi.org/10.1038/s41598-019-48981-w
23. Jungbluth N, Stucki M, Frischknecht R, Tuchschmid M, Doka G, Vollmer M (2009) Life-cycle inventories—photovoltaics. Swiss Centre for Life Cycle Inventories, Dübendorf, CH, pp 6–XII
24. Shockley W, Queisser HJ (1961) Detailed balance limit of efficiency of p-n junction solar cells. J Appl Phys 32(3):510–519. https://doi.org/10.1063/1.1736034
25. Bai F, Li M, Huang R, Li Y, Trevor M, Musselman KP (2014) A one-step template-free approach to achieve tapered silicon nanowire arrays with controllable filling ratios for solar cell applications. RSC Adv 4(4):1794–1798. https://doi.org/10.1039/C3RA45473H
26. Bhujel R, Rizal U, Agarwal A, Swain BS, Swain BP (2018) Synthesis and characterization of silicon nanowires by electroless etching. J Mater Eng Perform 27(6):2655–2660. https://doi.org/10.1007/s11665-018-3179-z
27. Zheng Y, Lake A, Rivas B, Klimeck (2004) Electronic properties of silicon nanowires. Electr Perform Electron Packag IWCE-04 52(6):82–83. https://doi.org/10.1109/IWCE.2004.1407334
28. Hasan M, Huq MF, Mahmood ZH (2013) A review on electronic and optical properties of silicon nanowire and its different growth techniques. SpringerPlus 2(1):151. https://doi.org/10.1186/2193-1801-2-151
29. Cao DT, Ngan LTQ, Viet TV, Anh CT (2013) Effect of $AgNO_3$ concentration on structure of aligned silicon nanowire arrays fabricated via silver-assisted chemical etching. Int J Nanotechnol 10(3/4):343. https://doi.org/10.1504/IJNT.2013.053147
30. Omar H, Jani AMM, Rusop M, Abdullah S (2016) The influence of H_2O_2 concentration to the structure of silicon nanowire growth by metal-assisted chemical etching. AIP Conf Proc 1733:020016. https://doi.org/10.1063/1.4948834
31. Rahmani M, Amdouni S, Zaïbi M, Meftah A (2021) Effect of etching duration on the morphological and opto-electrical properties of silicon nanowires obtained by Ag-assisted chemical etching. Silicon 13(1):179–187. https://doi.org/10.1007/s12633-020-00416-2

32. Fellahi O, Hadjersi T, Maamache M, Bouanik S, Manseri A (2010) Effect of temperature and silicon resistivity on the elaboration of silicon nanowires by electroless etching. Appl Surf Sci 257(2):591–595. https://doi.org/10.1016/j.apsusc.2010.07.039
33. Peng K-Q, Lee S-T (2011) Silicon nanowires for photovoltaic solar energy conversion. Adv Mater 23(2):198–215. https://doi.org/10.1002/adma.201002410
34. Gonchar KA, Kitaeva VY, Zharik GA, Eliseev AA, Osminkina LA (2019) Structural and optical properties of silicon nanowire arrays fabricated by metal assisted chemical etching with ammonium fluoride. Front Chem 6(Jan):1–7. https://doi.org/10.3389/fchem.2018.00653
35. Gonchar KA, Moiseev DV, Bozhev IV, Osminkina LA (2021) Influence of H_2O_2 concentration on the structural and photoluminescent properties of porous silicon nanowires fabricated by metal-assisted chemical etching. Mater Sci Semicond Process 125(December 2020):105644. https://doi.org/10.1016/j.mssp.2020.105644
36. Ge Z, Xu L, Cao Y, Wu T, Song H, Ma Z, Xu J, Chen K (2015) Substantial improvement of short wavelength response in n-SiNW/PEDOT:PSS solar cell. Nanoscale Res Lett 10(1):330. https://doi.org/10.1186/s11671-015-0998-9
37. Hutagalung SD, Fadhali MM, Areshi RA, Tan FD (2017) Optical and electrical characteristics of silicon nanowires prepared by electroless etching. Nanoscale Res Lett 12(1):425. https://doi.org/10.1186/s11671-017-2197-3
38. Sacconi F, Persson MP, Povolotskyi M, Latessa L, Pecchia A, Gagliardi A, Balint A, Fraunheim T, Di Carlo A (2007) Electronic and transport properties of silicon nanowires. J Comput Electron 6(1–3):329–333. https://doi.org/10.1007/s10825-006-0138-y
39. Leu PW, Shan B, Cho K (2006) Surface chemical control of the electronic structure of silicon nanowires: density functional calculations. Phys Rev B 73(19):195320. https://doi.org/10.1103/PhysRevB.73.195320
40. Rosli NN, Ibrahim MA, Ahmad Ludin N, Mat Teridi MA, Sopian K (2019) A review of graphene based transparent conducting films for use in solar photovoltaic applications. Renew Sustain Energy Rev 99(September 2018):83–99. https://doi.org/10.1016/j.rser.2018.09.011
41. Sahu S, Rout GC (2017) Band gap opening in graphene: a short theoretical study. Int Nano Lett 7(2):81–89. https://doi.org/10.1007/s40089-017-0203-5
42. Abdullah MF, Hashim AM (2019) Review and assessment of photovoltaic performance of graphene/Si heterojunction solar cells. J Mater Sci 54(2):911–948. https://doi.org/10.1007/s10853-018-2947-3
43. Mohan VB, Brown R, Jayaraman K, Bhattacharyya D (2015) Characterisation of reduced graphene oxide: Effects of reduction variables on electrical conductivity. Mater Sci Eng B 193(C):49–60. https://doi.org/10.1016/j.mseb.2014.11.002
44. De Silva KKH, Huang H-H, Joshi RK, Yoshimura M (2017) Chemical reduction of graphene oxide using green reductants. Carbon 119:190–199. https://doi.org/10.1016/j.carbon.2017.04.025
45. Sharma N, Sharma V, Jain Y, Kumari M, Gupta R, Sharma SK, Sachdev K (2017) Synthesis and characterization of graphene oxide (GO) and reduced graphene oxide (rGO) for gas sensing application. Macromol Symp 376(1):1700006. https://doi.org/10.1002/masy.201700006
46. Lei Z, Lu L, Zhao XS (2012) The electrocapacitive properties of graphene oxide reduced by urea. Energy Environ Sci 5(4):6391–6399. https://doi.org/10.1039/C1EE02478G
47. Aziz M, Abdul Halim FS, Jaafar J (2014) Preparation and characterization of graphene membrane electrode assembly. J Teknol 69(9):11–14. https://doi.org/10.11113/jt.v69.3388
48. Johra FT, Lee J-W, Jung W-G (2014) Facile and safe graphene preparation on solution based platform. J Ind Eng Chem 20(5):2883–2887. https://doi.org/10.1016/j.jiec.2013.11.022
49. Haghighi B, Tabrizi MA (2013) Green-synthesis of reduced graphene oxide nanosheets using rose water and a survey on their characteristics and applications. RSC Adv 3(32):13365. https://doi.org/10.1039/c3ra40856f
50. Roy I, Bhattacharyya A, Sarkar G, Saha NR, Rana D, Ghosh PP, Palit M, Das AR, Chattopadhyay D (2014) In situ synthesis of a reduced graphene oxide/cuprous oxide nanocomposite: a reusable catalyst. RSC Adv 4(94):52044–52052. https://doi.org/10.1039/C4RA08127G

51. Tene T, Usca GT, Guevara M, Molina R, Veltri F, Arias M, Caputi LS, Gomez CV (2020) Toward large-scale production of oxidized graphene. Nanomaterials 10(2):1–11. https://doi.org/10.3390/nano10020279
52. Zheng F, Xu W-L, Jin H-D, Hao X-T, Ghiggino KP (2015) Charge transfer from poly(3-hexylthiophene) to graphene oxide and reduced graphene oxide. RSC Adv 5(109):89515–89520. https://doi.org/10.1039/C5RA18540H
53. Velasco-Soto MA, Pérez-García SA, Alvarez-Quintana J, Cao Y, Nyborg L, Licea-Jiménez L (2015) Selective band gap manipulation of graphene oxide by its reduction with mild reagents. Carbon 93:967–973. https://doi.org/10.1016/j.carbon.2015.06.013
54. Kimiagar S, Abrinaei F (2018) Effect of temperature on the structural, linear, and nonlinear optical properties of MgO-doped graphene oxide nanocomposites. Nanophotonics 7(1):243–251. https://doi.org/10.1515/nanoph-2017-0030
55. Jin Y, Zheng Y, Podkolzin SG, Lee W (2020) Band gap of reduced graphene oxide tuned by controlling functional groups. J Mater Chem C 8(14):4885–4894. https://doi.org/10.1039/C9TC07063J
56. Sun F, Ghosh H, Wang J, Tan Z, Sivoththaman S (2020) Improving the electrical conductivity of reduced graphene oxide transparent electrodes for photovoltaics. In: 47th IEEE photovoltaic specialists conference (PVSC), June 2020, pp 2025–2028. https://doi.org/10.1109/PVSC45281.2020.9300537
57. Park H, Lim S, Nguyen DD, Suk JW (2019) Electrical measurements of thermally reduced graphene oxide powders under pressure. Nanomaterials 9(10):1387. https://doi.org/10.3390/nano9101387
58. Li X, Zhu H, Wang K, Cao A, Wei J, Li C, Jia Y, Li Z, Li X, Wu D (2010) Graphene-on-silicon Schottky junction solar cells. Adv Mater 22(25):2743–2748. https://doi.org/10.1002/adma.200904383
59. Song Y, Li X, Mackin C, Zhang X, Fang W, Palacios T, Zhu H, Kong J (2015) Role of interfacial oxide in high-efficiency graphene-silicon schottky barrier solar cells. Nano Lett 15(3):2104–2110
60. Rehman MA, Akhtar I, Choi W, Akbar K, Farooq A, Hussain S, Shehzad MA, Chun S-H, Jung J, Seo Y (2018) Influence of an Al_2O_3 interlayer in a directly grown graphene-silicon Schottky junction solar cell. Carbon 132:157–164. https://doi.org/10.1016/j.carbon.2018.02.042
61. Ma J, Bai H, Zhao W, Yuan Y, Zhang K (2018) High efficiency graphene/MoS 2/Si Schottky barrier solar cells using layer-controlled MoS 2 films. Solar Energy 160(September 2017):76–84. https://doi.org/10.1016/j.solener.2017.11.066
62. Mahala P, Gupta N, Singh S (2021) Silicon photovoltaic cell based on graphene oxide as an active layer. Microsyst Technol 27(11):4027–4033. https://doi.org/10.1007/s00542-020-04763-3
63. Han M, Ryu BD, Hyung J-H, Han N, Park YJ, Ko KB, Kang KK, Cuong TV, Hong C-H (2017) Enhanced thermal stability of reduced graphene oxide-Silicon Schottky heterojunction solar cells via nitrogen doping. Mater Sci Semicond Process 59(May 2016):45–49. https://doi.org/10.1016/j.mssp.2016.11.021
64. Hu X-G, Wei Q, Zhao Y-M, Hou P-X, Ren W, Liu C, Cheng H-M (2022) $FeCl_3$-functionalized graphene oxide/single-wall carbon nanotube/silicon heterojunction solar cells with an efficiency of 17.5%. J Mater Chem A 10(9):4644–4652. https://doi.org/10.1039/D2TA00217E
65. Fan G, Zhu H, Wang K, Wei J, Li X, Shu Q, Guo N, Wu D (2011) Graphene/silicon nanowire schottky junction for enhanced light harvesting. ACS Appl Mater Interfaces 3(3):721–725. https://doi.org/10.1021/am1010354
66. Yu L, Batmunkh M, Dadkhah M, Shearer CJ, Shapter JG (2018) Pyramid-textured antireflective silicon surface in graphene oxide/single-wall carbon nanotube-silicon heterojunction solar cells. Energy Environ Mater 1(4):232–240. https://doi.org/10.1002/eem2.12020
67. Fallahazad P, Naderi N, Eshraghi MJ (2020) Improved photovoltaic performance of graphene-based solar cells on textured silicon substrate. J Alloy Compd 834:155123. https://doi.org/10.1016/j.jallcom.2020.155123

68. Lee W-C, Tsai M-L, Chen Y-L, Tu W-C (2017) Fabrication and analysis of chemically-derived graphene/pyramidal Si heterojunction solar cells. Sci Rep 7(1):46478. https://doi.org/10.1038/srep46478
69. Zang Y, Li L, Chu Q, Pu H, Hu J, Jin H, Zhang Y (2018) Graphene as transparent electrode in Si solar cells: a dry transfer method. AIP Adv 8(6):065206. https://doi.org/10.1063/1.5030571
70. Abdullah MF, Hashim AM (2019) Improved coverage of rGO film on Si inverted pyramidal microstructures for enhancing the photovoltaic of rGO/Si heterojunction solar cell. Mater Sci Semicond Process 96(October 2018):137–144. https://doi.org/10.1016/j.mssp.2019.02.033
71. Jiao K, Wang X, Wang Y, Chen Y (2014) Graphene oxide as an effective interfacial layer for enhanced graphene/silicon solar cell performance. J Mater Chem C 2(37):7715–7721. https://doi.org/10.1039/c4tc00705k
72. Yang L, Yu X, Xu M, Chen H, Yang D (2014) Interface engineering for efficient and stable chemical-doping-free graphene-on-silicon solar cells by introducing a graphene oxide interlayer. J Mater Chem A 2(40):16877–16883. https://doi.org/10.1039/C4TA02216E
73. Yu L, Batmunkh M, Grace T, Dadkhah M, Shearer C, Shapter J (2017) Application of a hole transporting organic interlayer in graphene oxide/single walled carbon nanotube–silicon heterojunction solar cells. J Mater Chem A 5(18):8624–8634. https://doi.org/10.1039/C7TA01782K

Chapter 9
Energy Conversion Materials: An Electrolyte for Intermediate Temperature Solid Oxide Fuel Cell (IT-SOFCs) Applications

Somoju Ramesh

1 Introduction

A solid oxide fuel cell (SOFC) is an electrochemical device that converts chemical energy to electrical energy as long as the fuel is supplied. SOFC mainly consists of an anode, electrolyte, and cathode. As shown in Fig. 1, fuel is supplied to the anode, and the air is supplied to the cathode. Based on the operating temperature, SOFCs are classified into (a) High-temperature SOFCs (800–1000 °C), (b) Medium temperature SOFCs (500–700 °C), and lower temperature SOFCs (450–600 °C).

Due to the external workload, cathode side oxygen is reduced to oxide ions, passes through the electrolyte, and enters into the anode side where oxygen ions react with fuel; H_2 liberates electrons and water as a by-product. An electrolyte should allow only oxygen ions and block electrons; otherwise, short circuit might occur. SOFCs are clean energy conversion devices, unlike steam engines. SOFCs are more attractive than other fuel cells due to their efficiency (up to 80%), durability, non-restriction on metal electrodes [2–5]. Equation 1 is used to calculate the cell voltage.

$$V = E^o - iR - \eta_c - \eta_a \tag{1}$$

where R is the resistance of the cell, i is the current, η_c and η_a cathode and anode polarization losses, respectively, and E^o is the open circuit potential.

S. Ramesh (✉)
Department of Physics, GITAM School of Science, GITAM (Deemed to Be University), Telangana 502329, India
e-mail: ramesh.ou1@gmail.com

B. P. Swain (ed.), *Energy Materials*, Materials Horizons: From Nature to Nanomaterials, https://doi.org/10.1007/978-981-99-3866-7_9

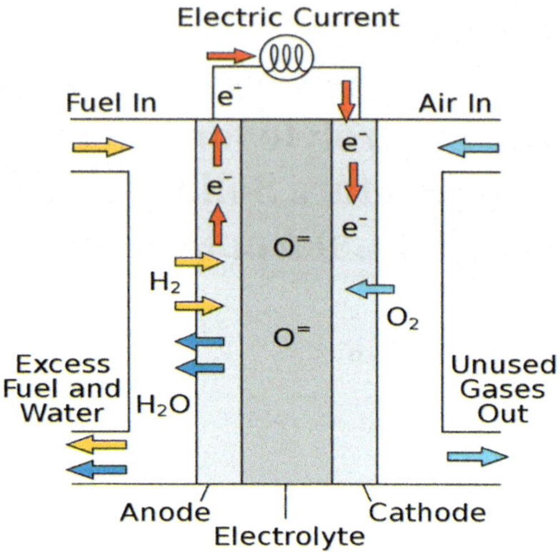

Fig. 1 Principle of SOFC [1]

2 Classification of Fuel Cells

Classification of fuel cells based on different electrolytes [2] is presented in Table 1. Among the various fuel cells, SOFCs drew more attention regarding efficiency and power generator. However, high operating temperature (>800 °C) is the constraint to taking it into commercialization—besides, higher operating temperature results in compatibility issues across the cell components, cracks, and chemical instability. Doped/co-doped CeO_2 is the alternate material to lower the operating temperature below 700 °C. Furthermore, nano-composite CeO_2 electrolyte materials are developed below 600 °C [2].

3 Synthesis Method

The sample synthesis method is important in nanoparticle preparation and achieving a relatively high density since an electrolyte material should be highly dense [7–12]. There are many reported chemical methods to produce nanoparticles over physical methods like the solid-state reaction method. The sample powder prepared through the solid-state reaction method resulted in non-uniform size particles, consequently requiring higher sintering temperatures (>1500 °C) to prepare dense ceramic samples (>95% TD). At higher sintering temperatures, uncontrolled grain growth occurs, which is an obstacle to electrical properties. However, chemical methods like sol–gel, precipitation, Pechini, polyvinyl, and modified sol–gel processes resulted in uniform nanoparticle size. Nanoparticles are active to heat treatment; higher density

Table 1 Classification of fuel cells [2, 6]

Fuel cell (FC) name	Electrolyte	Electrolyte type	Anode/fuel	Operating temperature (°C)	Overall efficiency	Suitable for
DFMC	Methanol	Liquid	Methanol + Deionized/methanol	30–130	30–40	Portable devices
PEMFC	Polymer membrane	Solid	Platinum/hydrogen	80–100	30–40	Cars, trucks, and buses
FAFC	Phosphoric acid	Liquid	Platinum/hydrogen–methanol	150–200	40–50	Buildings
AFC	Potassium hydroxide (KOH)	Liquid	Platinum or carbon/ hydrogen–ammonia	60–80	60–70	UPS
MCFC	Molten carbonate	Solid	Ceramic/methanol–coal–biogas	650	50	Utilities
SOFC	Yttria-stabilized zirconia (YSZ)	Solid	Steel or nickel/ hydrogen–coal–biogas–natural gas	800–1000	60–70	Power plants

DMFC—Direct methanol; *PEMFC*—Polymer electrolyte membrane; *PAFC*—Phosphoric acid; *AFC* Alkaline; *MCFC*—Molten carbonate; *SOFC*—Solid oxide fuel cell is achieved at a lower sintering temperature (< 1300 °C). Tables 2, 3 and 4 presents the different synthesis processes and sintering conditions maintained to reach higher density.

The sample $Ce_{0.8}Sm_{0.2}O_{2-\delta}$ synthesized by Sucrose-pectin modified sol–gel sintered at 1200 °C −6 h showed maximum electrical conductivity 1.0×10^{-2} S/cm at 600 °C [13]. Table 2 shows that electrical conductivity changed with grain size, i.e., grain size depends on dopant concentration and sintering temperature. Furthermore, it is noticed from Tables 2 and 3 that the electrical conductivity changed with different preparation methods; for example, the sample $Ce_{0.82}Dy_{0.155}Sr_{0.025}O_{2-\delta}$ exhibited the electrical conductivity 1.22×10^{-2} S/cm at 700 °C, and activation energy 0.73 eV prepared by modified sol–gel (sucrose-pectin) [23]. The sample $Ce_{0.83}Dy_{0.14}Ca_{0.03}O_{1.90}$ prepared by citrate–nitrate auto-combustion route showed electrical conductivity 1.45×10^{-2} S/cm at 600 °C and activation energy 0.94 eV [25]. The composition $Ce_{0.76}Pr_{0.08}Sm_{0.08}Gd_{0.08}O_{2-\delta}$ synthesized by sol–gel auto-combustion showed electrical conductivity 1.86×10^{-2} S/cm at 600 °C and activation energy 0.56 eV [32].

Table 2 Comparison of total electrical conductivities [13]

Electrolyte	Conductivity (S/cm)	Preparation method	Sintering conditions	Average grain size	References
$Ce_{0.8}Sm_{0.2}O_{2-\delta}$	1.0×10^{-2} at 600 °C	Sucrose-pectin modified sol–gel	1200 °C for 6 h	0.3 μm	[13]
$Ce_{0.83}Sm_{0.17}O_{2-X/2}$	5.73×10^{-3} at 600 °C	Hydrothermal	1400–1450 °C for 10 h	0.1–0.5 μm	[14]
$Ce_{0.8}Sm_{0.2}O_{1.9}$	5.00×10^{-3} at 600 °C	Sol–gel	1400–1450 °C for 10 h	–	[15]
$Ce_{0.8}Sm_{0.2}O_{2-\delta}$	2.30×10^{-2} at 700 °C	Citrate complexation	1300 °C for 6 h	0.52 μm	[16]
$Ce_{0.8}Sm_{0.2}O_{1.9\ 3}$	3.13×10^{-2} and 3.40 × 10 − 2 at 700 °C	Pechini	1400 °C for 6 h	0.78–0.85 μm	[17]
$Ce_{0.8}Sm_{0.2}O_{2-\delta}$	5.40×10^{-3} at 600 °C	Sol–gel	1300 °C for 10 h	0.85 μm	[18]
$Ce_{0.8}Sm_{0.2}O_{2-\delta}$	1.37×10^{-2} at 600 °C	Solid-state reaction	1550 °C for 4 h	0.77 μm	[19]
$Ce_{0.8}Sm_{0.2}O_{2-\delta}$	0.77×10^{-2} at 600 °C	Sol–gel auto-combustion	1300 °C for 4 h	0.60 μm	[20]
$Ce_{0.85}Sm_{0.15}O_{2-\delta}$	1.10×10^{-2} at 600 °C	Hydrothermal	1300 °C 4–8 h	0.78 μm	[21]
$Ce_{0.8}Sm_{0.2}O_{1.9}$	3.13×10^{-2} and 3.4×10^{-2} at 750 °C	Pechini	1400 °C for 6	0.85 μm	[22]

Table 3 Comparison of the total electrical conductivity reported in the literature [23]

Sample	Preparation method	Total conductivity, S/cm	Activation energy (eV)	References
$Ce_{0.82}Dy_{0.155}Sr_{0.025}O_{2-\delta}$ (sintered at 1200 °C)	Modified sol–gel (sucrose-pectin)	1.22×10^{-2} at 700 °C	0.73	[23]
$Ce_{0.82}Dy_{0.155}Sr_{0.025}O_{2-\delta}$ (sintered at 1250 °C)	Modified sol–gel (sucrose-pectin)	1.83×10^{-2} at 700 °C	0.75	[23]
$Ce_{0.82}Dy_{0.155}Sr_{0.025}O_{2-\delta}$ (sintered at 1300 °C)	Modified sol–gel (sucrose-pectin)	2.45×10^{-2} at 700 °C	0.76	[23]
$Ce_{0.8}Sm_{0.15}Sr_{0.05}O_{2-\delta}$	Citrate gel	2.21×10^{-4} at 750 °C	0.66	[24]
$Ce_{0.8}Dy_{0.20}O_{1.90}$	Citrate–nitrate auto-combustion route	1.04×10^{-3} at 600 °C	1.10	[25]
$Ce_{0.83}Dy_{0.14}Ca_{0.03}O_{1.90}$	Citrate–nitrate auto-combustion route	1.45×10^{-2} at 600 °C	0.94	[25]
(0.94SDC)(0.06Sr)	Ball milling	3.21×10^{-2} at 700 °C	0.71	[26]
$Ce_{0.82}Sm_{0.16}Sr_{0.02}O_{1.90}$	Citrate–nitrate auto-combustion	2.67×10^{-2} at 600 °C	0.66	[27]
$Ce_{0.8}Y_{0.18}Sr_{0.02}O_{2-\delta}$	Sol–gel	2.20×10^{-2} at 800 °C	0.89	[28]
$Ce_{0.8}Y_{0.18}Ca_{0.02}O_{2-\delta}$	Sol–gel	3.00×10^{-2} at 800 °C	1.06	[28]
$Ce_{0.8}Y_{0.15}Sr_{0.05}O_{2-\delta}$	Solid-state reaction	4.98×10^{-2} at 750 °C	0.76	[29]
$Sm_{0.1}Ca_{0.1}Ce_{0.8}O_{2-\delta}$	Cellulose templating	3.20×10^{-2} at 800 °C	0.82	[30]
$Ce_{0.8}Sm_{0.15}Ca_{0.05}O_{1.875}$	Sol–gel	1.69×10^{-2} at 600 °C	0.91	[31]
$Ce_{0.8}Sm_{0.15}Ca_{0.025}Sr_{0.025}O_{1.875}$	Sol–gel	1.51×10^{-1} at 600 °C	0.91	[31]

4 Doped Ceria Electrolyte Materials

Electrical conductivities of different doped ceria electrolyte materials are presented in Figs. 2 and 3.

It is noticed from Figs. 2 and 3 that the 15% Y_2O_3–ThO_2, 15% Cao–ZrO_2, and $La_{0.7}Ca_{0.3}AlO_3$ exhibited lower electrical conductivity than 8% Y_2O_3-stabilized zirconia. Ceria doped with different dopants, like divalent and trivalent dopants, showed much-improved electrical conductivity. Besides, $LSGMO_3$, $LSGMCO_3$, Bi_2O_3–Y_2O_3, and ZrO_2–Sc_2O_3 showed higher electrical conductivity.

Table 4 Comparison of $Ce_{0.76}Pr_{0.08}Sm_{0.08}Gd_{0.08}O_{2-\delta}$ with other reported ceria-based materials [32]

Sample name	Synthesis method	Sintering temp.–time	Conductivity (S/cm)	Activation energy (eV)	References
$Ce_{0.76}Pr_{0.08}Sm_{0.08}Gd_{0.08}O_{2-\delta}$	Sol–gel auto-combustion	1300 °C –4 h	1.86×10^{-2} at 600 °C	0.56	[32]
SDC20	Sol–gel auto-combustion	1300 °C –4 h	0.77×10^{-2} at 600 °C	0.60	[32]
$Ce_{0.7}Pr_{0.15}Sm_{0.075}Eu_{0.075}O_{2-\delta}$	Low temp. citrate auto-ignition	600 °C –6 h	3.05×10^{-4} at 500 °C	0.39	[33]
$Ce_{0.8}Sm_{0.12}Pr_{0.08}O_{2-\delta}$	Citric nitrate method	1300 °C –10 h	1.21×10^{-2} at 600 °C	0.77	[34]
$Ce_{0.84}Gd_{0.08}Pr_{0.08}O_2$	Sol–gel	1300 °C –10 h	1.71×10^{-2} at 600 °C	0.77	[35]
$Ce_{0.85}Gd_{0.1}Sm_{0.05}O_{1.925}$	Citrate method	1500 °C –14 h	4.6×10^{-2} at 700 °C	– -	[36]
$Ce_{0.85}Sm_{0.1}Nd_{0.05}O_{2-\delta}$	Conventional solid-state reaction	1600 °C –10 h	1.37×10^{-2} at 600 °C	0.98	[37]
$Ce_{0.8}Sm_{0.2}O_{2-\delta}$ (SDC20)	Sol–gel	1300 °C –10 h	5.4×10^{-3} at 600 °C	0.87	[18]
SDC20	Solid-state reaction	1550 °C –4 h	1.37×10^{-2} at 600 °C	0.77	[38]

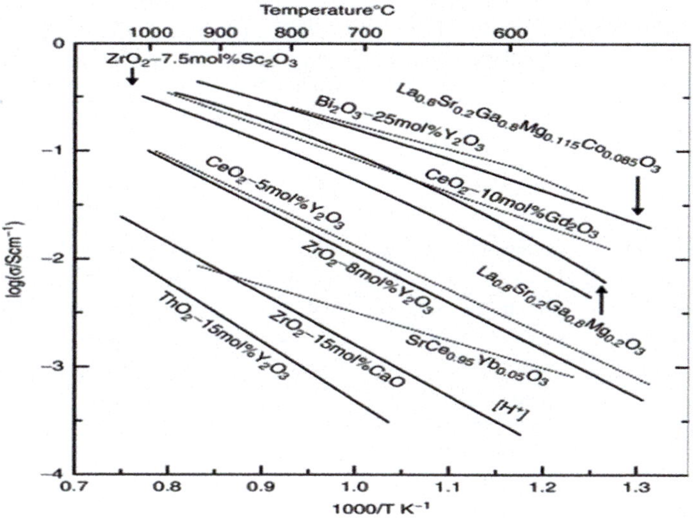

Fig. 2 Oxide-ion conductivity of different electrolytes [11]

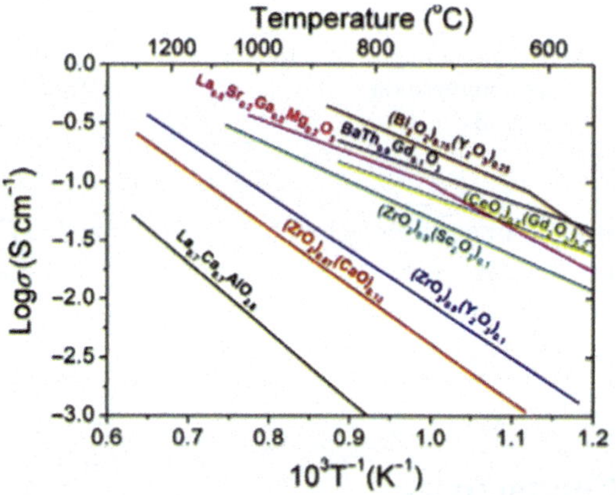

Fig. 3 Oxide-ion conductivity [12]

5 Doped Ceria-Based Composite Materials

Though ceria-based electrolyte materials significantly lowered the operating temperature of SOFCs from 1000 to 650 °C, their high electronic conductivity reduces their output power. This is the primary concern in the commercialization of SOFCs. To address this issue, doped ceria-based composite materials are the alternate new electrolytes, i.e., and they exhibit both ions (O^{2-}, H^+ ions conduction etc.), consequently

Table 5 Electrical conductivity of doped ceria-composite materials

Composite name	Synthesis method	Sintering temperature	Conductivity	References
$Ce_{0.76}La_{0.08}Pr_{0.08}Sm_{0.08}O_{2-\delta}$ (LPS)	Sol–gel auto-combustion	1300 °C	4.3 × 10^{-2} S/cm	[39]
LPS–LN	Solid-state mixing	700 °C	4.6 × 10^{-1} S/cm at 600 °C	[40]
(LPS-LN)—NiO anode	Solid-state mixing	700 °C	2.28 × 10^{-1} S/cm at 600 °C	[40]
Cathode-based symmetrical cell	Solid-state mixing	700 °C	4.2 × 10^{-2} S/cm	[40]
LSG–LN	Sol–gel auto-combustion technique followed by solid-state mixing methods	700 °C	4.2 × 10^{-1} S/cm at 600 °C	[41]
GDC–NiO	Conventional solid-state reaction method	–	2.8 × 10^{-4} S/cm at 973 K	[42]
GDC20–NiO	Mixing	1300 °C	7.5 × 10^{-2} S/cm at 600 °C	[43]
LSCF–CSGO	Glycine nitrate auto-combustion method	1673 K	4.3 × 10^{-4} S/cm at 973 K	[44]

lowering the operating temperatures below 550 °C. In doped ceria-based composite materials, carbonate content creates super-ionic paths for conduction. That is why, at lower temperatures, doped ceria-based composite materials exhibit higher ionic conductivity than doped ceria electrolytes [39–44] (Table 5).

6 Electrical Conductivity

Impedance spectroscopy helps study the transport properties (reactive and capacitive parts). The electrical and transport properties can be analyzed using Eqs. 2 and 3 [45]

$$Z^* = Z' + jZ'' \quad \text{complex impedance} \tag{2}$$

$$\sigma_g = \frac{L}{R_g A}, \sigma_{gb} = \frac{L}{R_{gb} A}, \sigma_t = \frac{L}{R_t A} \tag{3}$$

The total resistance R_t, the grain resistance R_g, and the grain boundary resistance R_{gb} are taken from the impedance spectra at different temperatures. The grain σ_b, grain boundary $\sigma_{gb,}$ and the total conductivity σ_t were then calculated using Eq., where L is the thickness, and A is the cross-sectional area of the sample.

Figure 4 presents the impedance spectra of the doped ceria electrolyte materials. The grain (R_g), grain boundary (R_{gb}), and electrode (R_e) contributions depend on the nature of the sample, grain growth, and experimental conditions like frequency range and temperature.

$$\sigma^*_{gb} = \frac{d_{gb}.L}{A.D.R_{gb}} \text{(without porosity correction)} \tag{4}$$

$$(\sigma_{gb} = \frac{L}{R_{gb}.A}; \sigma^*_{gb} = \sigma_{gb} \times \left(\frac{d_{gb}}{D}\right); \frac{C_g}{C_{gb}} = \frac{d_{gb}}{D}) \text{ [47]}$$

$$\sigma^*_{gb} = \frac{d_{gb}.L}{f_p.D.A.R_{gb}} \quad \text{(with porosity correction)} \tag{5}$$

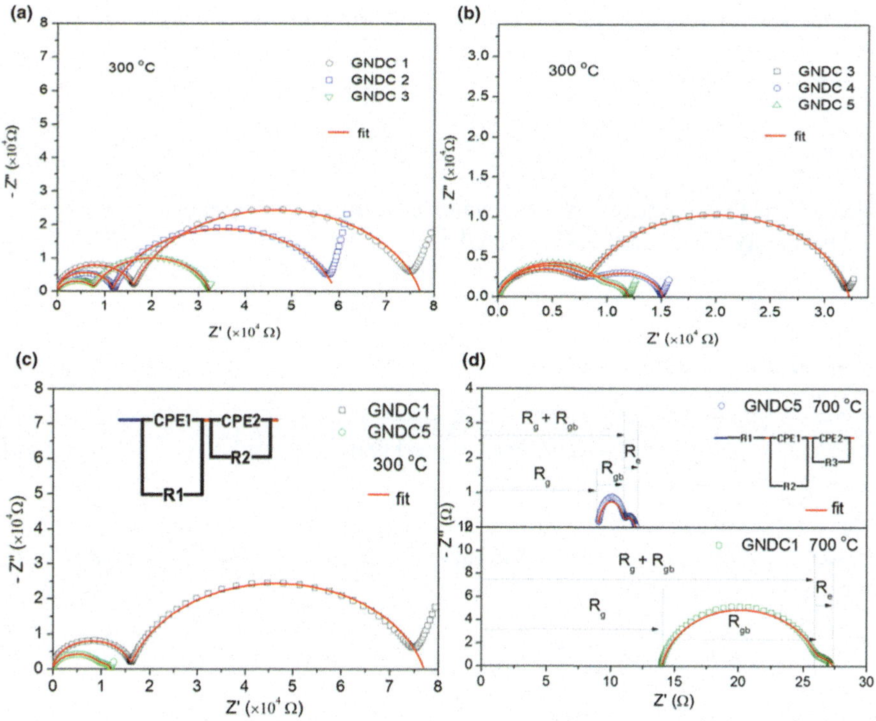

Fig. 4 Nyquist plots of **a** the GNDC1-GNDC3 samples at 300 °C, **b** the GNDC3-GNDC5 samples at 300 °C, **c** the impedance comparison of GNDC1 and GNDC5 samples and at 300 °C (inset Fig. represents equivalent circuit), and **d** the impedance comparison of the GNDC1 [46]

$$\sigma_g^* = \frac{L}{f_p.A.R_g} \text{ (with porosity correction)} \quad (6)$$

$$\sigma_g = \frac{L}{A.R_g} \text{ (without porosity correction)} \quad (7)$$

where D is the average grain size, d_{gb} is the grain boundary thickness, R_g is the grain resistance, R_{gb} is the grain boundary resistance, L is the thickness, A is the cross-sectional area, f_p sis the porosity factor (f_p = 1– density), C_g is the grain capacitance, and C_{gb} is the grain boundary capacitance.

Pérez-Coll [47] reported the porosity corrections when the samples' relative density was lower than 95%, using Eqs. 4–7. However, in the present study, the sample's relative density is more than 95%, and the different conductivities were calculated using Eqs. 4–7. [47, 48].

The electrical conductivity is given by

$$\sigma = ne\mu \quad (8)$$

n is the concentration of charge carriers, e is the electronic charge, and μ is the mobility

For oxide-ion conductivity [2]

$$\sigma_v = N_v q_v \mu_v \quad (9)$$

where N_v is the number of oxygen vacancies, and v is the vacancy

Diffusivity D and mobility μ related as

$$\mu = qB = qD/kT \quad (10)$$

where B is the absolute mobility,

$$D = a^2\vartheta_o \exp\left(\frac{\Delta S_m}{k}\right)\exp\left(-\frac{\Delta H_m}{kT}\right) \quad (11)$$

where 'a' is the vacancy jump distance, ϑ is the vibrational frequency, ΔS_m is the activation entropy of the migration, and ΔH_m is the activation enthalpy of the migration.

$$N_v = \left[V_O^{oo}\right]\left[1 - V_O^{oo}\right]N_o \quad (12)$$

where N_o represents the number of oxygen vacancies per unit volume

Using above equations, the conductivity can be written as

$$\sigma T = C'\left[V_O^{oo}\right]\left[1 - V_O^{oo}\right]xp\left(-\frac{\Delta H_m}{kT}\right) \quad (13)$$

where $C' = \left(\frac{4e^2}{k}\right)a^2\vartheta_o \exp\left(\frac{\Delta S_m}{k}\right)$

$$\sigma T = C'\left[V_O^{oo}\right]\exp\left(-\frac{\Delta H_m}{kT}\right)(\text{Since } V_O^{oo} \ll 1)$$

In case doped ceria has a large number of oxygen vacancies.
The conductivity equation is written as

$$\sigma T = \sigma_o \exp\left(-\frac{E_a}{kT}\right) \quad (14)$$

where σ_o is the pre-exponential factor, E_a is the activation energy, and T is the temperature. At lower temperatures, activation energy is the sum of association and migration energy; at higher temperatures, activation energy is solely equal to migration energy.

Figure 5 presents the SEM images of sintered and powder samples. The electrolyte material should be dense (more than 93% TD); surface view of SEM images showed grain growth with increased sintering temperature. The sample grain and grain boundary resistance contribution depends on grain size and temperature [23, 46], i.e., the grain resistance increased with grain size, whereas grain boundary resistance decreased. Hence, with higher-dense sample electrolyte materials, there is the possibility of higher gran resistance and lower grain boundary resistance. Besides, microstructure samples grain and grain boundary resistance depend on porosity [23, 46, 47]. The porosity corrections are presented in Eqs. 4–7 and presented in Fig. 6. Generally, total resistance depends on microstructure, density, and porosity factors [47]. The electrical conductivity values without porosity correction are lower than the porosity correction. The porosity correction to conductivity shows how important to prepare higher-dense electrolyte materials. Tables 6, 7 and 8 present the electrical conductivity values with and without porosity corrections.

The activation energy values are calculated using Eq. 14 and presented in Tables 6, 7 and 8. A suitable electrolyte material should have higher electrical conductivity and a minimum activation energy. The activation energy values changed with temperature, i.e., decreased with increasing temperature.

The fuel cell with the nanocrystalline structure GDC electrolyte (0.5 mm in thickness) can deliver a remarkable peak power density of 591.8 mW cm^{-2} at 550 °C, as presented in Fig. 7.

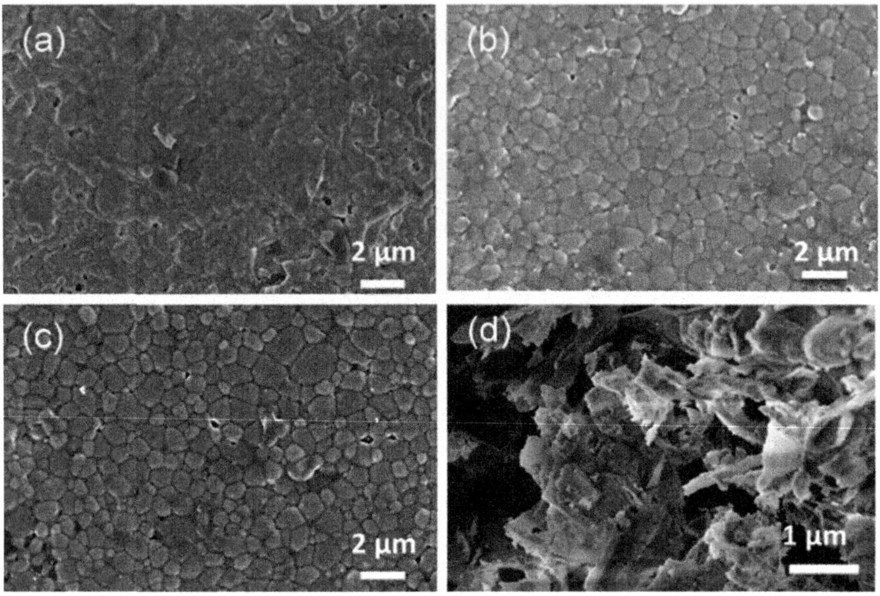

Fig. 5 SEM images of the SDDC samples sintered at **a** 1200 °C, **b** 1250 °C, **c** 1300 °C, and **d** powder calcined at 600 °C [23]

7 Electrolyte-Free Fuel Cell (EFFC) or Single-Component Cell

Three-component fuel cell (cathode, electrolyte, and anode) and single-component or electrolyte-free fuel cell (EFFC) is presented in Fig. 8. In a three-component fuel cell, the efficiency of the fuel cell is lowered due to high operating temperature (>650–700 °C). Due to high operating temperature, there is always a thermal expansion mismatch between cell components, chemical instability, and thermal degradation. To overcome this, the concept of an electrolyte-free fuel cell (EFFC) is helpful to optimize the efficiency further and decrease the operating temperature below 550 °C. In single-component fuel cells, both O^{2-} and H^+ ions are present, which further optimize the current–power and decrease the operating temperature compared with three-component fuel cells [49–55].

The comparison between three-component and single-component fuel cell's I-V and I-P properties are presented in Fig. 9. In comparison with a conventional SOFC device, a single LSCF-SCD (50:50) showed higher cell voltage, 0.95 V, and current density 1500 mA/cm^2. The performance of different single-component fuel cells over three-component fuel cells is presented in Table 9. The operating temperature was reduced to $< =$ 550 °C with improved current density.

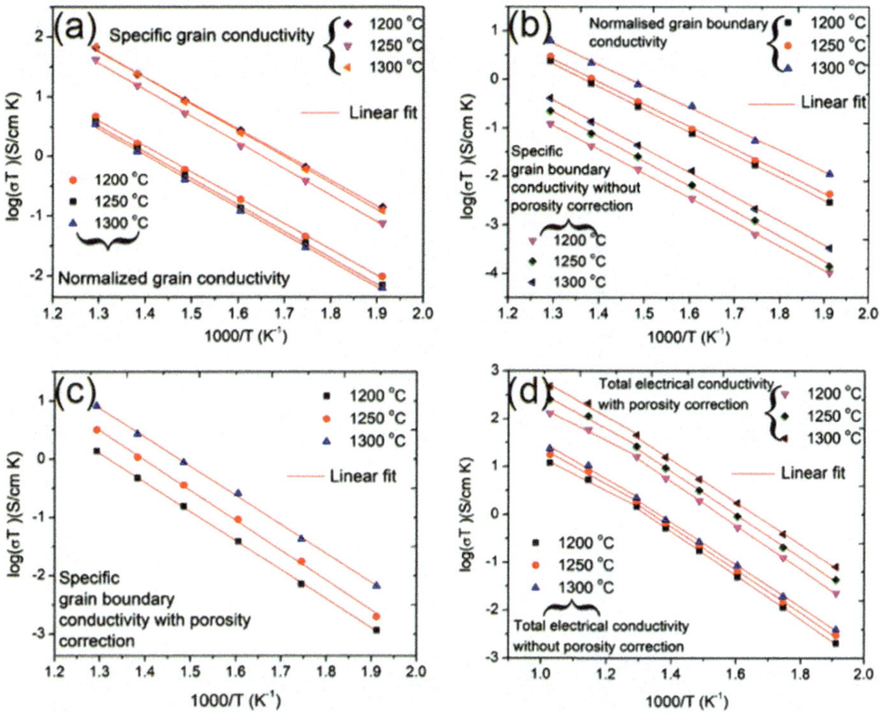

Fig. 6 Conductivity versus 1000/T plots for **a** normalized grain and specific grain conductivities, **b** normalized and specific grain boundary conductivity (without porosity correction), **c** specific grain boundary conductivity (with porosity correction), and **d** total electrical conductivity of the SDDC samples [23]

Table 6 Electrical conductivity values of the grain and the grain boundary of the SDDC ($Ce_{0.82}Dy_{0.155}Sr_{0.025}O_{2-\delta}$) sample [23]

	Electrical conductivity, σ S/cm					
	Grain			Grain boundary		
Temperature (°C)	1200	1250	1300	1200	1250	1300
300	7.99×10^{-5}	6.13×10^{-5}	5.25×10^{-5}	2.96×10^{-5}	3.68×10^{-5}	9.67×10^{-5}
400	8.97×10^{-4}	7.07×10^{-4}	6.13×10^{-4}	4.08×10^{-4}	5.25×10^{-4}	0.00119
500	0.00613	0.0049	0.00454	0.00306	0.00387	0.00817
Activation energy (eV)						
250–500	0.86	0.87	0.88	0.93	0.92	0.88

Table 7 Electrical conductivity values of specific grain and specific grain boundary of the SDDC ($Ce_{0.82}Dy_{0.155}Sr_{0.025}O_{2-\delta}$) sample [23]

	Electrical conductivity, σ S/cm								
	Specific grain (with porosity correction)			Specific grain boundary (without porosity correction)			Specific grain boundary (with porosity correction)		
Temperature (°C)	1200	1250	1300	1200	1250	1300	1200	1250	1300
300	6.8×10^{-4}	0.00114	0.00105	1.1×10^{-6}	2.1×10^{-6}	3.7×10^{-6}	1.2×10^{-5}	3.0×10^{-5}	7.4×10^{-5}
400	0.00786	0.01281	0.01225	2.0×10^{-5}	3.7×10^{-5}	6.5×10^{-5}	2.3×10^{-4}	5.3×10^{-4}	0.00131
500	0.05447	0.08754	0.09078	1.61×10^{-4}	2.9×10^{-4}	5.3×10^{-4}	0.00177	0.00416	0.01075
Activation energy (eV)									
250–500	0.88	0.87	0.86	0.99	1.03	0.99	0.99	1.02	0.99

Table 8 Total electrical conductivity values of the SDDC ($Ce_{0.82}Dy_{0.155}Sr_{0.025}O_{2-\delta}$) sample [23]

	Electrical conductivity, σ S/cm					
	Total (without porosity correction)			Total (with porosity correction)		
Temperature (°C)	1200	1250	1300	1200	1250	1300
300	2.01×10^{-5}	2.51×10^{-5}	3.40×10^{-5}	6.93×10^{-4}	0.00117	0.00113
400	2.59×10^{-4}	3.31×10^{-4}	4.04×10^{-4}	0.00809	0.01335	0.01356
500	0.00189	0.00237	0.00292	0.05623	0.09169	0.10152
600	0.00613	0.00919	0.01225	0.06775	0.13106	0.24391
700	0.01225	0.01838	0.02451	0.1355	0.25974	0.48782
Activation energy (eV)						
250–500	0.91	0.90	0.88	0.92	0.90	0.88
500–700	0.73	0.75	0.76	0.73	0.75	0.76

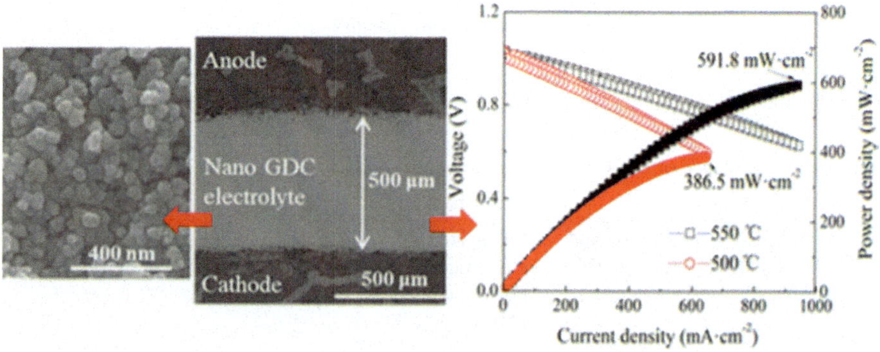

Fig. 7 Fuel cell power density [49]

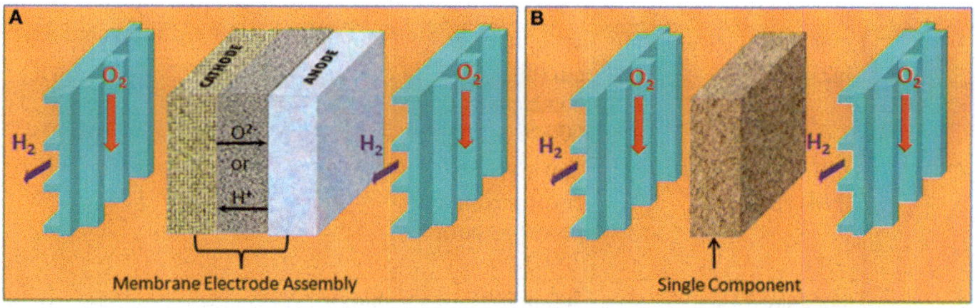

Fig. 8 Structures between **a** SOFC and **b** EFFC, or called single components [50]

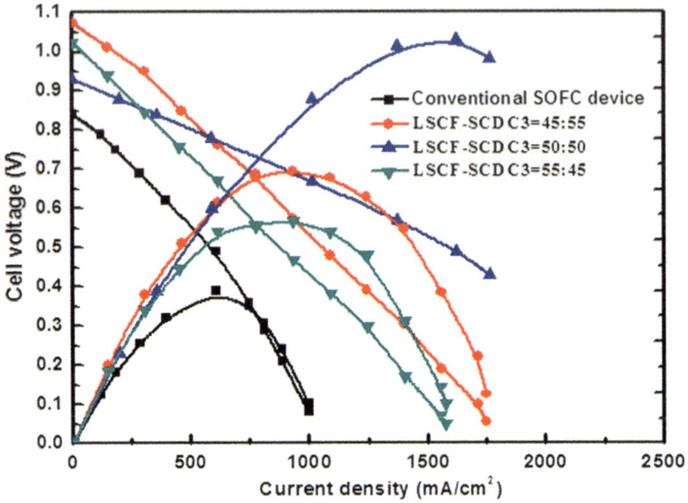

Fig. 9 Typical I–V and I–P characteristics for LSCF–SCDC fuel cells in comparison with the fuel cell using the ionic SCDC electrolyte [50]

Table 9 Performance of EFFC over the traditional SOFC [50]

S. No.	Materials	Performance	Layer	References
1	LiNiCuZnFe-oxide and Na2CO3–SDC (NSDC)	700 mW/ cm^2 (maximum power output), 550 °C	Single-component and three-component fuel cells	[51]
2	LiNiZn-oxide was mixed with GDC	450 mW/ cm^2 (maximum power output), 550 °C	Single	[52]
3	LiNiCuSr-oxide and MgZnDC	10×10^{-2} S/ cm (total conductivity), 600 °C	Single	[53]
4	Sr2Fe1.5Mo0.5O6 − δ (SFM) and NSDC	360 mW/ cm^2 (maximum power output), 750 °C	Single	[54]
5	Doped the oxidation–reduction catalyst Fe	700 mW/ cm^2 (maximum power output), 550 °C	Single	[55]

8 Conclusions

The concept of an electrolyte-free fuel cell (EFFC) is advantageous over the electrolyte fuel cell. The main concern of traditional fuel cells is high operating temperature and reduced output efficiency, which limits traditional SOFCs for large-scale production and commercialization. EFFCs are new devices that convert chemical energy into electrical energy. There is much scope for R&D in developing EFFC with high efficiency.

References

1. Solid oxide fuel cell (2022 In: Wikipedia. https://en.wikipedia.org/wiki/Solid_oxide_fuel_cell
2. Jaiswal N, Tanwar K, Suman R et al (2019) A brief review on ceria based solid electrolytes for solid oxide fuel cells. J Alloy Compd 781:984–1005
3. Vora SD (2000) Proceedings of the fourth European solid oxide fuel cell forum. In: McEvoy AJ (eds) Lucerne. Switzerland, pp 175, 3
4. Yokoyama H, Miyahara A, Veyo SE (1997) Proceedings of sixth international symposium on Solid oxide fuel cells, In: Stimming U, Singhal SC, Tagawa H, Lehnert W (eds) Aachen. Germany, pp 94, 4
5. Mori H, Omura H, Hisatome N, Ikeda K, Tomida K (1999) Proceedings of sixth international symposium on Solid oxide fuel cells (SOFC-VI), In: Singhal SC, Dokiya M (eds) Honolulu. Hawaii, pp 52
6. https://techetch.com/blog/the-different-types-of-fuel-cells/
7. Omer S, Wachsman ED, Nino JC (2008) Higher conductivity Sm^{3+} and Nd^{3+} co-doped ceria-based electrolyte materials. Solid State Ionics 178:1890–1897
8. Moure A, Moure C, Tartaj J (2011) A significant improvement of the processing and electric properties of CeO_2 co-doped with Ca and Sm by mechanosynthesis. J Power Sources 196:10543–10549
9. Omer S, Wachsman ED, Nino JC (2006) A co-doping approach towards enhanced ionic conductivity in fluorite-based electrolytes. Solid State Ionics 177:3199–3208
10. Pikalova EY, Murashkina AA, Maragou VI, Demin AK, Strekalovsky VN, Tsiakaras PE (2011) CeO_2 based materials doped with lanthanides for applications in intermediate temperature electrode
11. Ishihara T (2014) Oxide ion conductor. In: Kreysa G, Ota K, Savinell RF (eds) Encyclopedia of applied electrochemistry. Springer, New York. https://doi.org/10.1007/978-1-4419-6996-5_165
12. Ajith Kumar S, https://doi.org/10.1016/B978-0-12-817445-6.00005-3
13. Ramesh S, Rajitha G (2020) Structural characterization and electrical properties of Ce1-xSmxO2-δ by sucrose-pectin–assisted auto combustion process. Ionics 26(10):5089–5098
14. Kim DJ (1989) Lattice parameters, ionic conductivities, and solubility limits in fluorite-structure MO2 oxide [M $=Hf^{4+}$, Zr^{4+}, Ce^{4+}, Th^{4+}, U^{4+}] solid solutions. J Am Ceram Soc 72:1415–1421. https://doi.org/10.1111/j.1151-2916.1989.tb07663.x
15. Ismail A, Hooper J, Giorgi JB, Woo TK (2011) A DFT+U study of defect association and oxygen migration in samarium-doped ceria. Phys Chem Chem Phys 13:6116–6124. https://doi.org/10.1039/c0cp02062a
16. Kosinski MR, Baker RT (2011) Preparation and property–performance relationships in samarium-doped ceria nanopowders for solid oxide fuel cell electrolytes. J Power Sources 196:2498–2512. https://doi.org/10.1016/j.jpowsour.2010.11.041

17. Arabaci A, Serin Ö (2015) Characterization of Sm-doped ceria ceramics synthesized by two different method. J Mater Eng Perform 24:2730–2737. https://doi.org/10.1007/s11665-015-1563-5
18. Peng C, Wang Y, Jiang K, Bin BQ, Liang HW, Feng J, Meng J (2003) Study on the structure change and oxygen vacation shift for $Ce_{1-x}Sm_xO_{2-y}$solid solution. J Alloy Compd 349:273–278. https://doi.org/10.1016/S0925-8388(02)00903-9
19. Wu YC, Lin CC (2014) The microstructures and property analysis of aliovalent cations (Sm^{3+}, Mg^{2+}, Ca^{2+}, Sr^{2+}, Ba^{2+}) co-doped ceria-base electrolytes after an aging treatment. Int J Hydrog Energy 7988–8001. https://doi.org/10.1016/j.ijhydene.2014.03.063
20. Venkataramana K, Madhuri C, Suresh Reddy Y, Bikshamaiah G, Vishnuvardhan Rddy C (2017) Structural, electrical and thermal expansion studies of tri-doped ceria electrolyte materials for ITSOFCs. JALCOM 719:97–107. https://doi.org/10.1016/j.jallcom.2017.05.022
21. Gupta M, Shirbhate S, Ojha P, Acharya S (2018) Processing and conductivity behavior of La, Sm, Fe singly and doubly doped ceria: as electrolytes for IT-SOFCs. Solid State Ionics 320:199–209. https://doi.org/10.1016/j.ssi.2018.03.005
22. Arabacı A (2020) Effect of the calcination temperature on the properties of Sm-doped CeO_2. Emerg Mater Res 9:1–9. https://doi.org/10.1680/jemmr.18.00082
23. Ramesh S (2022) Effect of sintering temperature on microstructure and electrical properties of $Ce_{0.82}Dy_{0.155}Sr_{0.025}O_{2-\delta}$electrolyte. Process Appl Ceram 16(2):175–182
24. Sandhya K, Chitra Priya NS, Rajendran DN (2020) Enhancement of electrical properties $Ce_{0.8}$ $Sm_{0.2-x}O_{2-\delta}$ by Sr^{2+} doping. Appl Phys A 126:61
25. Tanwar K, Jaiswal N, Kumar D, Parkash O (2016) Synthesis and characterization of Dy and Ca Co-doped ceria based solid electrolytes for IT-SOFCs. J Alloys Compd 684:683–690
26. Zheng Y, He S, Ge L, Zhou M, Chen H, Guo L (2011) Effect of Sr on Sm-doped ceria electrolyte. Int J Hydrogen Energy 36:5128–5135
27. Jaiswal N, Upadhyay S, Kumar D, Parkash O (2014) Sm^{3+} and Sr^{2+} co-doped ceria prepared by citrate-nitrate autocombustion method. Int J Hydrogen Energy 39:543–551
28. Yang J, Ji B, Si J, Zhang Q, Yin Q, Xie J, Tian C (2016) Synthesis and properties of ceria based electrolyte for IT-SOFCs. Int J Hydrogen Energy 41:15979–15984
29. Zheng Y, Wu L, Gu H, Gao L, Chen H, Guo L (2009) The effect of Sr on the propertieos of Y-doped ceria electrolyte for IT-SOFCs. J Alloys Compd 486:586–589
30. Ozdemir H, Sarıboga V, Oksuzomer MAF, Gurkaynak MA (2012) Preparation and characterization of Ca-SmCe mixed oxides via cellulose templating method for solid oxide fuel cell applications. J Power Sources 219:155–162
31. Wu YC, Chen WY (2018) Fabrication and characterization of Ca^{2+}, Sr^{2+}, Ba^{2+}, Sm^{3+}, and La^{3+} co-doped ceria-based electrolyte powders for low-temperature anode-supported solid oxide fuel cells. Int J Hydrogen Energy 43:18463–18476
32. Venkataramana K, Madhuri C, Suresh Reddy Y, Bhikshamaiah G, Vishnuvardhan Reddy C (2017) Structural, electrical and thermal expansion studies of tri-doped ceria electrolyte materials for IT-SOFCs. J Alloy Compd. https://doi.org/10.1016/j.jallcom.2017.05.022
33. Anirban SK, Dutta A (2016) Microstructure and charge carrier dynamics in Pr-Sm-Eu triple-doped nanoceria. Solid State Ionics 295:48–56
34. Xiaomin LIN, Qiuyue L, Lili ZHU, Xiaomei LIU (2015) Synthesis and characterization of $Ce_{0.8}Sm_{0.2-x}Pr_xO_{2-\delta}$ (x= 0.02–0.08) solid electrolyte materials. J Rare Earths 33(4):411–416
35. Ramesh S, James Raju KC (2012) Preparation and characterization of $Ce_{1-x}(Gd_{0.5}Pr_{0.5})_xO_2$electrolyte for IT-SOFCs Int J Hydrogen Energy 37:10311–10317
36. Wang F-Y, Wan B-Z, Cheng S (2005) Study on Gd^{3+} and Sm^{3+} co-doped ceria-based electrolytes. J Solid State Electrochem 9:168–173
37. Rai A, Mehta P, Omar S (2014) Ionic conduction behavior in $Sm_xNd_{0.15-x}Ce_{0.85}O_{2-\delta}$. Solid State Ionics 263:190–196
38. Wu Y-C, Lin C-C (2014) The microstructures and property analysis of aliovalent cations (Sm^{3+}, Mg^{2+}, Ca^{2+}, Sr^{2+}, Ba^{2+}) co-doped ceria-base electrolytes after an aging treatment. Int J Hydrogen Energy 39:7988–8001

39. Madhuri C, Venkataramana K, Shanker J, Reddy CV (2020) Effect of La^{3+}, Pr^{3+}, and Sm^{3+} triple-doping on structural, electrical, and thermal properties of ceria solid electrolytes for intermediate temperature solid oxide fuel cells. J Alloy Compd 849:156636
40. Madhuri C, Venkataramana K, Ramesh S, Shanker J, Vishnuvardhan Reddy C (2022) J Electron Mater. https://doi.org/10.1007/s11664-022-09850-x
41. Venkataramana K, Madhuri C, Vishnuvardhan Reddy C (2020) Triple-doped ceria-carbonate ($Ce_{0.82}La_{0.06}Sm_{0.06}Gd_{0.06}O_{2-\delta}$—$(Li\text{-}Na)_2CO_3$) nano-composite solid electrolyte materials for LT-SOFC applications. Ceram Int 46:27584
42. Soman AK, Kuppusami P, Rabel AM (2017) Electrical conductivity of NiO-Gadolinia doped ceria anode material for intermediate temperature solid oxide fuel cells. Nano Hybrids Compos 17:224
43. Narsimha Reddy M, Vijaya Bhaskar Rao P, Bhoga SS, Bansod M, Sreehari Babu V (2020) Structural, mechanical and electrical properties of NiO—GDC20 composite anodes for low or intermediate temperature solid oxide fuel cells. J Phys Conf Ser 1495:012020
44. Ajith Kumar S, Kuppusami P, Vengatesh P (2018) AutoCombustion synthesis and electrochemical studies of $La_{0.6}Sr_{0.4}Co_{0.2}Fe_{0.8}O_{3-\delta}$—$Ce_{0.8}Sm_{0.1}Gd_{0.1}O_{1.90}$ nanocomposite cathode for intermediate temperature solid oxide fuel cells. Ceram Int 44:21188
45. Ramesh B, Ramesh S, Vijaya Kumar R, Lakshmipathi Rao M (2012) AC impedance studies on $LiFe_{5-x}Mn_xO_8$ ferrites. J Alloy Compd 513:289–293
46. Ramesh S (2022) Impedance and electrical properties of $Nd_x Gd_y Ce_{1-(x+y)}O_{2-\delta}$. J Electron Mater 51(10):5704–5716
47. Pérez-Coll D, Sánchez-López E, Mather GC (2010) Influence of porosity on the bulk and grain-boundary electrical properties of Gd-doped ceria. Solid State Ionics 181:1033
48. Ramesh S, Mahadevappa N, Jeevan V (2022) Synthesis, structure, microstructure, and electrical properties of $Ce_{0.81}Nd_{0.095}Sm_{0.095}O_{2-\delta}$. Bol Soc Esp Cerám Vidr 61(5):428–438
49. Chen G, Sun W, Luo Y, He Y, Zhang X, Zhu B, Li W, Liu X, Ding Y, Li Y, Geng S, Yu K (2019) Advanced fuel cell based on new nanocrystalline structure $Gd_{0.1}Ce_{0.9}O_2$ electrolyte. ACS Appl Mater Interfaces 11(11):10642–10650. https://doi.org/10.1021/acsami.8b20454
50. Lu Y, Zhu B, Cai Y, Kim J-S, Wang B, Wang J, Zhang Y, Li J (2016) Progress in electrolyte-free fuel cells. Front Energy Res 4:17. https://doi.org/10.3389/fenrg.2016.00017
51. Zhu B, Raza R, Qin H, Fan L (2011) Single component and three-component fuel cells. J Power Sources 196:6362–6365
52. Zhu B, Ma Y, Wang X, Raza R, Qin H, Fan L (2011) A fuel cell with a single component functioning. Electrochem Commun 13:225–227
53. Hu H, Lin Q, Zhu Z, Zhu B, Liu X (2014) Fabrication of electrolyte-free fuel cell with $Mg_{0.4}Zn_{0.6}O/Ce_{0.8}Sm_{0.2}O_{2-\delta}$–$Li_{0.3}Ni_{0.6}Cu_{0.07}Sr_{0.03}O_{2-\delta}$ layer. J Power Sources 248:577–581
54. Dong X, Tian L, Li J, Zhao Y, Tian Y, Li Y (2014) Single layer fuel cell based on a composite of $Ce_{0.8}Sm_{0.2}O_{2-\delta}$–$Na_2CO_3$ and a mixed ionic and electronic conductor $Sr_2Fe_{1.5}Mo_{0.5}O_{6-\delta}$. J Power Sources 249:270–276
55. Zhu B, Qin H, Raza R, Liu Q, Fan L, Patakangas J, Lund P (2011) A single component fuel cell reactor. Int J Hydrogen Energy 36:8536–8541

Chapter 10
Graphene-Based Materials in Energy Harvesting

Niranjan Patra and Gaddipati Bhavana

1 Introduction

Carbon has sp^3, sp^2, and sp hybridised orbitals, making it one of the most versatile materials. Anisotropy in some crystals of this material is possible due to the presence of anisotropic sp^2 carbon atoms in molecular form making it possible for carbon-based materials to be used in a variety of applications. The ability to precisely create nanocarbon crystals in the shape of sheets, needles, tubes, spheres, and many other morphologies has led to a surge in interest in carbon nanostructures.

Among different carbon nanostructures, Graphene is one of the most important atomic thick 2D carbon nanostructures consisting of sp^2 hybridised carbon atoms linked together in a highly regular "honeycomb" lattice as shown in Fig. 1. Graphene has a wide variety of potential uses, but its usage in energy harvesting and storage devices is one of the most intriguing of these sectors [1–8]. Graphene has a huge electrochemical window, large surface area (2600 m^2/g), high conductivity (100 S/cm^2), and excellent mobility of charge carriers (20 $m^2/(V\ s)$ [1]. Graphene's unique properties make it well-suited for use in energy production and harvesting. Graphene transparency is another one of its notable physical qualities; it is a million times thinner than paper, which is another notable difference [9]. This indicates that the thickness of 1 mm of graphene is only equivalent to a million sheets of paper. Graphene has an inter-layer spacing of roughly 0.335 nm. The transfer of graphene onto surfaces allows for the fabrication of transparent/semitransparent energy harvesting and storage systems.

Several methods exist for preparing graphene such as scotch-tape peeling, mechanical exfoliation of pyrolytic graphite, chemical vapour deposition (CVD) epitaxial growth on metal substrates, etc. Large-scale graphene production via CVD

N. Patra (✉) · G. Bhavana
Department of Engineering Chemistry, College of Engineering, Koneru Lakshmaiah Education Foundation, Greenfield, Vaddeswaram, Guntur 522502, Andhra Pradesh, India
e-mail: npatra@kluniversity.in

B. P. Swain (ed.), *Energy Materials*, Materials Horizons: From Nature to Nanomaterials,
https://doi.org/10.1007/978-981-99-3866-7_10

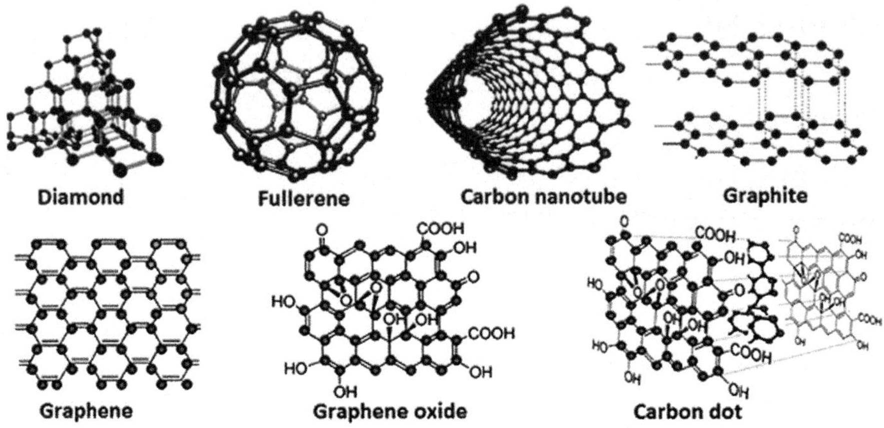

Fig. 1 Various forms of carbon allotrope [10]

could pave the way for its use in electronic applications like solar cells, thin-film transistors, and touch panels all of which call for graphene sheets with areas approximately tens of centimeters and most often necessitate transferring the material to a specific substrate. The CVD procedure is costly, just like the thermal degradation of the SiC technique. Modifying Hummer's approach yields graphene oxide (GO), which can be further reduced using $NaBH_4$ or hydrazine to yield pure graphene [9]. Adjusting the applied potential during electrochemical exfoliation makes for a sustainable product that is easy to maintain. Moreover, graphene may be prepared as a colloidal solution with variable solubility using this method, making it suitable for industrial-scale printed electronics.

2 Graphene-Based Materials in Energy Harvesting and Storage

2.1 Supercapacitors

Supercapacitors are energy storage device that works by combining the electrostatic properties of double-layer capacitance with the electrochemical properties of pseudocapacitance. Ultracapacitors, supercapacitors, and hybrid capacitors are all names given to charged carbon systems as well as those that combine carbon battery electrodes with conducting polymer electrodes [11]. Although the names anode and cathode are commonly used, they may not be strictly applicable in electrochemical capacitors (or supercapacitors) since energy is not necessarily transferred by redox reactions. So-called electrical double layers (EDLs) are generated and released by the orientation of electrolyte ions at the electrolyte/electrolyte interface; this leads to the parallel passage of electrons in the external wire and hence in the delivery of energy.

These capacitors have an unusually high energy density in comparison with traditional dielectric capacitors because of the inverse proportion between the energy stored and the thickness of the double layer. The amount of charge they can store is substantially greater than that of rechargeable batteries, and they can release that charge at much higher voltages. Supercapacitors are versatile energy storage devices that can be utilised as the sole source of power or in tandem with other power sources like batteries or fuel cells. Supercapacitors have many benefits over conventional energy storage methods, including high power capability, extended life, a wide thermal operating range, low weight, flexible packaging, and little to no maintenance [12]. Supercapacitor unique capacity to rapidly store and release energy makes them useful in a variety of other contexts, such as power balancing for utilities and as a backup supply in factories. Based on the mechanism used to store the electrical charge, supercapacitors can be divided into two distinct categories. Electrochemical double-layer capacitors (EDLCs) work because the ions at the electrode/electrolyte contact are drawn to the electrode due to pure electrostatic attraction which is often prepared using activated porous carbon. When utilising conducting polymers as an example, pseudo-capacitors are the second form of capacitor, and they function by transferring electrons to or from the valence bands of the redox cathode or anode reagent. In case both high energy and high poweroutput are required, supercapacitors are a useful addition since they prevent the battery from experiencing a voltage drop when under load.

EDLCs as having two porous carbon electrodes separated by a porous separator so that they do not come into electrical contact with one another [14, 15]. They are not like batteries in that a cathode and an anode can be defined, though the terms are often used that way for convenience. Energy-density-versus-power-density shows why EDLCs are gaining popularity; they bridge the gap between conventional capacitors and batteries as shown in Fig. 2 [13].

For the design of high performance supercapacitors, the development of innovative electrode materials is crucial. When it comes to filtering AC current in mobile electronics, graphene is typically employed in place of carbon as an electric double-layer capacitor [16]. Electrolytic aluminium capacitors have a maximum CV/volume value of only up to 0.14 FV/cm^3; however, these ECs can store up to 5.5 FV/cm^3 with organic electrolytes. In essence, traditional EDLCs are DC devices, with a full charge/discharge cycle of several seconds. However, their efficiency drops dramatically, and they begin to act more like resistors than capacitors at higher frequencies. This is because the electrodes, which are typically constructed of a high specific surface area conductive substance like activated carbon, are porous, hence increasing the device resistance. In order to solve this issue, EDLC with vertically oriented high specific surface area graphene electrodes that are not porous at all was developed [16]. The device improves upon commercial EDLCs by a factor of 105, increasing their working frequency to well over 5000 Hz. Furthermore, charging and discharging

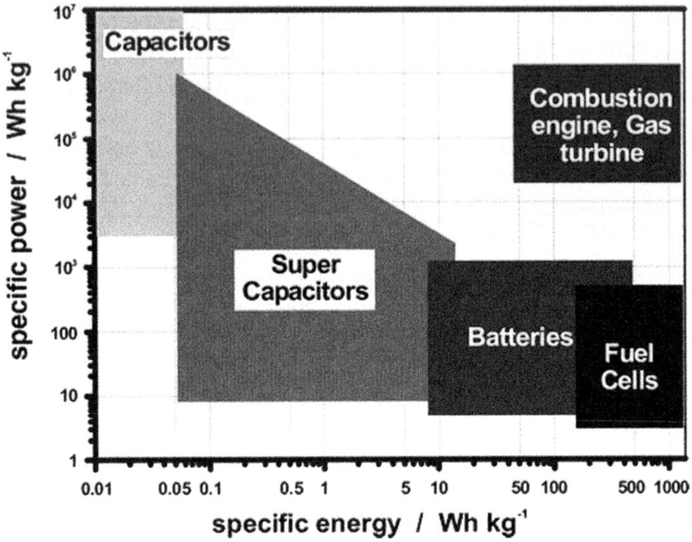

Fig. 2 Energy storage systems based on the Ragone strategy (adapted with permission from ACS [13])

may be done with great efficiency in times much shorter than 1 ms, and low-voltage aluminium electrolytic capacitors are much smaller. The electrical and ionic resistance of capacitors made with EDLCs fabricated from graphene can be reduced to less than 200 ms, while the resistive-capacitive (RC) time constant of conventional EDLCs is on the order of 1 s.

The integration of graphene EDLCs in EVs heralds a major step forward in transportation technology. To date, electric vehicles have been constrained due to the slow pace of charge/discharge provided by batteries, while the proposed usage of supercapacitors is hampered by their low specific energy. Supercapacitors based on curved graphene sheets have been shown to have a specific energy density comparable to that of contemporary Ni metal hydride batteries used in hybrid automobiles; they can also be recharged in less than 2 min, which is a significant advantage [16]. The mesoporous structure, with pore sizes between 2 and 25 nm, is stable because the curved shape prevents the graphene sheets from restacking.

Bai et al. [17] demonstrated a scalable approach to develop a hierarchical porous carbon nanofibers/graphene hybrid fibers (CNGFs) for use as electrodes in flexible supercapacitors to overcome the electrochemical performance of traditional graphene fibre electrodes made via wet spinning due to low specific surface area (SSA), mismatched pore size distribution, and significant interface resistanc. These hierarchical porous carbon nanofibers/graphene hybrid fibres (CNGFs) for use as electrodes in flexible supercapacitors exhibit both high specific capacitance and long-term cycle stability as shown in Fig. 3.

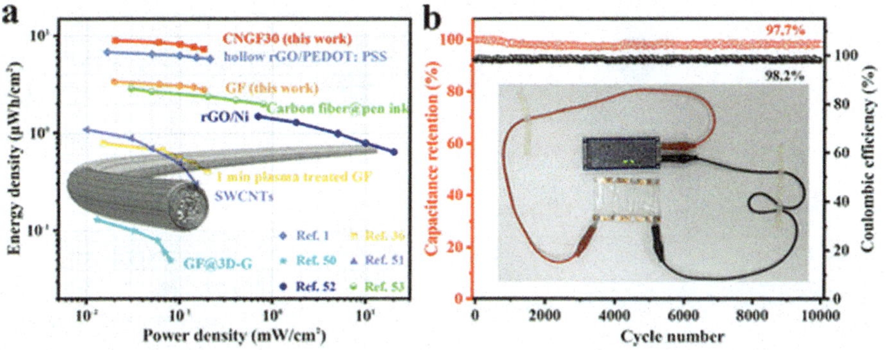

Fig. 3 (**a**) Energy density of different amalgamations. (**b**) Capacitance reflectance (adapted with permission from ACS [17])

Restocking graphene layers during synthesis is another hurdle in the electrode performance. Abdallah et al. [18] used electro-exfoliated graphene as the supercapacitor electrode for which the electrochemical properties were enhanced using a simple post-ultrasonication treatment to manage the electro-exfoliated graphene structure as shown in Fig. 4. Post-electro-exfoliation shows a maximum specific capacitance of 140.5 F g^{-1} at 0.5 A g^{-1}, which was higher than what was measured in a control sample subjected to the same amount of current but without ultrasonication. In addition, two-electrode measurements of the cycling performance of ultrasonicated EG showed that it retained the highest capacitance, 92.9%, after 10,000 cycles at 1 A g^{-1}, indicating that ultrasonication can increase the stability of exfoliated graphene compared to the sample without ultrasonication, which retained only 82.2% of its original capacitance.

Tsai et al. [18] investigated the scalability using supercapacitors with a high rate capability made from electrochemically exfoliated graphene (EG) and hybrid activated carbon (AC)/EG which provides a contrast between various rate capabilities of commercial high-power supercapacitors to those of pouch cells and large-scale EG or AC/EG composites system. The exfoliated graphene used in this study had a 9.6 at% oxygen content, a 9.36 C/O ratio, and a conductivity of 2.68 104 Sm^{-1}. By increasing the scan rate from 5 to 1 Vs^{-1}, the EEG-based supercapacitors' specific capacitance maintained above 80% of the maximum value. Incorporating EEG into the electrodes of an activated carbon (AC)-based supercapacitor can increase the device rate capabilities.

Hybrid CNT/graphene composites typically having a capacitance of 120 F/g have been created by incorporating carbon nanotube, suitable graphene-based material for dispersion, with a poly(ethyleneimine) (PEI) stabiliser that is charged and soluble [18]. Compared to it, this is a huge jump of the capacitances of vertically [19] and nonaligned CNT electrodes [20, 21].

Cheng et al. [22] developed a technique for making a hybrid composite film of single-walled carbon nanotubes (SWCNTs) and graphene as electrodes in supercapacitors that pack a lot of power into a small space. Through the use of a more

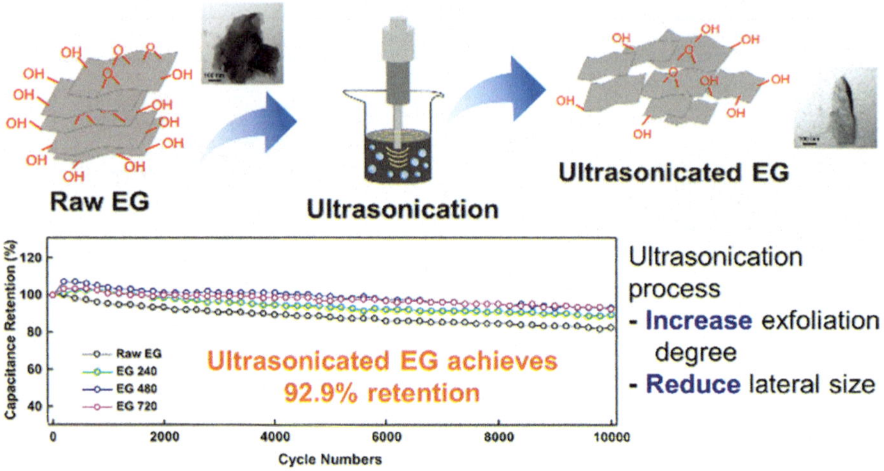

Fig. 4 Schematic explaining exfoliation of graphene using ultrasonication and the capacitance reflectance % (adapted with permission from ACS [18])

convenient two-electrode testing system, it has been observed that the specific capacitance of a single electrode in aqueous and organic electrolytes is 290.6 F g^{-1} and 201.0 F g^{-1}, respectively. Energy density in the organic electrolyte reached 62.8 Wh kg^{-1}, and power density reached 58.5 kW kg^{-1}. When compared to graphene electrodes, energy density was increased by 23% and power density by 31% which is due to the use of SWCNTs as shown in Fig. 5. The energy density of the graphene/CNT electrodes in ionic liquid at ambient temperature was 155.6 Wh kg^{-1}. In addition, after being cycled in ionic liquid 1000 times, the specific capacitance increased by 29%. In the CNT/graphene supercapacitors, SWCNTs served as a conductor, as well as a binder and a spacer.

2.2 *Batteries*

Batteries are nothing but electrochemical cells where the terminals are electrically linked. Redox reactions at the anode and cathode convert chemical energy into electrical energy in batteries as well as in fuel cells. The negative and positive electrodes are used because reactions at the anode often take place at lower electrode potentials than at the cathode. The anode is the more negatively charged electrode, whereas the cathode is the positive. The energy is stored and converted is where batteries and fuel cells diverge. The anode and the cathode are the "active masses" in a redox reaction, and the medium through which charges are transferred in a battery is a closed system. That is to say, the space where energy is stored and transformed is the same

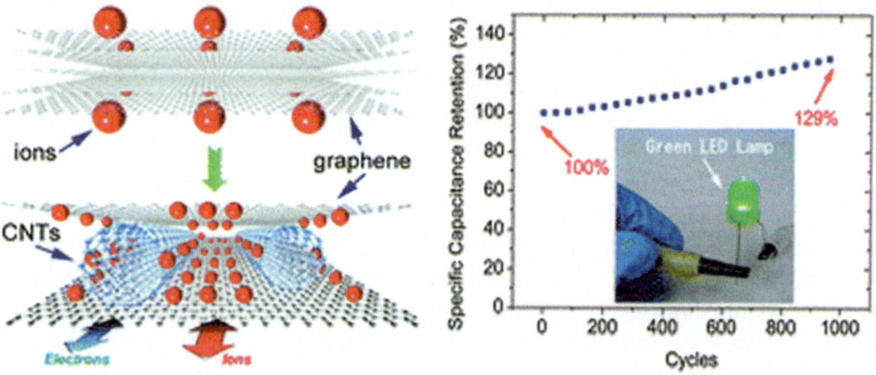

Fig. 5 Schematic of graphene and carbon nanotube composites where SWCNTs can be used to create a gap between the graphene nanosheets as well as the capacitance reflectance plot (adapted with permission from RSC [22])

space where the energy is used. Fuel cells are considered open systems because the reactive masses (such as oxygen from the air or hydrogen and hydrocarbon fuels) that undergo the redox reaction must be brought in from somewhere else.

Graphene's high electron mobility makes it suitable to be infused into a variety of different materials to help them store more charge in batteries, allowing them to power devices for longer [23]. Graphene anodes have been proposed as a means to increase battery capacity because of their superior charge retention (and specific energy) compared to graphite anodes [24]. Instead of using graphite for the anode in Li-ion batteries, graphene can be applied to the surface [25]. By subjecting the graphene to heat, defects can be introduced into the layers, creating channels through which Li^+ ions can flow and bind to the anode substrate [25]. The result is that recharge times for graphene-anode batteries will be lower than those of conventional Li-ion batteries with graphite anodes [26–28]. This could be used in electric and hybrid vehicles, where shorter charging times are essential for the vehicles to be competitive with conventional internal combustion engine vehicles [25].

In portable energy storage systems, graphene is shown to be a very intriguing and novel material. Graphene is the most frequent anode material in lithium-ion batteries which is called mesocarbon microbeads, although graphene has a far greater theoretical lithium storage capacity of 744 mAh g^{-1} assuming Li ions are bonded to both sides of the graphene sheet [29, 30]. The form and microstructure of graphene are reported to have significant effects on lithium storage. As an example, graphene with 1175, 1007, and 842 mAh g^{-1} reversible capacities, respectively, was demonstrated by graphene with single, triple, and quintuplicate layers as anode materials. Graphene-enhanced edge sites, flaws, and reduced size are all advantageous for better lithium storage [31, 32].

Porous graphene has gained a lot of attention as a potential material for energy storage devices, where their rapid diffusion of electrolytes and transit of electrons would be advantageous [33]. For instance, after 400 cycles, the specific capacity of highly ordered mesoporous graphene frame-structure with a large pore volume (1.8 $cm^3\ g^{-1}$) and high specific surface rea (1000 $m^2\ g^1$) exhibits a value of 520 mAh g^{-1}. Porous graphene sheets with high specific surface area, however, typically result in a sizable irreversible capacity, which significantly limits its practical employment in lithium-ion batteries.

Wang et al. [34] used an in-situ chemical approach to creating graphene-Si nanocomposite sheets that could stand alone. In the form of $Li_{4.4}Si$ alloys, silicon can reversibly host lithium, and graphene can limit the volume expansion that leads to powdering. After 100 cycles, the replacement electrode still had a capacity of roughly 708 mAh/g, according to electrochemical testing. The volume change that would ordinarily occur in Si electrodes is dampened by the vacuum spaces between graphene and Si, making these films remarkably long-lasting. Graphene can serve as a lithium storage medium and an electron conductor. More impressive results have been achieved by combining graphene with silicon nanoparticles. Silicon nanoparticles were found to be 21–22 nm in size, and the observed storage capacity was 2200 mAh/g after 50 cycles and 1500 mAh/g after 200 cycles [35]. The Si nanoparticles must be uniformly distributed throughout the graphene composite, and some of the graphene sheets must reconstruct graphite to create a continuous, highly conducting 3-D network to sandwich and trap the Si nanoparticles.

The uniformity of the nanoparticles and the sheet separation of the graphene are restricted when mechanical mixing is used. As a result, chemical strategies are now considered more viable. Graphene has SnO_2 incorporated into it using in-situ chemical techniques [36]. Through 100 cycles of cyclic voltammetry (CV), the SnO_2-graphene nanocomposite electrode retained its initial capacity of 520 mAh/g. CuO, another newly studied oxide, showed a reversible capacity of 600 mAh/g in a liquid solution-prepared CuO/graphene nanocomposite after 100 cycles as shown in Fig. 6 [37].

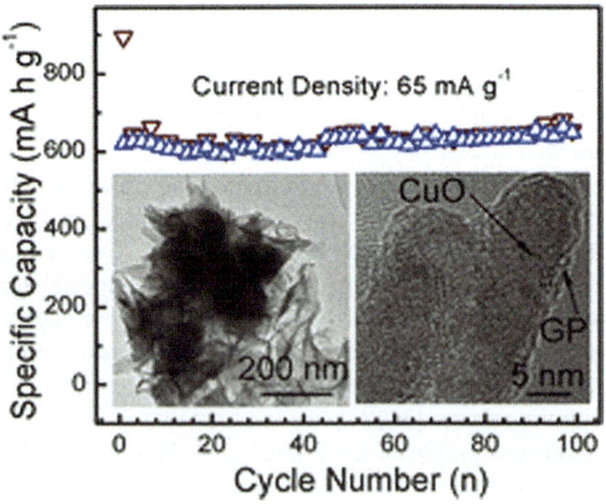

Fig. 6 TEM picture of CuO/graphene and its discharge–charge capacity up to 100 cycles at 65 mA g^{-1} (adapted with permission from RSC [37])

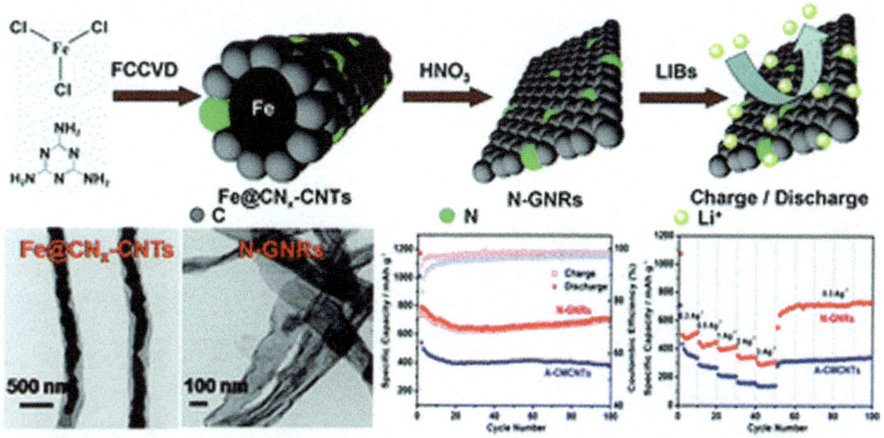

Fig. 7 Schematic representation of fabrication and N-GNRs and nitric acid treated commercial CNTs (A-CMCNTs) were cycled at a current density of 0.1 Ag^{-1} and Current densities measured at several rates to determine the rate capability of N-GNRs and A-CMCNTs: 0.3–3 A g^{-1} (adapted with permission from RSC [39])

It has been found that the rich edges of graphene nanoribbons, a quasi-one-dimensional version of graphene, make them ideal for Li^+ storage [38]. However, after 50 cycles, pure graphene nanoribbons only provide a meagre 250 mAh g^{-1} of capacity. N-doped GNRs, which make use of both N doping and the edge effect, can keep their capacity steady after 100 cycles at a current density of 0.1 A g^{-1} and exhibit outstanding rate capability at a high current density of 3 A g^{-1} [39] as shown in Fig. 7. By combining N-doped graphene nanoribbons into 3D aerogels, it is possible to create a micrometer-scale linked porous structure, which could increase the Li-ion storage capacity even further. Hence, at a current density of 0.5 A g^{-1}, a high capacity of 910 mAh g^{-1} was accomplished [40].

To improve battery performance, Cheng et al. [41] created honeycomb-like porous graphene sponge additions for use on both the anode and cathode materials. High rate cyclability and capacity retention after 100 charge/discharge cycles at an energy density of 162 Wh kg^{-1} were significantly enhanced. An improvement in electron conductivity was shown to be responsible for the enhanced charge/discharge rate capabilities as shown in Fig. 8. This improvement was achieved by increasing the adsorption of electrolytes and decreasing the charge transfer resistance of the active materials.

David et al. [42] created a free-standing anode material by using a molecular precursor to produce silicon oxycarbide glass particles and then encasing them in a chemically modified rGO matrix. The rGO matrix served as a robust mechanical support system for the efficient electron conductor and current collector, and amorphous silicon oxycarbide particles efficiently snagged Li^+ ions during charging. The

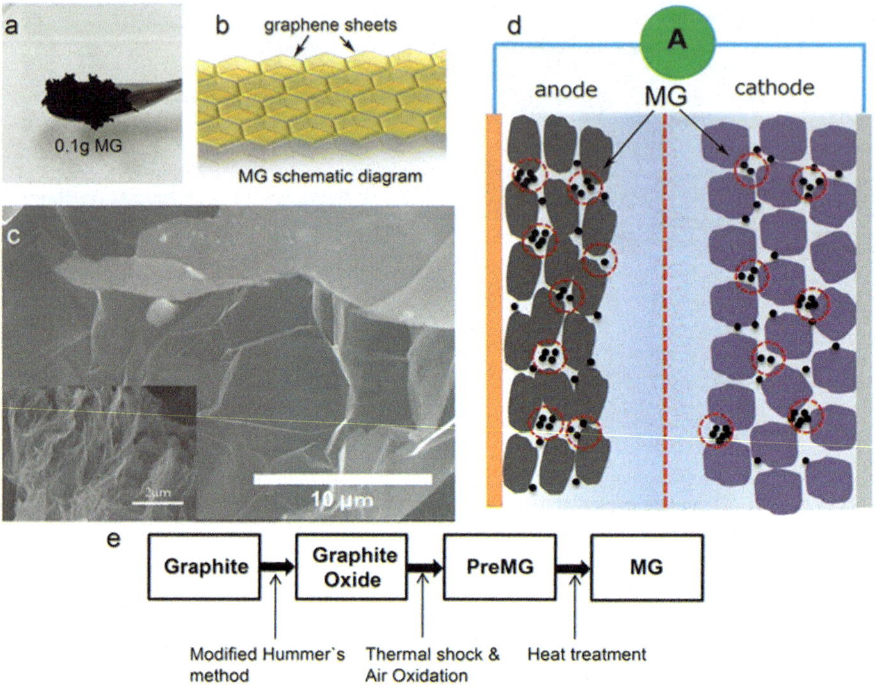

Fig. 8 (**a**) Digital image of graphene samples, (**b**) schematic image of a graphene sheet, (**c**) SEM image of a graphene sheet, (**d**) schematic of an anode cathode that makes up lithium-ion battery, (**e**) schematic of synthesis and heat treatment (adapted with permission from Nature [41])

paper electrode retained its charge capacity of around 588 mAh g^{-1} of the electrode after being subjected to 1020 charge/discharge cycles without showing any evidence of mechanical degradation or failure.

Holey-reduced graphene oxide (HRGO) was produced by Alsharaeh et al. [43] by depositing silver nanoparticles onto rGO sheets. The silver nanoparticles were subsequently removed using the nitric acid treatment and microwave irradiation to reveal a porous solid structure suitable for use as an anode in lithium-ion batteries. Using HRGO as the anode in an HRGO battery led to excellent cycling stability of over 100 charge/discharge cycles and a specific capacity of 423 mAh g^{-1} at 100 mA g^{-1}. They demonstrated a significant reversible capacity of roughly 400 mAh g^{-1} that was retained for up to 100 charge/discharge cycles demonstrating good electrochemical responsiveness and cyclic performance.

By flame-treating graphene oxide, Jiang et al. [43] were able to create a highly porous graphene foam for use as anodes in Li-ion batteries at a reasonable cost. In order to improve the conductivity of graphene oxide, it was subjected to a flame treatment to remove oxygen-containing groups from the graphene oxide sheets. Better intercalation kinetics and charge/discharge cycle stability are the results of flame treatment, which creates open pores via which Li^+ ions can access underlying graphene sheets. After 500 cycles, the flame-reduced graphene oxide battery had a specific capacity of 541 mAh g^{-1}, which was larger than the 186 mAh g^{-1} achieved by the chemically reduced graphene oxide battery. In addition to its use in conventional batteries, this material may find application in cutting-edge electrochemical energy storage devices like supercapacitors.

2.3 *Fuel Cells*

A fuel cell is a device that continuously converts chemical energy from the reaction of a liquid or gas fuel like methanol, hydrogen, hydrocarbons, or natural gas with an oxidising agent like oxygen or air into electrical energy [44, 45]. There are six basic categories of fuel cells based on the fuel they use [46]. Direct methanol fuel cells are a subset of the more widespread proton exchange membrane (PEM) fuel cell [47]. PEM fuel cells are similar to combustion engines in that they use hydrogen or methanol from a fuel tank to create energy through an oxidation reaction. The cathode side of the fuel cell is where the oxidising agent is added, while the anode side is where the fuel is placed. If hydrogen gas (H_2) is burned as fuel, water is produced [47].

The most fundamental distinction between a battery and a fuel cell is that the former requires a constant supply of fuel and an oxidising agent from an external source while the latter already contains both components. In contrast to batteries, which stop producing electricity once the limiting reactant is depleted, the fuel cell can continue to do so indefinitely, provided that active materials are continually added to the electrodes. Indeed, fuel cell electrodes don't directly contribute to the processes, but their catalytic qualities help speed up the anodic and cathodic reactions [46, 47].

Graphene is commonly used as catalyst support on fuel cell membranes, particularly proton exchange membrane fuel cells [48–50]. There are already a plethora of studies evaluating the viability of nitrogen-coordinated metal catalysts as a viable alternative to platinum catalyst [51]. As opposed to platinum catalysts, these catalysts have stability and activity concerns. Although active carbon is capable of addressing these concerns, it also has its own set of limitations. While active carbons have a huge surface area, the instability of their support presents significant concerns unless they are combined with another material. Graphene technological advancements have altered the status quo, with active carbons now being considered a more viable substitute for platinum due to their conductivity, high surface area, and ability to adhere to catalyst particles [48, 51]. Since graphene oxide has more functional groups, it can serve as a nucleation site and as a surface for attaching catalyst nanoparticles [52, 53].

The anode catalyst in fuel cells relies suitably on graphene, which is also utilised as a support material, as a replacement for the cathode catalyst, and as an electrolyte membrane in composite and standalone fuel cells. They employ it in the bipolar plates as well. In low-temperature fuel cells, hydrogen or other low-molecular-weight hydrocarbons, including methanol and ethanol are used as fuel [53–55]. Pt and Pt alloys are typically used as an active catalysts in the fuel cell electrodes. Pt, however, is not only rare and costly but it is also impacted by intermediates generated during the oxidation of various fuels [56]. At both the anode and the cathode efforts have been made to either lessen the amount of catalyst used or replace the Pt catalyst entirely with non-precious catalysts [56–58].

Graphene's enormous specific surface area makes it a superior dispersion for Pt nanoparticles than carbon black [48]. Increased Pt/graphene contacts, more Pt active sites, fewer defects on graphene (increasing graphene stability), and more ordered Pt surface morphology (introducing more active catalytic sites) are all hallmarks of DMFCs that employ the novel Pt/graphene catalyst. Graphene nanosheets have been studied for use in polymer electrolyte fuel cells due to their improved tolerance for carbon monoxide [49, 50].

Qu et al. [59] used chemical vapor deposition of methane in the presence of ammonia led to the synthesis of nitrogen-doped graphene (N-graphene) resulting from an N-graphene electrode for oxygen reduction via a four-electron pathway in alkaline fuel cells was shown to have superior electrocatalytic activity, long-term operating stability, and tolerance to crossover effect compared to platinum as shown in Fig. 9. This is the first report of graphene and its derivatives being used as metal-free catalysts for oxygen reduction, as far as we are aware. Since N-doping plays such a crucial part in the oxygen reduction reaction, it can be applied to other carbon materials to create metal-free, efficient Oxygen reduction reaction catalysts for fuel cell applications and perhaps even to create new catalytic materials with uses outside fuel cells. Similar behaviour to that of a Pt/C anode was seen with the novel electrode, and critically, it was found to be immune to CO poisoning. To boost electrical conductivity and enhance carbon-catalyst interaction, N-graphene can be employed in conjunction with Pt in proton exchange membrane fuel [60].

Methanol fuel cell catalysts made without metal can withstand gas poisoning and can tolerate the crossover effect. A metal-free electrocatalyst usable in methanol

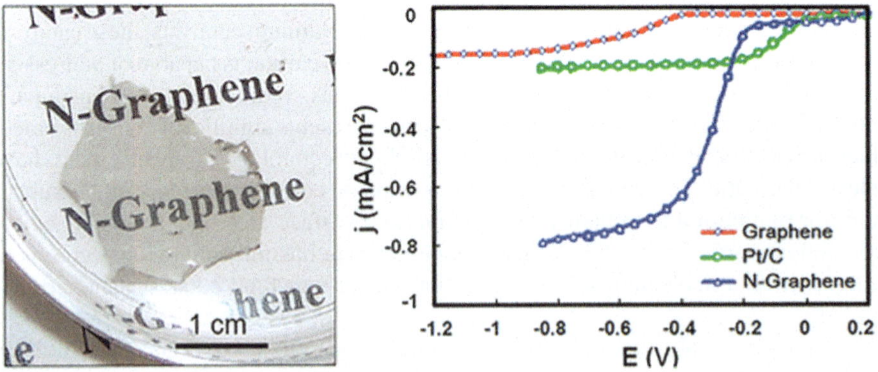

Fig. 9 Digital photo of a transparent N-graphene and voltammograms of the oxygen reduction reaction (adapted with permission from ACS [53])

fuel cells can be created by doping graphene with a nonmetal, such as nitrogen or phosphorus [61]. Vertical graphene nano-hills were fabricated by Akbar et al. [62] using a plasma-enhanced chemical vapour deposition method on a Si/SiO_2/planar graphene or Si/SiO_2/vertical graphene substrate. The electrode of a hydrazine fuel cell was also an electrocatalyst. Specifically, Akbar et al. [62] discovered that the electrocatalyst outperformed carbon-based electrocatalysts in terms of onset potential and current density by preventing N_2 gas bubbles from forming on the electrode surface during operation.

Electrocatalysts containing minute amounts of cobalt dispersed throughout nitrogen-doped graphene have been developed by Fei et al. [63] to speed up the hydrogen evolution reaction in aqueous (both acidic and basic) environments. The catalyst performed well as a catalyst, had a high efficiency in terms of atomic utility, and was inexpensive to prepare, all of which suggest it has potential as a substitute for platinum in water-splitting applications. Zhao et al. [64] developed a three-layered structured electrocatalyst composed of nickel, graphene, and a mixture of iron, manganese, and phosphorus that can split water into hydrogen and oxygen without the use of platinum. The metal phosphide speeds up the water-splitting reaction, the nickel increases the film's surface area, and the graphene prevents additional corrosion [65]. Electrocatalytic activity for the hydrogen evolution process and oxygen evolution reaction on the electrocatalyst was very high, making it a good candidate for application in fuel cells [64].

2.4 Solar Cells

Photovoltaic or more commonly known as solar cells convert sunlight directly into electricity. Photovoltaic cells have evolved via three distinct generations: Power conversion efficiency ranges from 15 to 20% for the first generation of silicon solar cells made from mono- or polycrystalline wafers [66]; from 10 to 15% for the second generation of solar cells made from thin films such as amorphous silicon, cadmium telluride, and copper indium gallium selenide [67, 68]; and from 5 to 10% for the third generation of solar cells made from nanostructures, organic materials, dyes, or other specialty materials [66, 69–72].

Dye-sensitised solar cells are third-generation and work as excited by sunlight, a dye molecule injects an electron into the conduction band (usually TiO_2). If an electron is injected into a transparent anode, it will go to the cathode via the external circuit. This will cause the current to flow. Nonetheless, the dye molecule will receive an electron from the iodine in the electrolyte, resulting in the creation of triiodide ions to compensate for the loss of an electron. When the triiodide ion diffuses to the counter electrode, it recovers the electron it lost in the external circuit [66, 68, 70, 72].

On the other hand, they have a number of limitations, such as the scarcity of indium as well as the expensive, low transparency in the near-infrared region, and the brittleness of indium tin oxide, which prevents it from being used in flexible organic photovoltaics [49, 73–76]. Some of these problems have been alleviated through the use of graphene and related materials. There is no difference in operation between conventional solar cells and those made from graphene or other graphene-based materials.

To mitigate the devastating effects of fossil fuel consumption on the environment, solar power has emerged as a viable alternative. This rapid expansion of the photovoltaic (PV) sector has led to a decrease in raw material supplies, even though economies of scale have reduced production prices which is true for indium, which is utilised in the electronic industry as a transparent conducting oxide, indium tin oxide (ITO). A transparent electrode, such as ITO or FTO, is required for all organic photovoltaic devices, whereas the other is typically aluminium, occasionally covered with MgO and LiF [9].

Organic photovoltaics (OPVs) have several advantages over traditional solar harvesting technologies, e.g., flexible and semitransparent, produced by covering vast areas in a continuous printing process, and cost-effective. Efficiency, longevity, and the availability of viable alternatives to ITO are the primary concerns. Graphene can replace ITO in OPVs because it is transparent to nearly all visible light when just one layer is present and becomes increasingly opaque when more layers are added [77]. Film sheet resistance is measured in terms of the material's resistance per unit area, or Ohms. While indium tic oxide has a transmittance of around 90% and a resistance of around ~20 Ω/sq, graphene employed in various devices has a resistance of ~6 k Ω/sq [78]. The biggest obstacle is figuring out methods for obtaining graphene sheets of the highest possible quality so that the sheet resistance can be reduced even more. Kalita et al. produced solar cell-using graphene from camphors [74]. By pyrolyzing camphor at 900 °C in argon, methyl carbons can be separated from the chemical structure, resulting in transparent graphene-structured carbon films of different thicknesses with an overall resistivity lower than that of ITO. Graphene obtained via chemical reduction of GO has a sheet resistance of 1 to 100 k Ω/sq at a transmittance below 80% [79].

High electrical conductivity was achieved in laser-scribed graphene electrodes produced by Thekkekara and Gu et al. [80]. That was made possible by shortening the ionic route in the electrolyte and increasing the active surface area of the new electrode. With a coulomb efficiency of 95%, the graphene electrode was able to store energy at a density of about 0.1 Wh cm^3, far higher than that of standard planar electrodes (0.001 Wh cm^3). The electrode was utilised to make thin-film on-chip energy storage integrated solar cells, which are flexible and small enough to be attached to smartphones, computers, automobiles, and buildings for autonomous energy generation [80].

Choi et al. [81] developed a graphene-based Schottky junction solar cell, in particular, has advanced quickly due to its very straightforward device architectures in comparison with those of traditional p–n junction type solar cells. Power conversion efficiencies of over 15% have been observed after several modifications were made, including chemical doping, antireflection coating, and interfacial oxide layer management. Interfacial charge recombination appeared to be greatly controlled through careful manipulation of the graphene integration process, leading to a noticeably increased device performance (from 0.8 to 12.5%) as shown in Fig. 10.

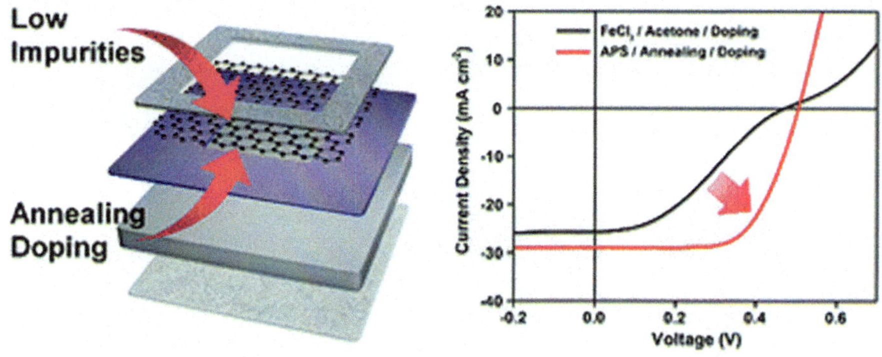

Fig. 10 Schematic of doping process and the current density Vs voltage (adapted with permission from RSC [81])

To show that transparent organic solar cells may function when built on flexible substrates like paper or plastic, Song et al. [72] used graphene that was itself transparent as both the anode and cathode electrodes. Power conversion efficiency was between 2.8 and 3.8% at optical transmittance in the range of 54 and 61% across the visible regime. This corresponds to a total visible spectrum optical transmittance of 61% and a value close to 69% at 550 nm wavelength. Even though the achieved power conversion efficiency was far lower than that of commercially available solar panels, further study of this concept has promise.

It was hypothesised by Tang et al. [82] that positively charged ions (Na^{+}, Ca^{2+}, and NH^{4+}) in rain may be separated from negatively charged ions using graphene layers, leading to a chemical reaction that could be used to generate energy. Raindrops can accomplish this because the water molecules within them contain different salts that can separate into positive and negative ions, much like the charges on the surface of cells. With a modified solar panel activated by either raindrops or sunlight, an efficiency of about 7% was reached in converting solar energy into electricity [77]. Graphene's potential to aid once again in this notion is promising, although it still falls short of the power conversion efficiency of conventional solar cells, which can reach up to 22% as shown in Fig. 11.

Dye-sensitised solar cells (DSSCs) are another type of flexible, cost-effective solar cell that offers an alternative to OPVs. In DSSCs, an electron is provided by a light-absorbing dye and is collected by the TiO_2 support layer before being sent to the external circuit. Efficiency levels of DSSCs are now on par with Si amorphous silicon cells. To simplify fabrication and reduce production costs, several works looked into replacing indium tin oxide and FTO with graphene as window electrodes [74, 75]. However, PCE is still much lower than that of conventional devices.

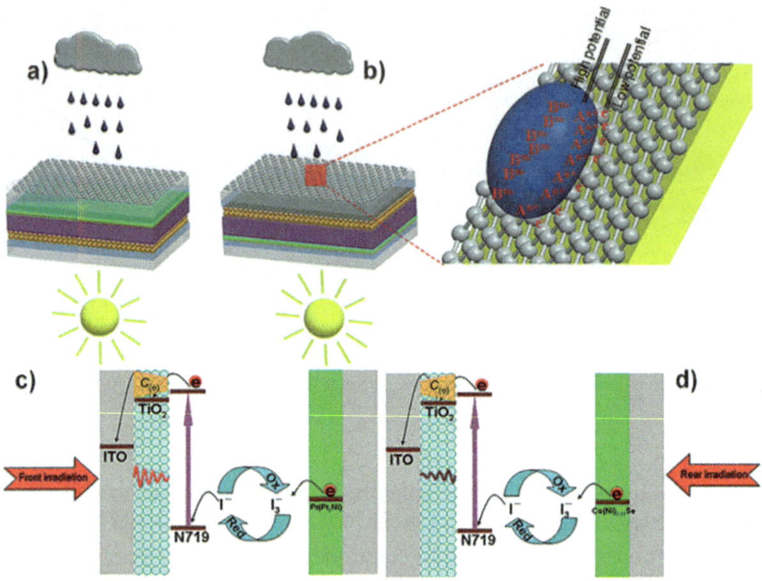

Fig. 11 (**a**, **b**) Schematic of the process and working principle of a flexible solar cell under natural light (**c**, **d**). On the PET side, the rGO film is deposited, and on the ITO side, the solar cell is built. Silver paint and ethylene vinyl acetate copolymer shield the two electrodes used to measure current and voltage in the presence of showers. To inject the droplets, a medical syringe is used, with the injection speed being regulated. (adapted with permission from Wiley [82])

3 Challenges and Opportunities

In this chapter, we have exclusively focused on graphene-based materials applications in supercapacitors, batteries, fuel cells, and solar cells. There have been significant advances in the aforementioned applications. Nonetheless, there are obstacles and additional studies into the aforementioned technologies are required to find solutions and improve efficiency. With the advent of graphene-based materials, the majority of the platinum in fuel cells has been diverted to other applications or replaced by graphene. While platinum serves as both the catalyst and the electrode in fuel cells, efforts to find a suitable substitute have centered on using graphene, either on its own or doped with other metals, in the anode and cathode. Hence, fuel cell technology will become more financially feasible. Doped graphene catalysts are promising, but their production still faces obstacles. This is because low-yield doped graphene is produced when a dopant element is introduced during the production process. Every one of these renewable energy industries could be hampered by low doped graphene yields because of the way that low yields alter the properties of doped graphene, making it less effective. Either pure graphene or doped with other metals, graphene nanotechnology has also been used to replace the window electrodes in solar cell technology. While progress has been slow, there is hope that graphene nanotechnology can overcome current inefficiencies. Graphene-based materials have also been successfully implemented in batteries. Graphene's success thus far bodes well for its application in sustainable power. Graphene will become a major factor in renewable energy, guaranteeing humanity's survival, once the obstacles are cleared.

4 Conclusions

Transparency, mechanical robustness, and thermal and chemical stability make graphene-based materials attractive for energy harvesting applications. Graphene is proving to be an invaluable substitute material for the ever-more-expensive metals that are in use, like platinum ad indium; furthermore, in order to get even higher performances in solar cells and fuel cells, functionalisation of graphene sheets may be a game-changer. As a rule, a material's conductivity decreases as its transparency increases, yet graphene defies this trend by becoming remarkably see-through without surrendering any of its conductivity. For applications such as flexible electronics cells, LEDs, touch screens, and electronic displays, this quality can be employed to replace the less versatile and durable ITO. By incorporating graphene into electrochemical supercapacitors, not only is the specific capacity increased, but the standard EDL capacitor is also given the ability to filter AC lines. Graphene used in lithium-ion batteries has the potential to aid in the creation of high-performance electrodes based on minerals like tin, and silicon and regulating the distance between graphene sheets could result in novel lithium storage features. Graphene implantation in battery electrodes has been shown to shorten charging times, which is good news for both the widespread use of electric vehicles and mobile devices.

References

1. Geimm AK, Novoselov KS (2007) The rise of graphene. Nature 6:183–191
2. Gabris MA, Ping J (2021) Carbon nanomaterial-based nanogenerators for harvesting energy from the environment. Nano Energy 90:106494. https://doi.org/10.1016/j.nanoen.2021.106494
3. Umar K, Ahmad H, Parveen T, Haseen U, Mohamad Ibrahim MN (2019) Metal-doped graphene nanocomposites and their application in energy storage. Elsevier Inc. https://doi.org/10.1016/B978-0-12-815811-1.00006-5
4. Fan W, Zhang L, Liu T (2017). Graphene-CNT hybrids for energy applications. https://doi.org/10.1007/978-981-10-2803-8_3
5. Zang X (2017) Graphene-based flexible energy storage devices. Elsevier Inc. https://doi.org/10.1016/B978-0-12-812651-6.00007-0
6. Zhuiykov S (2018). Composite graphene/semiconductor nano-structures for energy storage. https://doi.org/10.1016/b978-0-08-101919-1.00006-4
7. Horn M, Gupta B, MacLeod J, Liu J, Motta N (2019) Graphene-based supercapacitor electrodes: addressing challenges in mechanisms and materials. Curr Opin Green Sustain Chem 17:42–48. https://doi.org/10.1016/j.cogsc.2019.03.004

8. Balli B, Şavk A, Şen F (2018) Graphene and polymer composites for supercapacitor applications. Nanocarbon Compos Prep Prop Appl: 123–151. https://doi.org/10.1016/B978-0-08-102509-3.00005-5
9. Hummers Jr WSH (1958) Preparation of graphitic oxide 80:1937–1339. https://doi.org/10.1021/ja01539a017
10. Yan QL, Gozin M, Zhao FQ, Cohen A, Pang SP (2016) Highly energetic compositions based on functionalized carbon nanomaterials. Nanoscale 8:4799–4851. https://doi.org/10.1039/c5nr07855e
11. Conway BE (1999) Electrochemical supercapacitors: scientific fundamentals and technological applications
12. Ko R, Carlen M (2000) Principles and applications of electrochemical capacitors 45:2483–2498
13. Winter M, Brodd RJ (2004) What are batteries, fuel cells, and supercapacitors? Chem Rev 104:4245–4269. https://doi.org/10.1021/cr020730k
14. Stoller MD, Park S, Zhu Y, An J, Ruoff RS (2008) Graphene-based ultracapacitors: 6–10
15. Pandolfo AG, Hollenkamp AF (2006) Carbon properties and their role in supercapacitors. J Power Sour 157:11–27.https://doi.org/10.1016/j.jpowsour.2006.02.065
16. John Miller BCH, Outlaw RA (2010) Graphene double-layer capacitor with ac line-filtering performance. Science 329:1637–1639
17. Bing Bai YW, Qiu L (2023) Ultrafine carbon-nanofiber reinforced graphene fiber electrodes for flexible supercapacitor with high specific capacitance and durable cycle stability. ACS Appl Energy Mater 6:353–361
18. Abdillah OB, Floweri O, Irham MA (2022) Structural modulation of exfoliated graphene via a facile postultrasonication treatment toward enhanced electrochemical properties of supercapacitor electrode. Energy Fuels 36:14453–14463
19. Futaba DONN, Hata K, Yamada T, Hiraoka T, Hayamizu Y, Kakudate Y, Tanaike O, Hatori H, Yumura M, Iijima S (2006) Shape-engineerable and highly densely packed single-walled carbon nanotubes and their application as super-capacitor electrodes. Nat Commun 5:987–994. https://doi.org/10.1038/nmat1782
20. Lu W, Qu L, Henry K, Dai L (2009) High performance electrochemical capacitors from aligned carbon nanotube electrodes and ionic liquid electrolytes 189:1270–1277. https://doi.org/10.1016/j.jpowsour.2009.01.009
21. Frackowiak E (2001) Carbon materials for the electrochemical storage of energy in capacitors 39:937–950
22. Qian Cheng L-CQ, Tang J, Ma J, Zhang H, Shinya N (2011) Graphene and carbon nanotube composite electrodes for supercapacitors with ultra-high energy density. Phys Chem Chem Phys 13:17615–17624. https://doi.org/10.1039/c1cp21910c
23. Raji ARO, Villegas Salvatierra R, Kim ND, Fan X, Li Y, Silva GAL, Sha J, Tour JM (2017) Lithium batteries with nearly maximum metal storage. ACS Nano 11:6362–6369. https://doi.org/10.1021/acsnano.7b02731
24. Raccichini R, Varzi A, Passerini S, Scrosati B (2015) The role of graphene for electrochemical energy storage. Nat Mater 14:271–279. https://doi.org/10.1038/nmat4170
25. Liang M, Luo B, Zhi L (2009) Application of graphene and graphene-based materials in clean energy-related devices. Int J Energy Res 33:1161–1170. https://doi.org/10.1002/er.1598
26. Qi W, Shapter JG, Wu Q, Yin T, Gao G, Cui D (2017) Nanostructured anode materials for lithium-ion batteries: principle, recent progress and future perspectives. J Mater Chem A 5:19521–19540. https://doi.org/10.1039/c7ta05283a
27. Lithium-ion HA (2012) Photothermally reduced graphene as batteries. ACS Nano 6:7867–7878
28. Sarkar J, Bhattacharyya S (2012) Application of graphene and graphene-based materials in clean energy-related devices Minghui. Arch Thermodyn 33:23–40. https://doi.org/10.1002/er.1598
29. Scrosati B, Garche J (2010) Lithium batteries: status, prospects and future. 195:2419–2430. https://doi.org/10.1016/j.jpowsour.2009.11.048
30. Hsieh C, Lin J, Mo C (2011) Improved storage capacity and rate capability of Fe_3O_4—graphene anodes for lithium-ion batteries. Electrochim Acta 58:119–124. https://doi.org/10.1016/j.electacta.2011.09.008

31. Li X, Hu Y, Liu J, Lushington A, Li R, Sun X (2013) Structurally tailored graphene nanosheets as lithium ion battery anodes: an insight to yield exceptionally high lithium storage performance: 12607–12615. https://doi.org/10.1039/c3nr04823c
32. Tong X, Wang H, Wang G, Wan L, Ren Z, Bai J, Bai J (2011) State chemistry controllable synthesis of graphene sheets with different numbers of layers and effect of the number of graphene layers on the specific capacity of anode material in lithium-ion batteries. J Solid State Chem 184:982–989. https://doi.org/10.1016/j.jssc.2011.03.004
33. Wang D, Li F, Liu M, Lu GQ, Cheng H (2008) 3D aperiodic hierarchical porous graphitic carbon material for high-rate electrochemical capacitive energy storage: 373–376. https://doi.org/10.1002/anie.200702721
34. Wang J, Zhong C, Chou S, Liu H (2010) Flexible free-standing graphene-silicon composite fi lm for lithium-ion batteries. Electrochem Commun 12:1467–1470. https://doi.org/10.1016/j.elecom.2010.08.008
35. Lee JK, Smith KB, Hayner M, Kung HH (2010) Silicon nanoparticles—graphene paper composites for Li ion battery anodes. Chem Commun 46(12): 2025–2027. https://doi.org/10.1039/b919738a
36. Yao J, Shen X, Wang B, Liu H, Wang G (2009) In situ chemical synthesis of SnO_2—graphene nanocomposite as anode materials for lithium-ion batteries. Electrochem Commun 11:1849–1852. https://doi.org/10.1016/j.elecom.2009.07.035
37. Wang B, Wu X, Shu C, Guo Y, Wang C (2010) Synthesis of CuO/graphene nanocomposite as a high-performance anode material for lithium-ion batteries: 10661–10664. https://doi.org/10.1039/c0jm01941k.
38. Uthaisar C, Barone V, Peralta JE, Uthaisar C, Barone V, Peralta JE (2013) Lithium adsorption on zigzag graphene nanoribbons Lithium adsorption on zigzag graphene nanoribbons: 113715. https://doi.org/10.1063/1.3265431
39. Wang X (2014) Nitrogen-doped graphene nanoribbons for high-performance lithium ion batteries. J Mater Chem A Mater Energy Sustain 2:16832–16835. https://doi.org/10.1039/C4TA03531C
40. Liu Y, Wang X, Wan W, Li L, Dong Y, Zhao Z, Qiu J (2015). Multifunctional nitrogen-doped graphene nanoribbon aerogels for superior lithium storage and cell culture. https://doi.org/10.1039/c5nr05909g
41. Cheng Q (2017) Porous graphene sponge additives for lithium ion batteries with excellent rate capability. Sci Rep 7:1–11. https://doi.org/10.1038/s41598-017-01025-7
42. David L, Bhandavat R, Barrera U, Singh G (2016) Silicon oxycarbide glass-graphene composite paper electrode for long-cycle lithium-ion batteries. Nat Commun.7. https://doi.org/10.1038/ncomms10998
43. Alsharaeh E, Ahmed F, Aldawsari Y, Khasawneh M, Abuhimd H, Alshahrani M (2016) Novel synthesis of holey reduced graphene oxide (HRGO) by microwave irradiation method for anode in lithium-ion batteries. Sci Rep 6:1–13. https://doi.org/10.1038/srep29854
44. Rezk H, Taha E, Al-dhaifallah M, Obaid M (2019) Fuel cell as an effective energy storage in reverse osmosis desalination plant powered by photovoltaic system. Energy 175:423–433. https://doi.org/10.1016/j.energy.2019.02.167
45. Ali M, Hassan W, El Haj M, Allagui A, Cha SW (2019) On the technical challenges affecting the performance of direct internal reforming biogas solid oxide fuel cells. Renew Sustain Energy Rev 101:361–375. https://doi.org/10.1016/j.rser.2018.10.025
46. Cook B (2002) Introduction to fuel cells and hydrogen technology. Eng Sci Educ J 11:205–216. https://doi.org/10.1049/esej:20020601
47. Niakolas DK, Daletou M, Neophytides SG, Vayenas CG (2016) Fuel cells are a commercially viable alternative for the production of "clean" energy. Ambio 45:32–37. https://doi.org/10.1007/s13280-015-0731-z
48. Iwan A, Malinowski M, Pasciak G (2015) Polymer fuel cell components modified by graphene: electrodes, electrolytes and bipolar plates 49:954–967. https://doi.org/10.1016/j.rser.2015.04.093

49. Shaari N, Kamarudin SK (2017) Graphene in electrocatalyst and proton conductiong membrane in fuel cell applications: an overview. Renew Sustain Energy Rev 69:862–870. https://doi.org/10.1016/j.rser.2016.07.044
50. Ijaodola OS, Hassan ZE, Ogungbemi E, Khatib FN, Wilberforce T, Thompson J, Olabi AG (2019) Energy efficiency improvements by investigating the water flooding management on proton exchange membrane fuel cell (PEMFC). Energy 179:246–267. https://doi.org/10.1016/j.energy.2019.04.074
51. Liang Y (2011) CO_3O_4 nanocrystals on graphene as a synergistic catalyst for oxygen reduction reaction. Nat Mater 10:780–786. https://doi.org/10.1038/nmat3087
52. Bai H, Li C, Shi G (2011) Functional composite materials based on chemically converted graphene: 1089–1115. https://doi.org/10.1002/adma.201003753
53. Ali M, Yoshitoshi T, Tsujiguchi T, Nakagawa N (2010) Vertical operation of passive direct methanol fuel cell employing a porous carbon plate 195:1821–1828. https://doi.org/10.1016/j.jpowsour.2009.09.076
54. Ali M, Masdar MS, Tsujiguchi T, Nakagawa N, Taha E, Barakat NAM (2014) Elimination of toxic products formation in vapor-feed passive DMFC operated by absolute methanol using air cathode filter. Chem Eng J 240:38–44. https://doi.org/10.1016/j.cej.2013.11.043
55. Wu S, Zhang H, Ni M (2016) Performance assessment of a hybrid system integrating a molten carbonate fuel cell and a thermoelectric generator. Energy 112:520–527. https://doi.org/10.1016/j.energy.2016.06.128
56. Akhairi MAF, Kamarudin SK (2016) Catalysts in direct ethanol fuel cell (DEFC): an overview. Int J Hydrogen Energy: 1–15. https://doi.org/10.1016/j.ijhydene.2015.12.145
57. Jie Z, Shuihua T, Longyu L, Weifei YU (2013) Progress in non-platinum catalysts with applications in low temperature fuel cells, Chinese. J Catal 34:1051–1065. https://doi.org/10.1016/S1872-2067(12)60588-9
58. Sha L, Ye K, Wang G, Shao J, Zhu K, Cheng K, Yan J, Wang G (2019) Hierarchical $NiCO_2O_4$ nanowire array supported on Ni foam for efficient urea electrooxidation in alkaline medium. 412:265–271. https://doi.org/10.1016/j.jpowsour.2018.11.059
59. Qu L, Liu Y, Baek J, Dai L (2010) Nitrogen-doped graphene as efficient metal-free electrocatalyst for oxygen reduction in fuel cells. 4:1321–1326
60. Jafri RI, Rajalakshmi N, Ramaprabhu S (2010) Nitrogen doped graphene nanoplatelets as catalyst support for oxygen reduction reaction in proton exchange membrane fuel cell: 7114–7117. https://doi.org/10.1039/c0jm00467g
61. Zhang C, Mahmood N, Yin H, Liu F, Hou Y (2013) Synthesis of phosphorus-doped graphene and its multifunctional applications for oxygen reduction reaction and lithium ion batteries. Adv Mater 25:4932–4937. https://doi.org/10.1002/adma.201301870
62. Akbar K, Kim JH, Lee Z, Kim M, Yi Y, Chun SH (2017) Superaerophobic graphene nano-hills for direct hydrazine fuel cells. NPG Asia Mater 9:1–8. https://doi.org/10.1038/am.2017.55
63. Fei H, Dong J, Arellano-Jiménez MJ, Ye G, Dong Kim N, Samuel ELG, Peng Z, Zhu Z, Qin F, Bao J, Yacaman MJ, Ajayan PM, Chen D, Tour JM (2015) Atomic cobalt on nitrogen-doped graphene for hydrogen generation. Nat Commun 6:1–8. https://doi.org/10.1038/ncomms9668
64. Zhao Z, Schipper DE, Leitner AP, Thirumalai H, Chen JH, Xie L, Qin F, Alam MK, Grabow LC, Chen S, Wang D, Ren Z, Wang Z, Whitmire KH, Bao J (2017) Bifunctional metal phosphide FeMnP films from single source metal organic chemical vapor deposition for efficient overall water splitting. Nano Energy 39:444–453. https://doi.org/10.1016/j.nanoen.2017.07.027
65. Li F, Jiang X, Zhao J, Zhang S (2015) Graphene oxide: a promising nanomaterial for energy and environmental applications. Nano Energy 16:488–515. https://doi.org/10.1016/j.nanoen.2015.07.014
66. Razykov TM, Ferekides CS, Morel D, Stefanakos E, Ullal HS, Upadhyaya HM (2011) Solar photovoltaic electricity: current status and future prospects. Sol Energy 85:1580–1608. https://doi.org/10.1016/j.solener.2010.12.002
67. Xin X, He M, Han W, Jung J, Lin Z (2011) Low-cost copper zinc tin sulfide counter electrodes for high-efficiency dye-sensitized solar cells. Angew Chemie Int Ed 50:11739–11742. https://doi.org/10.1002/anie.201104786

68. Chen X, Tang Q, He B, Lin L, Yu L (2014) Platinum-free binary Co-Ni alloy counter electrodes for efficient dye-sensitized solar cells. Angew Chemie Int Ed 53:10799–10803. https://doi.org/10.1002/anie.201406982
69. Wang H, Sun K, Tao F, Stacchiola DJ, Hu YH (2013) 3D honeycomb-like structured graphene and its high efficiency as a counter-electrode catalyst for dye-sensitized solar cells. Angew Chemie Int Ed 52:9210–9214. https://doi.org/10.1002/anie.201303497
70. Balasingam SK, Jun Y (2015) Recent progress on reduced graphene oxide-based counter electrodes for cost-effective dye-sensitized solar cells. Isr J Chem 55:955–965. https://doi.org/10.1002/ijch.201400213
71. Pang B, Dong L, Ma S, Dong H, Yu L (2016) Performance of FTO-free conductive graphene-based counter electrodes for dye-sensitized solar cells. RSC Adv 6:41287–41293. https://doi.org/10.1039/c6ra02854c
72. Song Y, Chang S, Gradecak S, Kong J (2016) Visibly-transparent organic solar cells on flexible substrates with all-graphene electrodes. Adv Energy Mater 6:1–8. https://doi.org/10.1002/aenm.201600847
73. Kiziroglou ME, Yeatman EM (2012) Materials and techniques for energy harvesting. Woodhead Publishing Limited.https://doi.org/10.1533/9780857096371.4.539
74. Kalita G, Matsushima M, Uchida H, Wakita K, Umeno M (2010) Graphene constructed carbon thin films as transparent electrodes for solar cell applications: 9713–9717. https://doi.org/10.1039/c0jm01352h
75. Wang X, Zhi L, Mu K (2008) Transparent, conductive graphene electrodes for dye-sensitized solar cells
76. Spanggaard H, Krebs FC (2004) A brief history of the development of organic and polymeric photovoltaics 83:125–146. https://doi.org/10.1016/j.solmat.2004.02.021
77. Nair RR, Blake P, Grigorenko AN, Novoselov KS, Booth TJ, Stauber T, Peres NMR, Geim AK (2008) Fine structure constant defines defines visual transparency of graphene. Science 320:2–3. https://doi.org/10.1126/science.1156965
78. Tan Y, Zhang Y, Bolotin K, Zhao Y, Adam S, Hwang EH, Das Sarma S, Stormer HL, Kim P (2007) Measurement of scattering rate and minimum conductivity in graphene 246803:10–13. https://doi.org/10.1103/PhysRevLett.99.246803
79. Eda G, Fanchini G, Chhowalla M (2008) Large-area ultrathin films of reduced graphene oxide as a transparent and flexible electronic material 3:1–5. https://doi.org/10.1038/nnano.2008.83
80. Thekkekara LV, Gu M (2017) Bioinspired fractal electrodes for solar energy storages. Sci Rep 7:1–9. https://doi.org/10.1038/srep45585
81. Choi Y, Lee J, Seo J, Jung S, Kim U, Park H (2017) The effect of the graphene integration process on the performance of graphene-based Schottky junction solar cells. J Mater Chem A 5:18716–18724. https://doi.org/10.1039/c7ta05481e
82. Tang Q, Wang X, Yang P, He B (2016) A solar cell that is triggered by sun and rain. Angew Chemie Int Ed 55:5243–5246. https://doi.org/10.1002/anie.201602114

Chapter 11
Cathode Materials in Lithium Ion Batteries as Energy Storage Devices

Swetapadma Praharaj and Dibyaranjan Rout

1 Introduction

With the growing population and modern societal needs, scarcity of energy and high rates of pollution has been the ever-mounting problems of the twenty-first century. To compromise such situations, efforts are being made to introduce green energy sources i.e. solar, wind and hydroelectric power. However, most of them suffer from intermittent and uncontrollable nature which is a demerit for energy storage and limits their widespread use. Again, optimum performance from green technologies can only be expected using sophisticated structures with the proper choice of materials. For instance, poor charge mobility in polymer/organic semiconductors cap the maximum energy conversion efficiency of organic photovoltaics to below 6%. While the conversion efficiency of thermoelectric is below 3%, portable electric power sources provide low power density owing to poor mass and charge transport properties. Under such circumstances, novel nanostructured materials which are chemically modified through structural and microstructural engineering exhibit significantly improved properties to serve the new generation high energy density storage and conversion devices. In this regard, electrochemical sources can be considered as the near-term solution because of their high energy storage and conversion efficiency. Lithium ion batteries or LiBs are a prototypical electrochemical source for energy storage and conversion. Presently, LiBs are quite efficient, extremely light and rechargeable power sources for electronic items such as digital cameras, laptops, smartphones and smartwatches. Besides, these are being extensively in electric vehicles (EVs) and hybrid electric vehicles (HEVs) requiring high power and energy densities. However, the power densities offered by these batteries are relatively low compared to their counterparts such as electrochemical double-layer capacitors and

S. Praharaj · D. Rout (✉)
Department of Physics, School of Applied Sciences, KIIT Deemed to be University, Bhubaneswar, Odisha 24, India
e-mail: droutfpy@kiit.ac.in

B. P. Swain (ed.), *Energy Materials*, Materials Horizons: From Nature to Nanomaterials,
https://doi.org/10.1007/978-981-99-3866-7_11

ultracapacitors. Such fewer power densities are primarily due to the large polarization caused by the sluggish motion of the Li ions with the active material resulting in an increase in the resistance of the electrolyte during high charge–discharge rates. One of the effective ways of addressing this issue is to engineer novel materials with large surface areas and short diffusion paths for electronic conduction and ionic transport.

Nanostructured materials are known to exhibit exotic electrical, mechanical, optical and electrochemical properties owing to their dimensions. The overall response of these materials is an amalgamation of surface as well as bulk properties [1]. Therefore, nanomaterials have found their place in potential applications such as sensors, field effect transistors, nanocables and nanoprobes [2–5]. It is expected that nanomaterials having different morphologies, structures and good electrochemical properties can improve the efficiency of LiB cathodes by accelerating the Faradaic reactions at the electrode–electrolyte interface, shortening the mass and charge transport path, and enhancing the intercalation of Li ion and discharge. Hence, this chapter aims to capture the recent progress in different types of cathode materials used in LiBs for energy storage and the challenges faced in this direction.

2 Working Principle of Li Ion Batteries

Generally, a battery transforms chemical potential into electrical energy through Faradaic reactions. The whole mechanism can be realized by a thorough understanding of its three basic components i.e. anode, cathode and electrolyte. Primarily, the heterogeneous or Faradaic reactions at the surface of electrodes promote charge and mass transport through the electrodes [6]. In batteries, the preferred electrodes are intercalation type which serves as a host solid to accommodate the guest species from the electrolyte. Intercalation refers to the reversible insertion and reinsertion of mobile guest species including atoms, ions or molecules into a host lattice which is usually an interconnected system of voids/ vacant spaces of suitable size [7]. The working principle of LiBs revolves around this mechanism. A schematic diagram showing the working mechanism of Li ion batteries is shown in Fig. 1.

A negative Li intercalation compound called as 'anode' in combination with another Li intercalation material with higher positive redox potential i.e. 'cathode' constitutes a lithium ion transfer cell. Both the electrodes are separated by a lithium ion conductor called as 'electrolyte'. When the battery is assembled, it is in a discharged state and all the Li ions are on the cathode. On charging, lithium ions get released from the cathode host lattice, solvate and migrate through the electrolyte and intercalate into the anode. In the meantime, the motion of the electrons in the outer circuit is also from the cathode to the anode. Besides, the chemical potential of Li ion is more in the anode than in the cathode due to which charges are stored in the form of electrochemical energy. The reverse process takes place on discharging and electrochemical energy is released in the form of charges. Often a 'separator' i.e. a microporous membrane separates the cathode and anode, which permits the movement of electrolyte but restricts the shorting of electrodes. The electrolyte is so

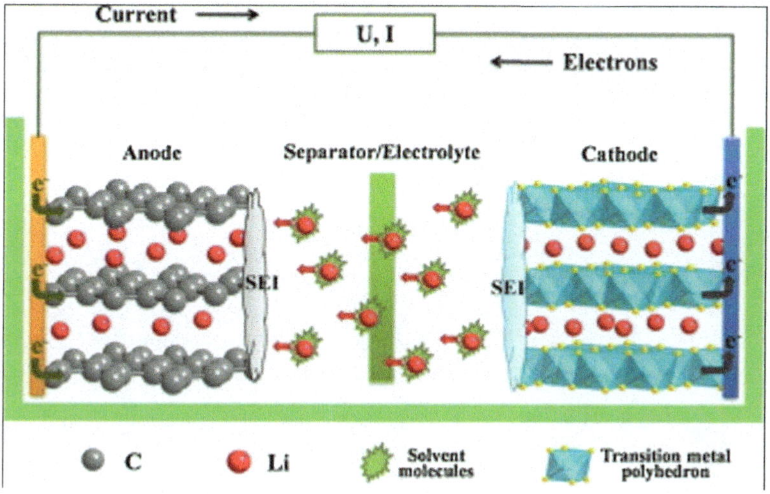

Fig. 1 Schematic diagram showing the working mechanism of lithium ion batteries [8]

chosen that it is electronically insulating but ionically conducting. Further, during the first cycle, solid electrolyte interphase (SEI) layer is formed on the interface owing to the decomposition of organic solid electrolyte at an extreme voltage (usually less than 1.2 V or greater than 4.6 V). The present LiB technology is highly dependent on the cathode materials (structure and morphology) so as to manipulate the cell voltage and capacity. Hence, the development of cathode materials is extremely essential and is receiving significant attention in recent times.

3 Cathode Materials in LiBs

The choice of materials for cathode depends on various crucial factors such as rate capability, energy density, cyclic stability, cost-effectiveness and safety. Rate capability and cyclic stability are mainly regulated by the electronic and ionic transport rates along with morphology of the particles and anisotropy in the structure of the materials. On the other hand, energy density is decided by the intrinsic chemistry of materials, for instance, effective redox couple and maximum Li concentration [8]. Considering these factors, material optimization and design are carried out on two basic aspects—(1) alterations in the intrinsic chemistry and (2) modifications in surface morphology and particle size. Figure 2 displays a comparison between different cathode materials based on theoretical and practical gravimetric energy densities. To date, two major categories of cathode materials are known. One of them includes layered anionic close-packed lattice structures with transition metal ions occupying the space between alternate anionic sheets. Li ions finally intercalate into the remaining vacant spaces between the layers; for example $LiCoO_2$,

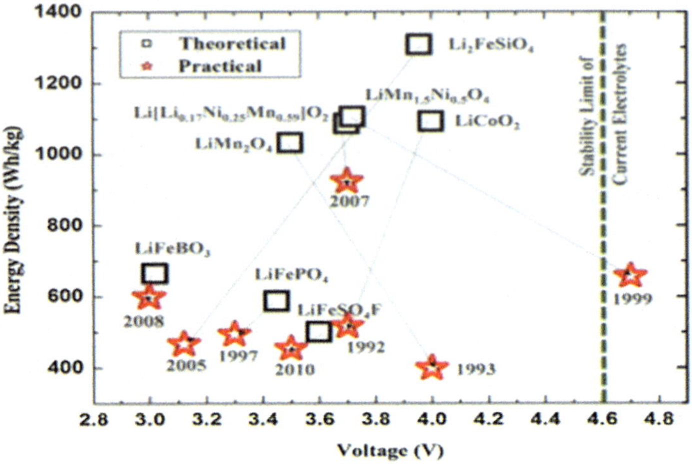

Fig. 2 Schematic diagram showing the variation of theoretical and practical energy densities of various cathode materials [8]

$LiTiS_2$, $LiNi_{1-x}Co_xO_2$, $LiNiMn_xCo_{1-2x}O_2$, spinels with transition metal ions ordered in layers. The advantage of the above category materials is their higher energy density due to compact lattices. The second class of materials possesses open structures like the V_2O_5 family, tunnel compounds of Mn-based oxides and transition metal phosphates (e.g. olivine $LiFePO_4$). Some advantages of this group of compounds include superior safety and cost-effectiveness in comparison to the former.

3.1 Layered Compounds with General Formula $LiMO_2$ (M is a Metal Atom)

Figure 3 represents the archetypal structure of $LiMO_2$ layers which consists of a close-packed *fcc* lattice of oxygen ions with cations placed at the octahedral sites. Further, the metal oxide (MO_2) and lithium layers are alternatively stacked [9]. Among the layered oxides, $LiCoO_2$ is most widely used in portable electronics [10] and has been commercialized for more than 20 years as LiBs. However, $LiCoO_2$ can deliver a specific capacity of 140 mAh/g which is only half of its theoretical capacity. This is because of the structural instability caused when almost half of the Li ions are removed from the compound during the charging of LiB. Again, cobalt ions in $LiCoO_2$ is not only an environmental hazard, but also increase the cost of batteries. To overcome the shortcomings of $LiCoO_2$, the development of new materials was targeted. $LiNiO_2$ is one such compound that is not only isostructural to $LiCoO_2$ but also exhibits higher reversible capacity owing to the fact that more amount of Li can be extracted during the redox cycles as compared to $LiCoO_2$. This feature allows the specific capacity to reach above 150 mAh/g with reasonable cyclability [11]. Besides,

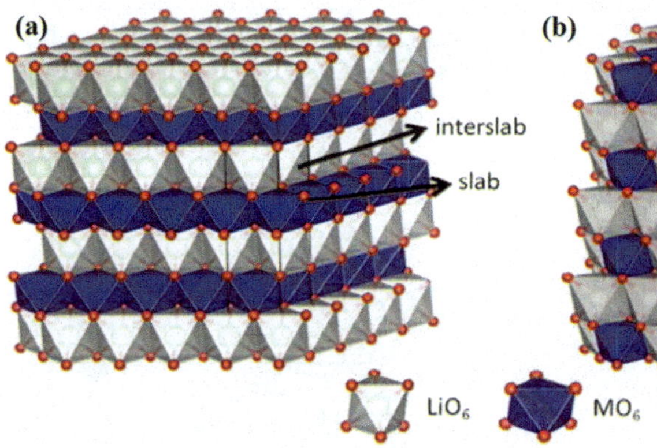

Fig. 3 Structure of layered $LiMO_2$ compounds [9]

the crystal structure of $LiNiO_2$ is more complicated due to the occupancy of the Li sites by additional Ni atoms which are a deviation from the original stoichiometry [12, 13]. Such inherent non-stoichiometry makes the synthesis of a stoichiometric composition (all probable Li sites filled with Li) rather difficult.

The third most widely accepted cathode material is $LiMn_2O_4$ which has the advantages of less toxicity than the other two and abundant availability. Moreover, $Li_xMn_2O_4$ permits the intercalation of Li ions in the range $0 < x < 2$ [14]. For $1 \leq x \leq 2$, the compound can exist in two different states tetragonal at the surface and cubic in the bulk. This may be explained as the continuous intercalation of Li ions into the $Li_xMn_2O_4$ layers decreases the average valance of Mn ions leading to a marked cooperative Jahn–Teller effect. in this effect, the cubic spinel changes to a distorted tetragonal with c/a $\approx$1.16 along with an increase in volume by 6.5%. Finally, such a distortion leads to a low capacity of 120–125 mAh/g and prominent capacity degradation at moderate temperatures [15]. Efforts are undertaken by different researchers to improve the rate capability of Li metal oxides in terms of doping or designing solid solutions. Following this Ohuzuku in 2001 [16] constructed a layered compound $LiNi_{0.5}Mn_{0.5}O_2$ by combining $LiNiO_2$ and $LiMnO_2$ in the ratio 1:1. This layered system exhibited good electrochemical properties as shown in Fig. 4 [17]. The cyclic behaviour studied within 2.5–4.4 V at a current rate of 21.7 mA/g predicted a reversible capacity of about 190 mAh/g with negligible capacity loss even after 100 cycles. Rate capability tests depicted a stable capability of 120 mAh/g even at a high rate of 6 C. These satisfactory results were interpreted by theoretical scientists from first principle calculations. They suggested a structural pattern with alternate cationic layers of $Li_{11/12}Ni_{1/12}$ and $Li_{1/12}Ni_{5/12}Mn_{6/12}$ in the octahedral sites of the close-packed oxygen frame. The cations arrange themselves in a flower-like motif with lithium ions enclosed in a hexagonal cage of manganese which is gain surrounded by a bigger hexagon of Ni.

Those orderings were predicted using Monte Carlo simulations and were seen to be in good agreement with the experimental results. Based on this model, it was

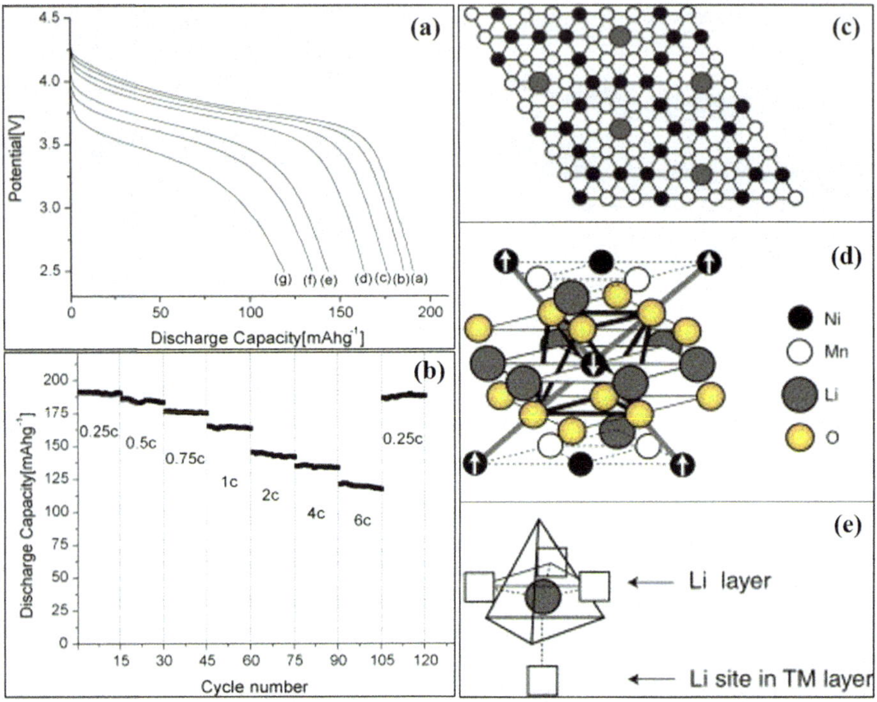

Fig. 4 **a** Potential vs discharge capacity curves; **b** discharge capacity vs number of cycles of $LiNi_{0.5}Mn_{0.5}O_2$ at different current densities [17]; **c** floral ordering pattern of Mn layers (small white circles) with Li (large grey circles) and Ni (black circles); **d** Ni in the Li plane and anti-ferromagnetic ordering of 180° Ni–O–Ni bonds (arrows indicate the magnetic moment of the electron); and **e** four face sharing octahedral site available for Li occupancy, face sharing octahedral sites are vacant of $Li_xNi_{0.5}Mn_{0.5}O_2$ [18]

explained that during the early stages of the charging cycle, lithium ions present in the flower pattern are removed which leaves the tetrahedral sites vacant to be occupied by Li ions again [18]. This material structure was found to be stable up to a temperature of 300 °C. Above this temperature, oxygen starts releasing which leads to the decomposition of the material.

Considering the degradation of $LiNi_{0.5}Mn_{0.5}O_2$ at high temperatures, further manipulations were made to improve the rate capability and capacity. One such composition series is $LiCo_xNi_yMn_{1-x-y}O_2$. Although in 2001 Ohzuku et al. [19] fabricated $LiCo_{1/3}Ni_{1/3}Mn_{1/3}O_2$ with satisfactory electrochemical performance, the significance of this series was also realized by many other researchers owing to the presence of Co ions which was found to reduce the Ni defect in lithium layer. Different synthesis methods such as sol–gel, co-precipitation, combustion, self-template and hydrothermal, were adopted by different groups to bring about diverse morphology of the compound to enhance its rate capability and specific capacity. A three-dimensional nanoflower structure was successfully synthesized by Hua et al. [20] using a fast co-precipitation method. The pristine compound in the form of nanoflower (Fig. 5) facilitated easy pathways for the diffusion of Li ions and electrons due to which it exhibited an outstanding charge–discharge rate (20 C) of 126 mAh/g. Similarly, hierarchical hollow peanut-like mesoporous structures were developed by Fang et al. [21] employing a simple self-template solid-state reaction technique. The sample calcined at 850 °C for 10 h demonstrated an initial discharge capacity of 155.1 mAh/g and maintains 107.8 mAh/g after 500 cycles at 1C which may be attributed to the mesoporous structure and short ion transport path. Besides, the introduction of extra Li ions into the MO_2 layers was also found to improve the material capacity. In this regard, a series of Li layered compounds, $Li[Li_{1/3-2x/3}Ni_xMn_{2/3-x/3}]O_2$, exhibited higher theoretical specific capacity (~300 mAh/g) beyond the limitation of single lithium ion per MO_2. Moreover, some permutations in this structure could greatly increase its experimental capacity also. In 2010, Wei et al. [22] utilized the idea that Li ions intercalate following a preferable path in a direction parallel to the Li layers. Hence in their report, they developed a crystal habit-tuned nanoplate material of $Li(Li_{0.17}Ni_{0.25}Mn_{0.58})O_2$ in which (010) nanoplates were significantly increased to provide high rate performance. The designed material could deliver high reversible capacity along with excellent cyclic stability. The reversible capacity was measured to be 200 mAh/g at 6C and reduced negligibly t 186 mAh/g after 50 cycles. Zhou et al. [23] studied $xLi_2MnO_3.(1-x)LiMn_{1/3}Ni_{1/3}Co_{1/3}O_2$ synthesized by simple hydrothermal method and observed that lithium content is one of the important factors in affecting the electrochemical performance of the compounds.

Furthermore, applying coating materials including different metal oxides like Nb_2O_5, Ta_2O_5, Al_2O_3, ZnO; fluorides AlF_3 and polyanionic compounds such as $AlPO_4$, $CoPO_4$ are also instrumental in improving the capacity of cathode materials. A systematic review of the functionality of oxide coatings on cathode materials was given by Shobana [24] in 2019. Before that Myung et al. [25] studied the effect of different metal oxide coatings such as Al_2O_3, Ta_2O_5, ZrO_2, Nb_2O_5 and ZnO on Li-rich cathode material. They observed that Al_2O_3 coating yields the best electrochemical result and gets transformed into AlF_3. This fluoride layer was effective against HF attack during cycling. Additionally, this coating has the ability to suppress the oxygen loss occurring in active materials and turn, improving the thermal stability of materials. Another important approach includes the design of core–shell structure particles with a gradient from surface to bulk. Many reviews in the past have presented the significance of core–shell architectures in building LiB cathodes [26–28]. In the core–shell structure, active core materials maintain performance while the less active shell acts as a buffer layer and helps to enhance active materials' performance. Recently in 2020, Chen et al. [29] used the residual Li ions on the surface of spherical $LiNi_{0.5}Co_{0.2}Mn_{0.3}O_2$ to construct the core–shell structure of Li_3PO_4 (LPO)@ $LiNi_{0.5}Co_{0.2}Mn_{0.3}O_2$. At an optimal composition of 0.5 wt% of LPO, the thickness of the shell was found to be 12 nm. Again, it exhibits both a longer life cycle and a better structural stability. Not only that it protects the cathode

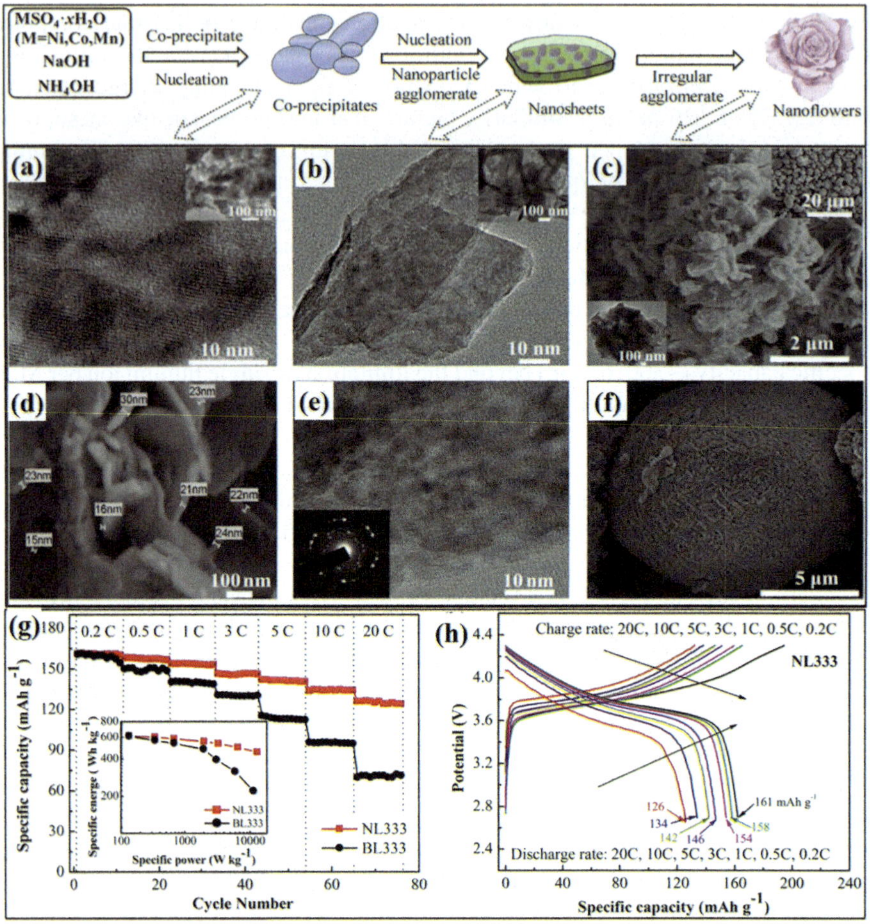

Fig. 5 Schematic illustration showing the development of $LiCo_{1/3}Ni_{1/3}Mn_{1/3}O_2$ at different stages of co-precipitation **a** 0.5 h, **b** 1.5 h, **c** 3.0 h; **d** high-resolution SEM micrograph; **e** TEM and corresponding SEAD pattern of 3.0 h synthesized $LiCo_{1/3}Ni_{1/3}Mn_{1/3}O_2$; **f** SEM image of bulk $LiCo_{1/3}Ni_{1/3}Mn_{1/3}O_2$; **g** rate performance and Ragone plot; **h** charge–discharge profile of $LiCo_{1/3}Ni_{1/3}Mn_{1/3}O_2$ [20]

from eroding and also prevents volume loss. The performance of this electrode is summarized in Fig. 6. Apart from the above-discussed layered category of materials, other prospective cathode materials are important and are discussed below.

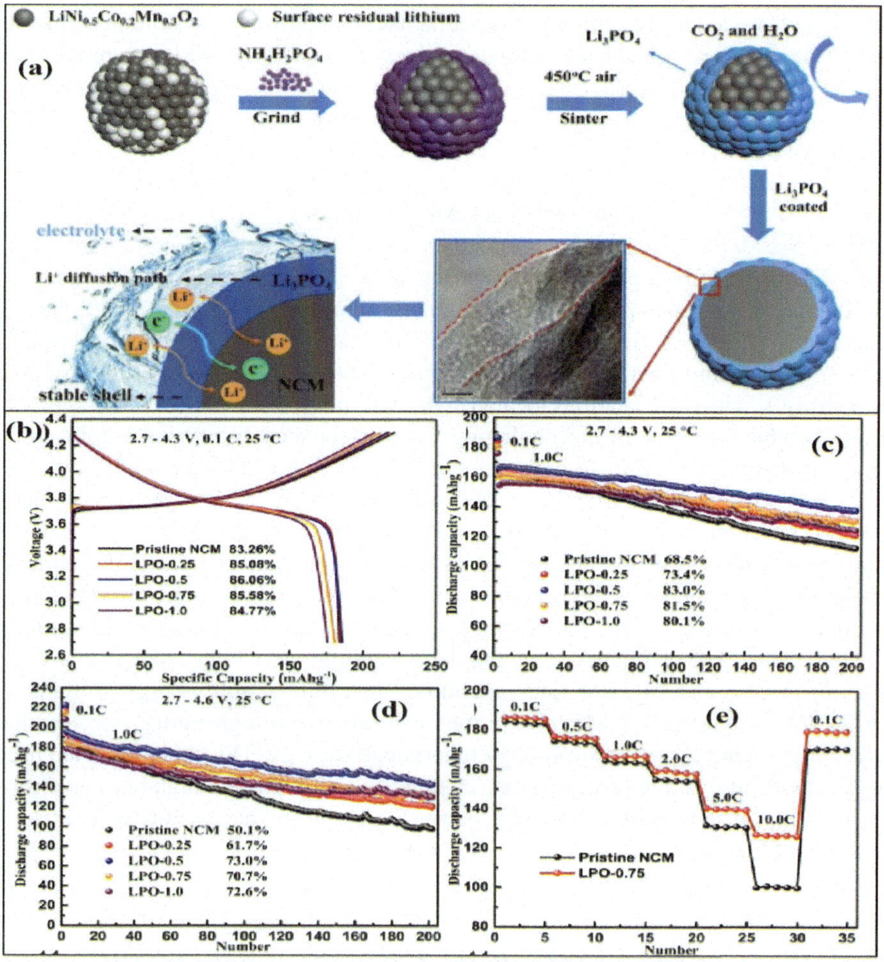

Fig. 6 **a** Schematic diagram showing the synthesis of LPO@$LiNi_{0.5}Co_{0.2}Mn_{0.3}O_2$; electrochemical properties of pristine $LiNi_{0.5}Co_{0.2}Mn_{0.3}O_2$ and LPO@$LiNi_{0.5}Co_{0.2}Mn_{0.3}O_2$, **b** voltage vs specific capacity, **c** and **d** cyclic performance in the voltage range 2.7–4.3 V and 2.7–4.6 V, respectively; **e** rate capability [29]

3.2 Layered Spinel Compounds with General Formula LiM_2O_4 (M is a Metal Atom)

Spine compounds with the formula LiM_2O_4 possess almost a similar type of structure as layered $LiMO_2$. The only difference lies in the occupancy of metal ions which not only sit at the octahedral sites but also ¼ of them settle at the tetrahedral sites of the Li layer. Such an arrangement leaves vacant octahedral sites in the M layer. Moreover, the structure can be said to be based on a three-dimensional MO_2 host along with vacancies to promote three-dimensional pathways for lithium diffusion. In the late eighties, spinel $LiMn_2O_4$ became very popular as a cathode material when it was first proposed by Thackery et al. [30]; however, it faced problems of capacity fading. The possible reasons might be the dissolution of Mn^{2+} into the electrolyte produced due to the disproportionate reaction:

$$2Mn^{3+} \rightarrow Mn^{4+} + Mn^{2+}$$

and generation of new phases while cycling and related microphases.

In this regard, replacing Mn with other metal atoms Al, Mg, Zn, Cr, Ti, Cr, Nd and La is one of the innovative approaches to improve the overall electrochemical performance of the compound. $LiNi_{0.5}Mn_{1.5}O_4$ is one such spinel that mimics the structure of $LiMn_2O_4$ with the Ni cations located at the Mn sites [31]. Various synthesis procedures producing different morphologies and particle sizes have been employed by many researchers to achieve high-rate capabilities and cyclic stability. Zhou et al. [32] 2012 presented a morphology-controlled fabrication of $LiNi_{0.5}Mn_{1.5}O_4$ hollow microspheres and microcubes using an impregnation technique followed by a solid-state reaction. One of the advantages of the impregnation method is the uniform distribution of reagents at the nanolevel, as a result of which the ionic migration path reduces significantly to the nanometer level and the undesirable growth of particles is suppressed. The resultant compound demonstrates a very high discharge capacity of 120 mAh/g and superior rate capability and cyclic stability (Fig. 7). Later in 2013, $P4_332$-type $LiNi_{0.5}Mn_{1.5}O_4$ porous nanorods with high purity were synthesized employing a novel morphology inheritance strategy. This material when used as a cathode in LiB demonstrated long cyclability and extremely high rate capability. In addition, it delivered a discharge capacity of 109 mAh/g at 20C and capacity retention of 99% up to 300 cycles at 5C [33].

3.3 *Olivine Compounds with the General Formula $LiMPO_4$ (M is a Metal Atom)*

Despite rigorous work in the direction of cathode materials in LiBs, less attention was paid to polyanionic compounds. However, the decade has seen a proportionate rise in interest in these materials owing to the inherent stability that can delay or minimize the oxygen loss occurring in layered/spinel oxides. Metal ions such as Mn, Co and Fe can form olivine compounds, out of which $LiFePO_4$ has attracted much attention due to its cost-effectiveness, low toxicity, good thermal stability and eco-friendly nature. Being discovered by Good enough and his associates for the first time in 1997 [34], $LiFePO_4$ consists of slightly distorted hcp oxygen anion arrays with almost 1/2 of the octahedral sites filled by Fe and 1/8th by Li ions. Although this olivine compound has superior cycling performance, the major drawback comes in terms of low energy density. These limitations can be overcome by certain techniques

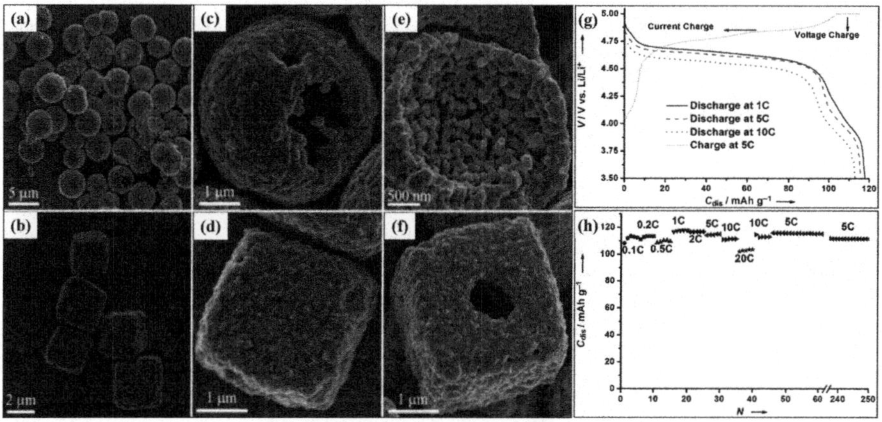

Fig. 7 FESEM micrographs showing **a–c** hollow microsphere and **d–f** hollow microcubes of $LiNi_{0.5}Mn_{1.5}O_4$; **g** charge–discharge curves; and **h** cyclic performance at different current densities in $LiNi_{0.5}Mn_{1.5}O_4$ sample [32]

such as coating with a conductive film, doping with other metal ions at the Li or Fe sites and size modulations. Other than that, compositing $LiFePO_4$ with carbon materials is one of the efficient ways of improving battery performance. Carbon materials not only serve as conducting agent for increasing electrical conductivity but as increase the porosity produced shorten the diffusion paths, thereby enhancing the overall battery efficiency [35]. Very recently, Jia et al. [36] regenerated spent $LiFePO_4$ employing environmentally benign ethanol. They attempted to improvise the cyclic stability of the regenerated material by elevating the d-band centre of Fe atoms by developing a heterogeneous interface between the compound and N_2-doped carbon. The elevated d-band suppressed the movement of Fe ions during cycling and hence strengthened the Fe–O bond. Again the lithium ion diffusion was improved in the regenerated material which led to excellent reversibility of phase transition and capacity retention of 80% even after 1000 cycles at 10 C. The electro properties of regenerated $LiFePO_4$ are summarized in Fig. 8. The other olivine compounds are $LiMnPO_4$, $LiNiPO_4$ and $LiCoPO_4$. Wang et al. [37] reported cluster and rod-like olivine $LiMnPO_4$ synthesized by a simple solvothermal method in the water–organic solvent mixture. Further, carbon was coated on the surface of $LiMnPO_4$ by chemical vapour deposition (source: methylbenzene) and then ball-milled using acetylene black. The hierarchical nanoplates demonstrated much superior rate capability and specific capacity than nanorods owing to wider interfacial exposure for lithium ion diffusion and electrochemical reactions. Moreover, it was also observed that carbon-coated (CVD) electrodes exhibited better electrochemical performance than ball-milled ones. On the other hand, doping with zinc was also established to be effective in enhancing the battery performance of $LiMnPO_4$ [38]. Similar studies were also carried out in the direction of $LiNiPO_4$ and $LiCoPO_4$.

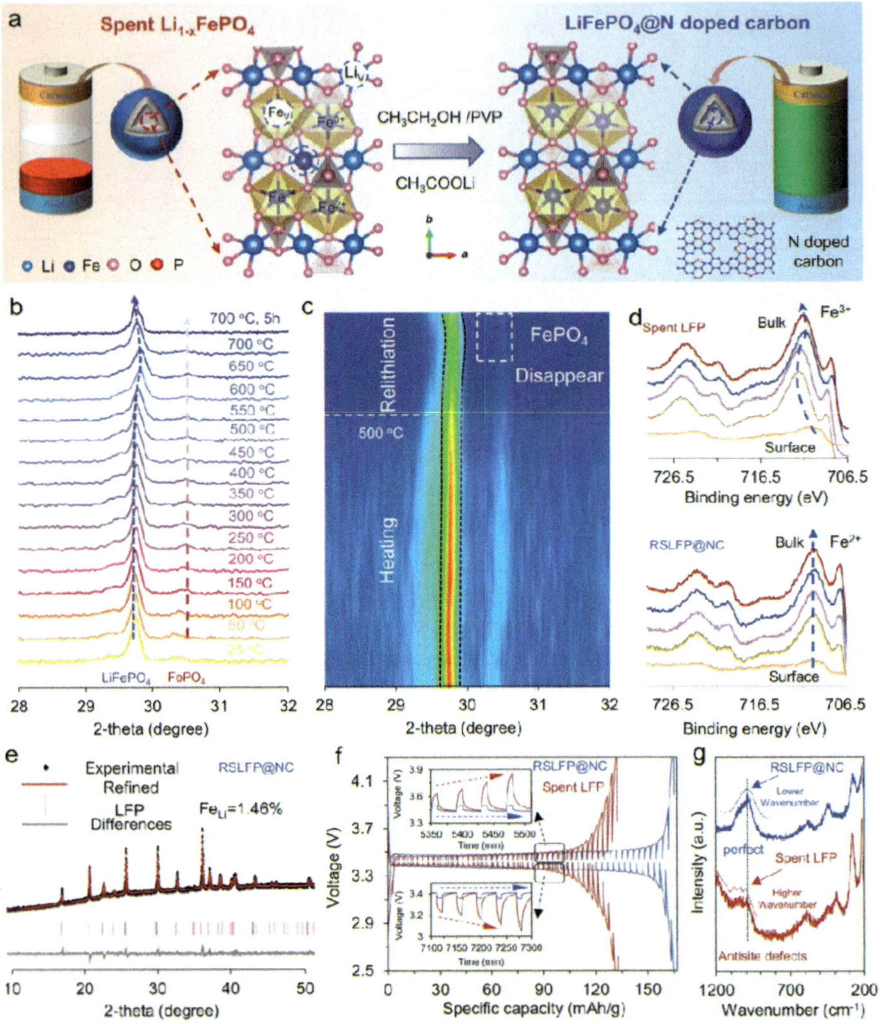

Fig. 8 **a** Schematic illustration showing the regeneration of spent $LiFePO_4$; **b** and **c** Changes observed in in-situ XRD data and corresponding contour maps during heat treatment; **d** XPS spectra of Fe 2p; **e** Rietveld refinement of the diffraction pattern of $LiFePO_4$@N-doped carbon; **f** Charge–discharge profiles; and **g** FTIR spectra of spent $LiFePO_4$ and regenerated $LiFePO_4$@N-doped carbon [36]

3.4 *Silicate Compounds with the General Formula Li_2MSiO_4 (M is a Metal Atom)*

Silicate intercalation compounds are a relatively newer class of materials that were first pioneered by Anton Nyten in 2005 [39]. In the beginning, he investigated the extraction of two lithium ions per unit formula in Li_2FeSiO_4. Structurally, the transition metal and silicate tetrahedral with corner-sharing form a layered structure in which a two-dimensional zigzag transport path is existing for Li ion to intercalate. Details on the dependence of the electrochemistry of Li_2FeSiO_4 on structure are estimated by Sirisopanaporn et al. [40]. With such a unique structure, the silicates offer a capacity (theoretical) of up to 166 mAh/g for one extracted Li ion while the removal of two Li ions delivers 333 mAh/g of capacity. However, achieving such high capacities is difficult. Many researchers employed different means to improve the maximum achievable capacity. Therefore, soon after A. Nyten, R. Dominko and his associates [41] studied the effect of three independent synthesis techniques i.e. hydrothermal method, Pechini and modified Pechini method on the morphology and electrochemical performance of Li_2FeSiO_4/carbon composites. In another approach, $Li_2FeSi_{0.98}M_{0.02}O_4$/carbon with M = Zn, Mg, Co, Ni, and Mn were considered for the LiB cathode. The results indicate optimum initial discharge capacity for Mg- and Zn-doped samples. Moreover, it was found that the discharge capacity of $Li_2FeSi_{0.99}Mg_{0.01}O_4$/C and $Li_2FeSi_{0.98}Zn_{0.02}O_4$/C delivered discharge capacities of 125 mAh/g and 166.2 mAh/g, respectively, and the capacity retention of 94.6% after 10 cycles [42]. A comparison of the electrochemical performance of the above-discussed compounds is given in Fig. 9.

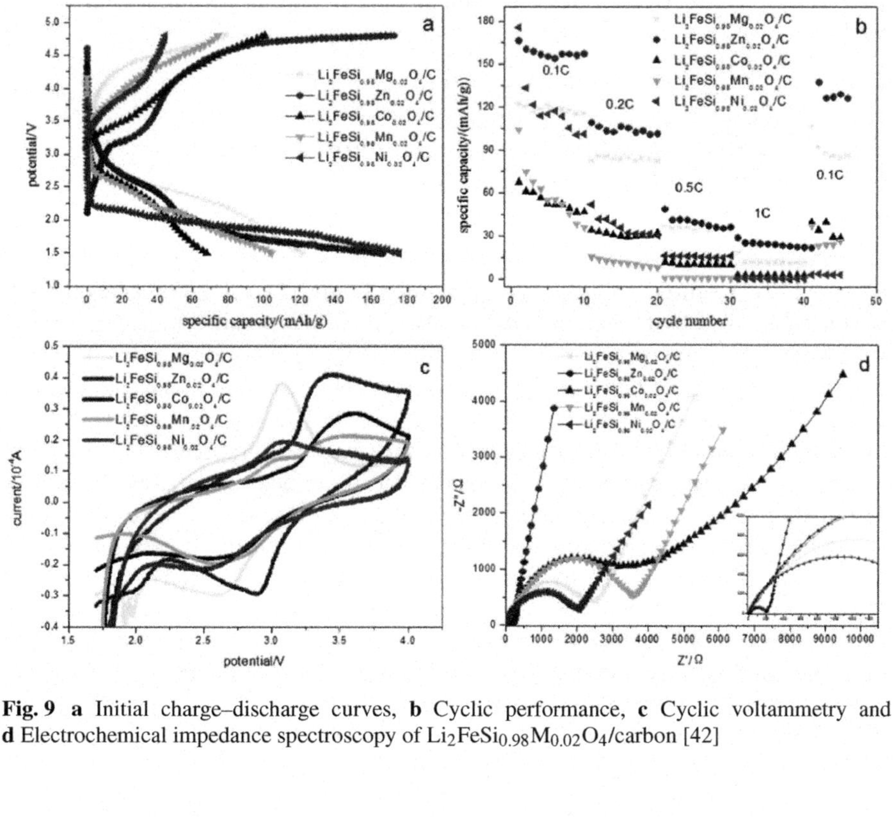

Fig. 9 **a** Initial charge–discharge curves, **b** Cyclic performance, **c** Cyclic voltammetry and **d** Electrochemical impedance spectroscopy of $Li_2FeSi_{0.98}M_{0.02}O_4$/carbon [42]

Again in 2006, another silicate compound i.e. Li_2MnSiO_4 was synthesized using an advanced Pechini sol–gel method by Dominko et al. [43]. Soon they realized that this compound was not easy to synthesize and met with failures while trying methods such as hydrothermal, sol–gel, etc. However, carbon coating was found to be prospective in improving the cathode performance of this compound. There were many reports on these composites. One of the works in 2014 is dedicated to Li_2MnSiO_4/C nanocomposites with hierarchical macroporosity. These composites are prepared with poly(methyl methacrylate) (PMMA) colloidal crystals as a sacrificial hard template and water-soluble phenol formaldehyde (PF) resin as a carbon source. These nanostructured materials demonstrate a very high reversible discharge capacity of 200 mAh/g at C/10 (voltage range—1.5–4.8 V at 45 °C). Although operating at higher temperatures, the capacity became high, but it also faded away very fast [44]. Later in 2018, Peng et al. [45] proposed a rational design of nanostructures to overcome the inherent defect and engineer superior electrochemical performance of the LiB cathodes.

They synthesized CNT (carbon nanotube)@Li_2MnSiO_4@C core–shell heterostructures. Such architecture not only provided an electron superhighway and mechanical flexibility but also provided short diffusion paths to facilitate short transport paths for Li intercalation. In this way, the intrinsic limitation of Li_2MnSiO_4 i.e. low ionic/electronic conductivity and capacity fading at a fast rate is overcome. The search for silicate material is still on since they exhibit premium intercalation qualities.

3.5 *Tavorite Compounds with the General Formula $LiMPO_4F$ (M is a Metal Atom)*

Tavorite or $LiMPO_4F$ group of materials are a derivative of olivine structure and possess many features similar to olivine family. In this class of compounds, Li ions are circumscribed by transition metal octahedra and phosphate tetrahedra. Though strong phosphate and oxygen bonds provide better thermal stability, they suffer from poor energy density. In that case, fluorine converts the one-dimensional path to multidimensional paths for Li diffusion [46]. Barker et al. [47] in 2005 fabricated $LiVPO_4F$ that exhibits a structured voltage response for the Li extraction process by a two-phase reaction mechanism centred at 4.2 V versus Li. It yielded a discharge capacity of 123 mAh/g and good capacity retention over the first 300 cycles. The results proved $LiVPO_4F$ is a better cathode material than $LiCoO_2$. In another work, the electrochemical performance of $LiVPO_4F$/C nanocomposites was enhanced by coating with MoS_2 nanosheets. The coatings were deposited using the solution method followed by low-temperature calcination. These nanocomposites exhibited good rate capability and significantly improved cyclic stability as compared to the bare ones. The sample with 1.75 wt % of MoS_2 offered a capacity retention of 91.7% in 100 cycles at 2C and a room temperature specific capacity of 112 mAh/g at 8C [48]. The synthesis technique and electrochemical properties of the same are summarized in Fig. 10. $LiFePO_4F$ is another closely studied class of tavorite compounds whose intercalation behaviour is somewhat different from others. It starts from the charged state and is subsequently discharged to Li_2FePO_4F. As a member of the tavorite family, it possesses good ionic conductivity owing to the plenty of diffusion pathways and this makes $LiFePO_4F$ electrochemically active. Again, the absence of OH groups during the synthesis of $LiFePO_4F$ is pivotal for electrochemical reversibility. Besides, 100% extraction of Li from this compound is not easy due to the large redox potential of Fe. Therefore, only one Li ion per unit formula is possible. Despite these limitations, this compound shows significant capacity retention of 150 mAh/g over 40 cycles even at high temperatures [49]. In a similar investigation by Chen et al. [50], the initial discharge capacity of 128 mAh/g and 82 mAh/g was obtained at 1 C and 2 C, respectively. However, the discharge capacity of up to 1000 cycles at high rates of 2 C and 5 C was less but the reversibility is good. Nearly monodisperse nanoparticles of size 500 nm of $LiFePO_4F$ were synthesized by Zhang et al. [51]. Voltammetric measurements of the sample provide a clear signature of Fe^{3+}/Fe^{2+} redox couple involving a two-phase transition. In addition to that, it demonstrates an initial discharge capacity of 110.2 mAh/g at 0.5 C and retains up to 104 mAh/g even after 200 cycles. The uniform nanosphere-like morphology is advantageous in terms of the short transport path of electrons and ions.

3.6 Borate Compounds with the General Formula $LiBO_3$ (M is a Metal Atom)

Borates, an important class of compounds, have attracted much attention due to the presence of the lightest polyanionic group BO_3 which confirms better theoretical energy density as compared to other polyanionic cathode materials. The electrochemical behaviour of this group of compounds was first reported by Legagneur et al. [52] in which only 0.04 Li ions per unit chemical formula could intercalate. In this regard, $LiFeBO_3$ is one of the most commonly studied cathode materials. The 3D structure of $FeBO_3$ is constructed from FeO_5 bipyramids and BO_3 trigonal planer. These FeO_5 bipyramids form long and single chains by sharing edges along the $[\bar{1}01]$ plane, while BO3 joins with three chains by corner-sharing. Li ions also form long chains along [001] direction by locating themselves at two tetrahedral sites sharing a common edge. Further, the structural and electrochemical properties of $LiFeBO_3$ were studied by Dong et al. [53] and Tao et al. [54] in detail. Dong et al. [53] synthesized the samples using the solid-state reaction method which showed an initial discharge capacity of 125.8 mAh/g (current density of 5 mA/g). On decreasing the current density to 50 mA/g, the specific capacity degrades to 88.6 mA. To further refine the performance of the cathodes, the compounds are coated with carbon which facilitates a short diffusion path for the lithium diffusion. A discharge capacity of 158.3 mAh/g was achieved for the carbon-coated $LiFeBO_3$ which was reasonably

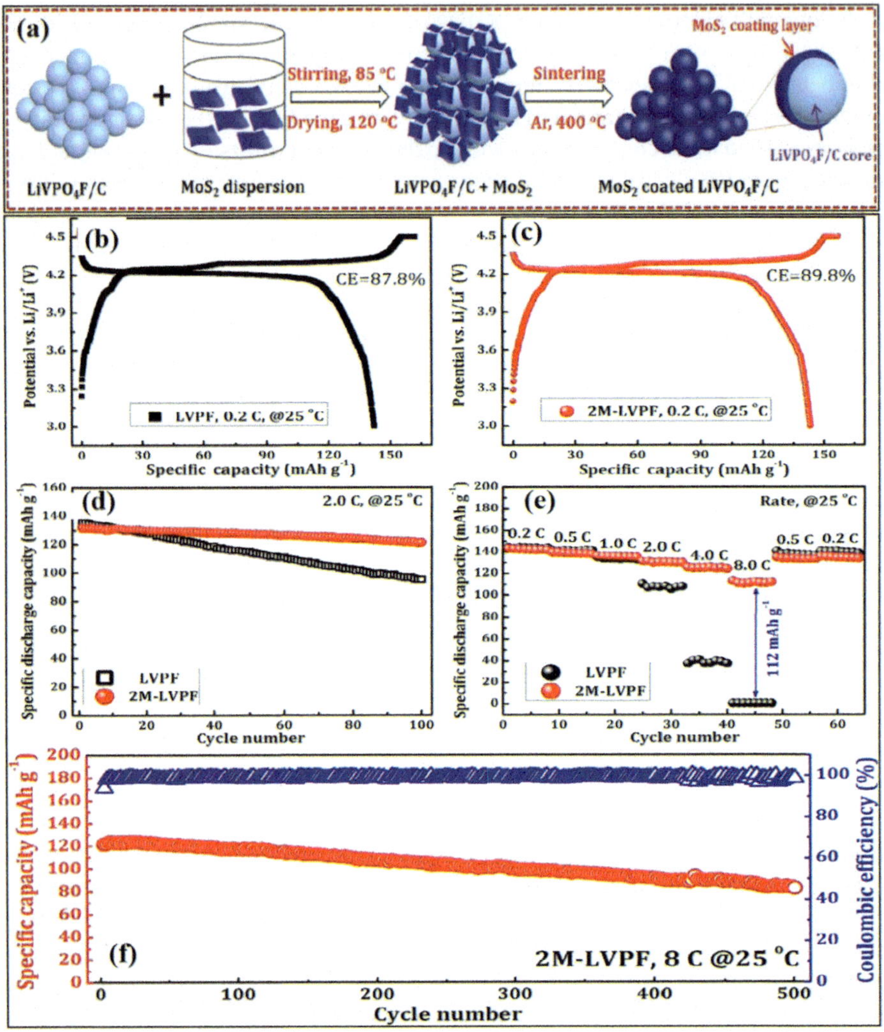

Fig. 10 **a** Schematic diagram showing the synthesis of MoS2-coated $LiVPO_4F/C$; **b** Initial charge–discharge curves of $LiVPO_4F/C$; **c** Initial charge–discharge curves of MoS_2-coated $LiVPO_4F/C$; **d** Cyclic stability and **e** Rate capability of $LiVPO_4F/C$ and MoS_2-coated $LiVPO_4F/C$; **f** Long-term cycle performance of MoS_2-coated $LiVPO_4F/C$ [48]

higher than the bare compound. In a similar work, carbon-coated $LiFeBO_3$ yielded a first discharge capacity of 210 mAh/g within a 1.5–4.5 V window (rate C/20, 55 °C) [54]. Recently, Son et al. [55] developed fluorine-substituted $LiFeBO_3$ with carbon coating ($LiFeBO_3/C$) by solid-state reaction method. The synergistic effect of F substitution and carbon coating improved the overall electrochemical performance of the electrode with a robust initial discharge capacity of 367 mAh/g and 231.7 mAh/g at 0.05 C and 1.0 C, respectively. The cyclic stability of the system

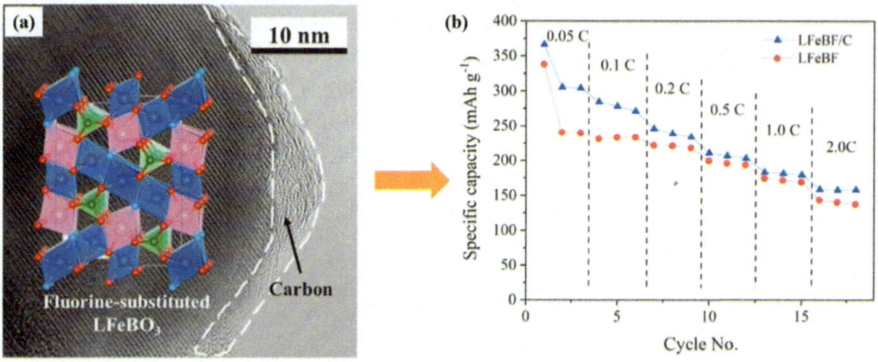

Fig. 11 Schematic figure relating the structure of $LiFeBO_3$ with its cyclic performance [55]

is displayed in Fig. 11. Apart from $LiFeBO_3$, Mn-based borates are also widely studied in the last few years. It exists in two polymorphic forms—hexagonal and monoclinic. While the hexagonal phase generated an initial discharge capacity of 75.5 mAh/g within a voltage window of 1.0–4.8 V, the monoclinic phase was not known to show any electrochemical behaviour till 2011. However, second discharge of 100 mAh/g with capacity retention over many cycles on carbon-coated $LiMnBO_3$ was found. In 2013, hexagonal $LiMnBO_3$ (h- $LiMnBO_3$) nanorods were synthesized by Afyon et al. [56] using the sol–gel method. In situ carbon coating was carried out by carbonization of propionic acid fabrication of h- $LiMnBO_3$. The nanorods are partially covered by carbon, while the remaining carbon particles disperse among the crystallites affecting the electron transport positively. An initial discharge capacity of 136 mAh/g was obtained within 4.7–1.7 V at C/20. Further improvements were achieved by forming a composite reduced graphene oxide (rGO). h- $LiMnBO_3$/rGO delivered an initial discharge capacity of 145 mAh/g at C/20 with a retention of 111 mAh/g after 10 cycles. Borates being the relatively newest class of compounds studied for their intercalation properties need a lot of improvement in synthesis methodologies and surface modifications.

3.7 Conclusion

Lithium ion batteries are taking a pivotal role in operating devices starting from small consumer electronics to electric vehicles. Under such circumstances, a lot of research is taking place worldwide in order to enhance the energy density of LiBs. Among the different components of a battery, cathode materials are significantly important for improving their overall electrochemical performance. Here, in this chapter, we have made an attempt to collage the progress made in the direction of cathode materials towards high power and energy densities; longer cycle life and better safety. Six different classes of Li intercalation compounds along with their advancements are discussed in detail. It was identified that two aspects are important in improving the performance of the cathode: good conductivity and surface area of exposure. Conductivity could be increased by coating with conducting materials such as carbon while reducing the particle size led to more area of exposure and porosity. The experimental methods along with computational techniques are expected to bring a paradigm shift in this field of research.

References

1. Cao G (2004) Nanostructures and nanomaterials: synthesis, properties and applications. Imperial college press
2. Comini E, Faglia G, Sberveglieri G, Pan Z, Wang ZL (2002) Stable and highly sensitive gas sensors based on semiconducting oxide nanobelts. Appl Phys Lett 81(10):1869–1871
3. Heo YW, Tien LC, Kwon Y, Norton DP, Pearton SJ, Kang BS, Ren F (2004) Depletion-mode ZnO nanowire field-effect transistor. Appl Phys Lett 85(12):2274–2276
4. Shen L, Yuan C, Luo H, Zhang X, Xu K, Zhang F (2011) In situ growth of $Li_4Ti_5O_{12}$ on multi-walled carbon nanotubes: novel coaxial nanocables for high rate lithium ion batteries. J Mater Chem 21(3):761–767
5. Elnathan R, Barbato MG, Guo X, Mariano A, Wang Z, Santoro F, Shi P, Voelcker NH, Xie X, Young JL, Zhao Y (2022) Biointerface design for vertical nanoprobes. Nat Rev Mater 1–21
6. Winter M, Brodd RJ (2004) What are batteries, fuel cells, and supercapacitors? Chem Rev 104(10):4245–4270
7. Wang Y, Cao G (2008) Developments in nanostructured cathode materials for high-performance lithium-ion batteries. Adv Mater 20(12):2251–2269
8. Xu B, Qian D, Wang Z, Meng YS (2012) Recent progress in cathode materials research for advanced lithium ion batteries. Mater Sci Eng R Rep 73(5–6):51–65
9. Tang D, Liu D, Liu Y, Yang Z, Chen L (2014) Investigation on the electrochemical activation process of $Li_{1.20}Ni_{0.32}Co_{0.004}Mn_{0.476}O_2$. Prog Nat Sci Mater Int 24(4):388–396
10. Manthiram A (2011) Materials challenges and opportunities of lithium ion batteries. J Phys Chem Lett 2(3):176–184
11. Ohzuku T, Ueda A, Nagayama M (1993) Electrochemistry and structural chemistry of $LiNiO_2$ (R3m) for 4 volt secondary lithium cells. J Electrochem Soc 140(7):1862
12. Dahn JR, von Sacken U, Michal CA (1990) Structure and electrochemistry of $Li_{1\pm y}NiO_2$ and a new Li_2NiO_2 phase with the $Ni(OH)_2$ structure. Solid State Ionics 44(1–2):87–97
13. Rougier A, Gravereau P, Delmas C (1996) Optimization of the composition of the $Li_{1-z}Ni_{1+z}O_2$ electrode materials: structural, magnetic, and electrochemical studies. J Electrochem Soc 143(4):1168
14. Kim DK, Muralidharan P, Lee HW, Ruffo R, Yang Y, Chan CK, Peng H, Huggins RA, Cui Y (2008) Spinel $LiMn_2O_4$ nanorods as lithium ion battery cathodes. Nano Lett 8(11):3948–3952
15. Thackeray MM (1995) Structural considerations of layered and spinel lithiated oxides for lithium ion batteries. J Electrochem Soc 142(8):2558
16. Ohzuku T, Makimura Y (2001) Layered lithium insertion material of $LiNi_{1/2}Mn_{1/2}O_2$: a possible alternative to $LiCoO_2$ for advanced lithium-ion batteries. Chem Lett 30(8):744–745
17. Meng X, Dou S, Wang WL (2008) High power and high capacity cathode material $LiNi_{0.5}Mn_{0.5}O_2$ for advanced lithium-ion batteries. J Power Sources 184(2):489–493
18. Van der Ven A, Ceder G (2004) Ordering in $Li_x(Ni_{0.5}Mn_{0.5})O_2$ and its relation to charge capacity and electrochemical behavior in rechargeable lithium batteries. Electrochem Commun 6(10):1045–1050
19. Ohzuku T, Makimura Y (2001) Layered lithium insertion material of $LiCo_{1/3}Ni_{1/3}Mn_{1/3}O_2$ for lithium-ion batteries. Chem Lett 30(7):642–643

20. Hua WB, Guo XD, Zheng Z, Wang YJ, Zhong BH, Fang B, Wang JZ, Chou SL, Liu H (2015) Uncovering a facile large-scale synthesis of $LiNi_{1/3}Co_{1/3}Mn_{1/3}O_2$ nanoflowers for high power lithium-ion batteries. J Power Sources 275:200–206
21. Fang Y, Huang Y, Tong W, Cai Y, Wang X, Guo Y, Jia D, Zong J (2018) Synthesis of hollow peanut-like hierarchical mesoporous $LiNi_{1/3}Co_{1/3}Mn_{1/3}O_2$ cathode materials with exceptional cycle performance for lithium-ion batteries by a simple self-template solid-state method. J Alloy Compd 743:707–715
22. Wei GZ, Lu X, Ke FS, Huang L, Li JT, Wang ZX, Zhou ZY, Sun SG (2010) Crystal habit-tuned nanoplate material of $LiLi_{1/3-2x/3}Ni_xMn_{2/3-x/3}]O_2$ for high-rate performance lithium-ion batteries. Adv Mater 22(39):4364–4367
23. Zhou S, Liu G, Wen L, Luo H (2021) Preparation and electrochemical properties of uniform spherical lithium rich materials $xLi_2MnO_3{\bullet}(1-x)$ $LiMn_{1/3}Ni_{1/3}Co_{1/3}O_2$. J Alloy Compd 874:159885
24. Shobana MK (2019) Metal oxide coated cathode materials for Li ion batteries–a review. J Alloy Compd 802:477–487
25. Myung ST, Izumi K, Komaba S, Yashiro H, Bang HJ, Sun YK, Kumagai N (2007) Functionality of oxide coating for $Li[Li_{0.05}Ni_{0.4}Co_{0.15}Mn_{0.4}]O_2$ as positive electrode materials for lithium-ion secondary batteries. J Phys Chem C 111(10):4061–4067
26. Su L, Jing Y, Zhou Z (2011) Li ion battery materials with core–shell nanostructures. Nanoscale 3(10):3967–3983
27. Shen L, Li H, Uchaker E, Zhang X, Cao G (2012) General strategy for designing core–shell nanostructured materials for high-power lithium ion batteries. Nano Lett 12(11):5673–5678
28. Lu W, Guo X, Luo Y, Li Q, Zhu R, Pang H (2019) Core-shell materials for advanced batteries. Chem Eng J 355:208–237
29. Chen H, Xiao L, Chen H, Zhu Y, Xiang K, Liao H (2020) Construction of core-shell $Li_3PO_4@LiNi_{0.5}Co_{0.2}Mn_{0.3}O_2$ cathode with improved cycling stability for lithium ion batteries. Electrochim Acta 344:136142
30. Goodenough JB, Thackeray MM, David WF, Bruce PG (1984) Lithium insertion/extraction reactions with manganese oxides. Revue de Chim Minerale 21(4):435–455
31. Santhanam R, Rambabu B (2010) Research progress in high voltage spinel $LiNi_{0.5}Mn_{1.5}O_4$ material. J Power Sources 195(17):5442–5451
32. Zhou L, Zhao D, Lou X (2012) $LiNi_{0.5}Mn_{1.5}O_4$ hollow structures as high-performance cathodes for lithium-ion batteries. Angew Chem Int Ed 51(1):239–241
33. Zhang X, Cheng F, Yang J, Chen J (2013) $LiNi_{0.5}Mn_{1.5}O_4$ porous nanorods as high-rate and long-life cathodes for Li-ion batteries. Nano Lett 13(6):2822–2825
34. Padhi AK, Nanjundaswamy KS, Masquelier C, Okada S, Goodenough JB (1997) Effect of structure on the redox Fe3+/Fe2+ in iron phosphate. J Electrochem Soc 144(5):1609–1613
35. Eftekhari A (2017) $LiFePO_4$/C nanocomposites for lithium-ion batteries. J Power Sources 343:395–411
36. Jia K, Ma J, Wang J, Liang Z, Ji G, Piao Z, Gao R, Zhu Y, Zhuang Z, Zhou G, Cheng HM (2023) Long-life regenerated $LiFePO_4$ from spent cathode by elevating the d-band center of Fe. Adv Mater 35(5):2208034
37. Wang F, Yang J, Gao P, NuLi Y, Wang J (2011) Morphology regulation and carbon coating of $LiMnPO_4$ cathode material for enhanced electrochemical performance. J Power Sources 196(23):10258–10262
38. Fang H, Yi H, Hu C, Yang B, Yao Y, Ma W, Dai Y (2012) Effect of Zn doping on the performance of $LiMnPO_4$ cathode for lithium ion batteries. Electrochim Acta 71:266–269
39. Nytén A, Abouimrane A, Armand M, Gustafsson T, Thomas JO (2005) Electrochemical performance of Li_2FeSiO_4 as a new Li-battery cathode material. Electrochem Commun 7(2):156–160
40. Sirisopanaporn C, Masquelier C, Bruce PG, Armstrong AR, Dominko R (2011) Dependence of Li_2FeSiO_4 electrochemistry on structure. J Am Chem Soc 133(5):1263–1265
41. Dominko R, Conte DE, Hanzel D, Gaberscek M, Jamnik J (2008) Impact of synthesis conditions on the structure and performance of Li_2FeSiO_4. J Power Sources 178(2):842–847

42. Li L, Han E, Qiao S, Liu H, Shi Y, Yuan W (2019) Synthesis characterization and improved electrochemical performance of Li_2FeSiO_4/C as cathode for lithium-ion battery by metal doping. Prog Nat Sci Mater Int 29(2):111–118
43. Dominko R, Bele M, Gaberšček M, Meden A, Remškar M, Jamnik J (2006) Structure and electrochemical performance of Li_2MnSiO_4 and Li_2FeSiO_4 as potential Li-battery cathode materials. Electrochem Commun 8(2):217–222
44. He G, Manthiram A (2014) Nanostructured Li_2MnSiO_4/C cathodes with hierarchical macro-/mesoporosity for lithium-ion batteries. Adv Func Mater 24(33):5277–5283
45. Peng T, Guo W, Zhang Q, Zhang Y, Chen M, Wang Y, Yan H, Lu Y, Luo Y (2018) Uniform coaxial CNT@Li_2MnSiO_4@C as advanced cathode material for lithium-ion battery. Electrochim Acta 291:1–8
46. Gover RKB, Burns P, Bryan ASMY, Saidi MY, Swoyer JL, Barker J (2006) LiVPO4F: a new active material for safe lithium-ion batteries. Solid State Ionics 177(26–32):2635–2638
47. Barker J, Gover RKB, Burns P, Bryan A, Saidi MY, Swoyer JL (2005) Structural and electrochemical properties of lithium vanadium fluorophosphate, $LiVPO_4F$. J Power Sources 146(1–2):516–520
48. Liu Z, Peng W, Shih K, Wang J, Wang Z, Guo H, Yan G, Li X, Song L (2016) A MoS_2 coating strategy to improve the comprehensive electrochemical performance of $LiVPO_4F$. J Power Sources 315:294–301
49. Ramesh TN, Lee KT, Ellis BL, Nazar LF (2010) Tavorite lithium iron fluorophosphate cathode materials: phase transition and electrochemistry of $LiFePO_4F$–Li_2FePO_4F. Electrochem Solid-State Lett 13(4):A43
50. Chen D, Shao GQ, Li B, Zhao GG, Li J, Liu JH, Gao ZS, Zhang HF (2014) Synthesis, crystal structure and electrochemical properties of $LiFePO_4F$ cathode material for Li-ion batteries. Electrochim Acta 147:663–668
51. Zhang Y, Liang Q, Huang C, Gao P, Shu H, Zhang X, Yang X, Liu L, Wang X (2018) Nearly monodispersed $LiFePO_4F$ nanospheres as cathode material for lithium ion batteries. J Solid State Electrochem 22:1995–2002
52. Legagneur V, An Y, Mosbah A, Portal R, La Salle ALG, Verbaere A, Guyomard D, Piffard Y (2001) $LiMBO_3$ (M = Mn, Fe, Co): synthesis, crystal structure and lithium deinsertion/insertion properties. Solid State Ionics 139(1–2):37–46
53. Dong YZ, Zhao YM, Shi ZD, An XN, Fu P, Chen L (2008) The structure and electrochemical performance of $LiFeBO_3$ as a novel Li-battery cathode material. Electrochim Acta 53(5):2339–2345
54. Tao L, Rousse G, Chotard JN, Dupont L, Bruyere S, Hanžel D, Mali G, Dominko R, Levasseur S, Masquelier C (2014) Preparation, structure and electrochemistry of $LiFeBO_3$: a cathode material for Li-ion batteries. J Mater Chem A 2(7):2060–2070
55. Son Y, Nguyen DT, Lee Y (2022) Preparation and characterization of fluorine-substituted $LiFeBO_3$ with carbon coating to enhance electrochemical performance and stability as a cathode material for Li-ion batteries. J Power Sources 533:231395
56. Afyon S, Kundu D, Krumeich F, Nesper R (2013) Nano $LiMnBO_3$, a high-capacity cathode material for Li-ion batteries. J Power Sources 224:145–151

The manufacturer's authorised representative in the EU is Springer Nature Customer Service Centre GmbH, Europaplatz 3, 69115 Heidelberg, Germany. If you have any concerns regarding our products, please contact ProductSafety@springernature.com

Printed and bound by CPI Group (UK) Ltd, Croydon, CR0 4YY
01/10/2026
02225900-0006